LINKING SPECIES
& ECOSYSTEMS

LINKING SPECIES
& ECOSYSTEMS

edited by

Clive G. Jones
Institute of Ecosystem Studies

John H. Lawton
National Environment Research Council for Population Biology

CHAPMAN & HALL

 I⟨T⟩P   An International Thomson Publishing Company

New York • Albany • Bonn • Boston • Cincinnati • Detroit • London • Madrid • Melbourne • Mexico City
Pacific Grove • San Francisco • Singapore • Tokyo • Toronto • Washington

Copyright © 1995 by Chapman & Hall

Printed in the United States of America

For more information contact:

Chapman & Hall
115 Fifth Avenue
New York, NY 10003

Chapman & Hall
2-6 Boundary Row
London SE1 8HN
England

Thomas Nelson Australia
102 Dodds Street
South Melbourne, 3205
Victoria, Australia

Chapman & Hall GmbH
Postfach 100 263
D-69442 Weinheim
Germany

Nelson Canada
1120 Birchmount Road
Scarborough, Ontario
Canada M1K 5G4

International Thomson Publishing Asia
221 Henderson Road #05-10
Henderson Building
Singapore 0315

International Thomson Editores
Campos Eliseos 385, Piso 7
Col. Polanco
11560 Mexico D.F.
Mexico

International Thomson Publishing - Japan
Hirakawacho-cho Kyowa Building, 3F
1-2-1 Hirakawacho-cho
Chiyoda-ku, 102 Tokyo
Japan

2 3 4 5 6 7 8 9 XXX 01 00 99 98 97 96

Library of Congress Cataloging-in-Publication Data

Linking species and ecosystems / edited by Clive G. Jones and John H. Lawton.
         P.   Cm.
    Papers from the Fifth Cary Conference, held at the Institute of Ecosystem Studies, Millbrook, NY, May 8--12, 1993.
    Includes bibliographical references and index.
    ISBN 0-412-04801-9
    1. Biotic communities--Congresses. 2. Species--Congresses. 3. Ecology--Congresses. I. Jones, Clive G., 1951- II. Lawton, John H., 1943-- . III. Cary Conferences (5th : 1993 : Institute of Ecosystem Studies)
    QH540.L56   1994   94-8281
    574.5'247--dc20   CIP

    Chapter 15 is written by an employee of the government of the United States of America and may not be copyrighted.

Visit Chapman & Hall on the Internet http://www.chaphall.com/chaphall.html

To order this or any other Chapman & Hall book, please contact **International Thomson Publishing, 7625 Empire Drive, Florence, KY 41042.** Phone (606) 525-6600 or 1-800-842-3636. Fax: (606) 525-7778. E-mail: order@chaphall.com.

For a complete listing of Chapman & Hall titles, send your request to **Chapman & Hall, Dept. BC, 115 Fifth Avenue, New York, NY 10003.**

CONTENTS

APPROACHES

CONTEXT

FORWARD *(SIC)*

Gene E. Likens

Because ecology is an integrative science, it has great potential for generating broad and fundamental understanding about complex natural phenomena. Such ecological understanding can provide relevant and critical guidance to decision makers seeking management solutions to complicated environmental problems. It is my belief that the ultimate challenges for ecology are not only to do nontrivial, fundamental research, but also to integrate the ecological information available from all levels of inquiry into an understanding that is useful to managers and decision makers.

Unfortunately, recently there has been a fractionation of ecology into its various specialities, thus, potentially at least, reducing the power of the field as a whole to integrate and synthesize to the fullest extent (Likens, 1992). It is not surprising that the field is becoming fractionated, given its complexity and the natural desires of creative scientists for identity and visibility for their work. Nevertheless, this partitioning is counterproductive given the overriding need for synthesis of ecological information.

There are divergent views about the seriousness of this problem. Some think that the problem really is unimportant or doesn't exist. Others believe that the problem is characterized by a deep schism, often including disdain among the different approaches to the subject. In an attempt to provide a new, over-arching definition of ecology the scientific staff of the Institute of Ecosystem Studies has proposed the following: "Ecology is the scientific study of the processes influencing the distribution and abundance of organisms, the interactions among organisms, and the interactions between organisms and the transformation and flux of energy and matter." This definition describes a robust and dynamic field of inquiry, a unified science dedicated to understanding the ecology of the Earth (Likens, 1992).

One of the problems of fractionation in ecology is related to the often naive perceptions that one group has about the goals, approaches, and activities of another. For example, many ecologists believe that ecosystem ecologists are interested only in energy flow or biogeochemical cycling. My view as an ecosystem ecologist is much broader (see Likens, 1992). But, the point is, as usual, stereotyping can result in false reality. In reality, there are numerous critical linkages between the approaches, concepts, and results of population ecology and ecosystem ecology, two areas of ecology that conceptually often are viewed to be at the boundaries of the field and separated by an artificial schism.

The Cary Conference that gave rise to this book was organized specifically to address such linkages between population ecology and ecosystem ecology. The Cary Conferences were conceived in general to be forums for such philosophical discussions in ecology. This Conference was no exception. The overviews, reviews, and case studies herein provide conceptual frameworks to help integrate the field, with a wide range of examples of how this integration can be achieved. The Conference sparked a renewed sense of commitment among the participants toward integration and synthesis in ecology. It is hoped that this book will do the same among its readers.

ACKNOWLEDGMENTS

This book had its intellectual genesis at the Fifth Cary Conference, "Linking Species and Ecosystems," held at the Institute of Ecosystem Studies, Millbrook, New York, May 8–12, 1993. We are deeply indebted to all who made the conference a success. In particular we thank the conference participants and the steering committee—James Brown, Gene Likens, Michael Pace, Moshe Shachak, and Peter Vitousek—for their insights, ideas, and the enthusiasm with which they all tackled the problems of integrating species and ecosystem perspectives. The conference could not have happened, or have run so smoothly, without Jan Mittan, Conference Coordinator, the conference staff—James Baxter, Martin Christ, Christopher Tripler, and D. Alexander Wait—and many of the staff at the Institute of Ecosystem Studies. We thank the National Science Foundation (DEB-9311600), the Mary Flagler Cary Charitable Trust, the Royal Society of London, and the NERC Centre for Population Biology (Core Funding) for the financial support and travel grants that made the conference possible.

We are very grateful to all the authors for producing their manuscripts (mainly!) on time, and for tolerating our harassment, and to colleagues at the Institute of Ecosystem Studies, the NERC Centre for Population Biology,

Imperial College at Silwood Park, and at many other institutions for their rapid, insightful, and critical review of manuscripts. Last but not least, we thank Sharon Okada for help with graphics, Heather Rolland Fischer for tireless typing and reference checking, Marielle Reiter for prompt production, and Greg Payne, Life Sciences Editor at Chapman & Hall. This is a contribution to the program of the Institute of Ecosystem Studies.

CLIVE G. JONES
JOHN H. LAWTON

CONTRIBUTORS

Gören I. Ågren, Department of Ecology and Environmental Research, Swedish University of Agricultural Sciences, Box 7072, S-750 07, Uppsala, Sweden.

J.M. Anderson, Rothamsted International, Rothamsted Experimental Station, Harpenden, AL5 2JQ, England, U.K.

S. Archer, Texas A & M University, College Station, TX 77843, U.S.A.

Gary T. Banta, Institute of Biology, Odense University, DK-5230, Odense M, Denmark.

Jan Bengtsson, Department of Ecology and Environmental Research, Swedish University of Agricultural Sciences, Box 7072, S-750 07, Uppsala, Sweden.

Niall Broekhuizen, Department of Statistics and Modelling Science, University of Strathclyde, Glasgow, G1 1XH, Scotland, U.K.

James H. Brown, Department of Biology, University of New Mexico, Albuquerque, NM 87131, U.S.A.

V.B. Brown, Natural Resource Ecology Laboratory, Colorado State University, Fort Collins, CO 80523, U.S.A.

Charles D. Canham, Institute of Ecosystem Studies, Box AB, Millbrook, NY 12545, U.S.A.

Stephen R. Carpenter, Center for Limnology, 680 North Park Street, University of Wisconsin, Madison, WI 53706, U.S.A.

S.L. Coale, Institute of Marine Sciences, University of California, Santa Cruz, CA 95064, U.S.A.

Donald L. DeAngelis, Environmental Sciences Division, Oak Ridge National Laboratory, Oak Ridge, TN 37831, U.S.A.

Heather E. Erickson, Center for Streamside Studies, AR-10, University of Washington, Seattle, WA 98195, U.S.A.

James A. Estes, U.S. Fish and Wildlife Service, Institute of Marine Sciences, University of California, Santa Cruz, CA 95064, U.S.A.

Kenneth H. Foreman, Boston University Marine Program, Woods Hole, MA 02543, U.S.A.

Jerry F. Franklin, College of Forest Resources, AR-10, University of Washington, Seattle, WA 98195, U.S.A.

Thomas M. Frost, Center for Limnology, 680 North Park Street, University of Wisconsin, Madison, WI 53706, U.S.A.

Anne E. Giblin, The Ecosystems Center, Marine Biological Laboratory, Woods Hole, MA 02543, U.S.A.

Nancy B. Grimm, Department of Zoology, Arizona State University, Tempe, AZ 85287-1501, U.S.A.

William S.C. Gurney, Department of Statistics and Modelling Science, University of Strathclyde, Glasgow, G1 1XH Scotland, U.K.

K.A. Hibbard, Texas A & M University, College Station, TX 77843, U.S.A.

Heinrich D. Holland, Department of Earth and Planetary Sciences, Harvard University, Cambridge, MA 02138, U.S.A.

Robert D. Holt, Museum of Natural History, Department of Systematics and Ecology, The University of Kansas, Lawrence, KS 66045, U.S.A.

Nancy Huntly, Department of Biological Sciences and Center for Ecological Research and Education, Idaho State University, Pocatello, ID 83209-8007, U.S.A.

Anthony R. Ives, Department of Zoology, 430 Lincoln Drive, University of Wisconsin, Madison, WI 53706, U.S.A.

Carol A. Johnston, University of Minnesota, Center for Water and the Environment, Duluth, MN 55812, U.S.A.

Clive G. Jones, Institute of Ecosystem Studies, Box AB, Millbrook, NY 12545, U.S.A.

Timothy K. Kratz, Center for Limnology, 680 North Park Street, University of Wisconsin, Madison, WI 53706, U.S.A.

John H. Lawton, Centre for Population Biology, Imperial College at Silwood Park, Ascot, Berkshire, SL5 7PY, England, U.K., and Institute of Ecosystem Studies, Box AB, Millbrook, NY 12545, U.S.A.

Jeffrey Levinton, Department of Ecology and Evolution, State University of New York, Stony Brook, NY 11794, U.S.A.

Gene E. Likens, Institute of Ecosystem Studies, Box AB, Millbrook, NY 12545, U.S.A.

Jane Lubchenco, Department of Zoology, Oregon State University, 3029 Cordley Hall, Corvallis, OR 97331-2914, U.S.A.

C.P. Lund, Department of Biological Sciences, Stanford University, Stanford, CA 94305, U.S.A.

Neo D. Martinez, Bodega Marine Laboratory, University of California, Davis, Bodega Bay, CA 94923-0247, U.S.A.

Robert J. Naiman, Center for Streamside Studies, AR-10, University of Washington, Seattle, WA 98195, U.S.A.

Robert V. O'Neill, Environmental Sciences Division, Oak Ridge National Laboratory, P.O. Box 2008, Oak Ridge, TN 37831-6038, U.S.A.

Stephen W. Pacala, Department of Ecology and Evolutionary Biology, Princeton University, Princeton, NJ 08544, U.S.A.

Michael L. Pace, Institute of Ecosystem Studies, Box AB, Millbrook, NY 12545, U.S.A.

Robert W. Parmelee, Department of Entomology, 1735 Neil Avenue, Ohio State University, Columbus, OH 43210, U.S.A.

John Pastor, University of Minnesota, Center for Water and the Environment, Duluth, MN 55812, U.S.A.

Tryggve Persson, Department of Ecology and Environmental Research, Swedish University of Agricultural Sciences, Box 7072, S-750 07, Uppsala, Sweden.

C.H. Pilskaln, Monterey Bay Aquarium Research Institute, Pacific Grove, CA 93950, U.S.A.

Gilles Pinay, CNRS, Centre d'Ecologie, 29 rue Jeanne Marvig, 31055 Toulouse, France.

Michael M. Pollock, Center for Streamside Studies, AR-10, University of Washington, Seattle, WA 98195, U.S.A.

Mary E. Power, Department of Integrative Biology, University of California, Berkeley, CA 94720, U.S.A.

Edward B. Rastetter, The Ecosystems Center, Marine Biological Laboratory, Woods Hole, MA 02543, U.S.A.

Alex H. Ross, NIWAR Fresh Water, P.O. Box 8602, Christchurch, New Zealand

David S. Schimel, National Center for Atmospheric Research, P.O. Box 3000, Boulder, CO 80307-3000, U.S.A.

D.W. Schindler, Department of Zoology, CW 312 Biological Sciences Building, University of Alberta, Edmonton, T6G 2E9, Canada.

Moshe Shachak, The Jacob Blaustein Institute for Desert Research, Mitrani Center for Desert Ecology, Ben-Gurion University of the Negev, Sede Boker Campus, 84990 Israel, and Institute of Ecosystem Studies, Box AB, Millbrook, NY 12545, U.S.A.

Gaius R. Shaver, The Ecosystems Center, Marine Biological Laboratory, Woods Hole, MA 02543, U.S.A.

M.W. Silver, Institute of Marine Sciences, University of California, Santa Cruz, CA 95064, U.S.A.

Lawrence B. Slobodkin, Department of Ecology and Evolution, State University of New York, Stony Brook, NY 11794, U.S.A.

Patricia A. Soranno, Center for Limnology, 680 North Park Street, University of Wisconsin, Madison, WI 53706, U.S.A.

D.K. Steinberg, Institute of Marine Sciences, University of California, Santa Cruz, CA 95064, U.S.A.

Robert W. Sterner, Gray Freshwater Biological Institute, University of Minnesota, Box 100, County Roads 15 and 19, Navarre, MN 55392, U.S.A.

Monica G. Turner, Environmental Sciences Division, Oak Ridge National Laboratory, P.O. Box 2008, Oak Ridge, TN 37831-6038, U.S.A.

David A. Wedin, Department of Botany, University of Toronto, 25 Willocks Street, Toronto, Ontario, M5S 3B2, Canada.

David Wei Zheng, Department of Ecology and Environmental Research, Swedish University of Agricultural Sciences, Box 7072, S-750 07, Uppsala, Sweden.

LINKING SPECIES
& ECOSYSTEMS

ISSUES

THE NEED FOR INTEGRATION IN ECOLOGY

The vigorous growth in ecology from its origins in the early years of this century has been accompanied by the creation of numerous subdisciplines (McIntosh, 1985). Although specialization may be inevitable, it also causes problems. Gaps in understanding arise at the interfaces of subdisciplines (Slobodkin, 1985; Swanson, 1987) that are ignored because they do not fit into convenient pigeonholes. It is also difficult, many would say well-nigh impossible, for individual scientists to keep abreast of the explosion in ecological information outside their own immediate fields of interest (Root, 1987). So the paradigms, assumptions, definitions, lexicon, methods, and approaches in one area risk becoming incomprehensible to practitioners in another (Ziman, 1985). As a result, the conceptual frameworks in each area tend to become increasingly divergent over time, hampering communication across the subject as a whole (Clark, 1985; Ehrlich, 1986; Root, 1987; Slobodkin, 1992a; Pickett et al., 1994).

We all have great aspirations for the advancement and utility of ecology, particularly in the face of human-accelerated environmental change (Likens,

1991; Lubchenco et al., 1991; Steffen et al., 1992; Kareiva et al., 1993). The history of science shows that integration across disciplines is a critical ingredient of scientific progress (Cohen, 1985). Integration forces us to ask new questions, fills gaps in our understanding, and helps information to flow across the entire subject (Pickett et al., 1994). It follows that because ecology is one of the most complex, multifaceted sciences (Slobodkin, 1985; Likens, 1992)—embracing areas of study as diverse as biogeochemistry, biogeography, population dynamics, ecophysiology, genetics, and evolution—integration is particularly important for the health and vigor of the discipline (e.g., Reiners, 1986; Huston et al., 1988). In the words of Paul Ehrlich: "Ecologists have gone a long way toward explaining different aspects of the extremely complex systems they investigate. Now they are beginning to tie the pieces together. Integration cannot come too soon" (Ehrlich, 1986; p. 16).

ECOSYSTEM ECOLOGY VS. POPULATION AND COMMUNITY ECOLOGY

The need for integration is nowhere more apparent than in two of the major subdisciplines of ecology—population/community ecology and ecosystem ecology. These two subdisciplines have quite different approaches to understanding nature. One major distinction is a focus on organisms versus a focus on materials and energy (O'Neill et al., 1986; Pickett et al., 1994) clearly seen in ecological textbooks (e.g., Odum, 1971; Begon and Mortimer, 1986; Begon et al., 1990). Questions in population and community ecology usually address the determinants of the distribution and abundance of species and their assemblages (e.g., Krebs, 1985; Fenchel, 1987); questions in ecosystem ecology commonly focus on material and energy fluxes (e.g., Odum, 1971; Schlesinger, 1991). It sometimes seems that chalk and cheese have more in common than these two areas of ecology.

There is, of course, nothing wrong with having different perspectives and goals, underpinned by different assumptions and methods (Slobodkin, 1985; Pickett et al., 1994); we have no doubt that major advances in ecological understanding will continue to be made in each of these subdisciplines. However, a truly integrative ecology should deal effectively with organismal interactions, material and energy fluxes, and their relationships. In the broadest sense, ecology is *"the scientific study of the processes influencing the distribution and abundance of organisms, the interaction among organisms, and the interactions between organisms and the transformation and flux of energy and matter"* (Likens, 1992).

This definition is not simply an umbrella that covers and justifies independent research in each subdiscipline. Explicit recognition of the interrelationships among organisms, their assemblages, and material and energy fluxes means that integration is a prerequisite for deeper theoretical understanding and the practical applications of ecology (MacMahon et al., 1978; Pastor and Post, 1986). We share Ehrlich's view that "ecologists who study populations do not talk as much as they should to those who are interested in the complexities of ecosystems" (Ehrlich, 1986; p. 16). Current division between these subdisciplines did not always exist. Lotka's classic *Elements of Physical Biology* (Lotka, 1925) embraced population processes, nutrient dynamics, and energy fluxes in a pioneering attempt to understand the natural world; and Tansley (1935) formulated and defined the concept of ecosystem in an essay primarily devoted to the nature of species assemblages and succession.

Of course, some ecologists have always found the division between ecosystem ecology and population/community ecology an artificial one; they may not even be aware of the split! Many of the contributors to this book clearly belong in this category, but they are exceptions. We sense that the tide is now turning, with "the natural intellectual links between population and ecosystem ecology . . . receiving renewed attention by ecologists after many years of relative neglect" (Carpenter et al., 1994). Books by Schulze and Mooney (1993), devoted to the effects of biodiversity on ecosystem function, and by Kareiva et al. (1993), addressing impacts of environmental change from physiology to landscape-level processes, are but two examples.

THE ORGANIZATION OF THE BOOK

Individual organisms, species populations, and communities inhabit ecosystems, and by definition must be affected by ecosystem processes—nutrient fluxes, productivity, and the physical environment. Conversely, ecosystem processes must be affected by organisms; there can be no primary production without plants, and no nitrogen cycle without microbes. Yet, in general, we do not understand how to link organismal activities, population dynamics, and community assemblages to ecosystem processes, and there is little general theory that relates ecosystem properties to the activities, dynamics, and assemblages of species. Against this background, the aims of this book are to redress the balance, by exploring the problems of linkage, both empirically and conceptually.

As well as providing numerous, well-studied examples of when and how particular species affect ecosystem processes and vice versa, this book highlights general approaches and ways of thinking about the problems that others may find helpful, and it seeks to identify general principles. It is decidedly not the final word on many, indeed most of the issues. Individual chapters raise as many questions as they provide answers, and authors frequently disagree about approaches to the problems of linkage. Nor is there, in Lubchenco's words (Ch. 28), any Grand Unifying Theory (GUT). But on many issues, progress is being made.

In the opening chapter, Grimm (Ch. 1) gives an overview of the range of problems encountered by linking species and ecosystem perspectives, with an emphasis on the populations and processes found in freshwater streams. This introductory chapter highlights interaction strength and spatially based studies as ways of combining the expertise of ecosystem and population ecologists. Brown (Ch. 2) offers a more radical approach, arguing that because all species and interactions are unique, *a priori* predictions about the role of species in ecosystems are extremely difficult, if not impossible. He believes that deeper understanding will emerge only by treating organisms and ecosystems as "Complex Adaptive Systems" (CASs), with energy as the critical common currency.

Later in the book (Ch. 14—27), other authors return to the problem of how to develop general principles and approaches that link species and ecosystem perspectives; none are as radical as Brown's suggestions! In between (Ch. 3–13), we present a range of case histories and examples of linkage between population and ecosystem perspectives. The book closes with four chapters (Ch. 28–31) that address practical issues that emerge from the need to integrate the two subdisciplines. Our hope is that it will help to accelerate a change in the way we teach, practice, and apply ecological understanding.

1

WHY LINK SPECIES AND ECOSYSTEMS? A PERSPECTIVE FROM ECOSYSTEM ECOLOGY

Nancy B. Grimm

SUMMARY

Population, community, and ecosystem ecologists historically have asked different kinds of questions about nature, and as a result have defined domains of study quite differently. In this chapter, the general question "Why link species and ecosystems?" will be explored from an ecosystem ecologist's point of view. I begin by considering the subdisciplinary distinctions in ecology: how they are reflected in questions asked, what theories underlie them, and what areas are ripe for integration.

Examples drawn largely from stream ecology are presented in a discussion of redundancy and keystone species. The primary conclusion is that these phenomena occur in nature, are applicable to ecosystem processes as well as community structure, and should be predictable from ecosystem and species characteristics. Although available data do not always allow prediction of *which* species is likely to affect ecosystem functioning, use of variables such as biomass or production as indicators of a given species' status may lead to misinterpretation. Concepts of interaction strength are applicable to ecosystem studies.

Species occur in nature as members of interactive assemblages, and their interactions may affect ecosystem functioning in either subtle or dramatic ways. Given the attention that has been paid by community ecologists to biotic interactions, does an understanding of these processes contribute to understanding ecosystem functioning? A complex stream system and conceptual model are described for the purpose of highlighting questions relevant to integrating biotic interactions, disturbance, and ecosystem functioning.

Finally, some loose ideas on the themes of disturbance and stability, patchiness, and scale as potential starting points for linking species and ecosystems are given in the final section. To facilitate interaction between population/community and ecosystem/landscape ecology, units of study should be spatially based.

INTRODUCTION

In one of the final paragraphs of his book on the history of ecosystem ecology, historian of biology Joel Hagen wrote,

> "Populations are important, but an evolutionary ecology worthy of the name must come to grips with the question of how large communities and ecosystems are structured. How this might be done remains unclear, for evolutionary and ecosystem ecologists have been talking past one another for almost a generation." (Hagen, 1992).

Hagen refers to a split in ecology that he traced to the publication of George Williams' book, *Adaptation and Natural Selection* (Williams, 1966). With respect to the topic of this book, Hagen's statement encapsulates the view that the endpoint or goal of ecological investigation is an understanding of the *structure* of populations, communities, and ecosystems. Structure refers to the number and kinds of component parts, and a thorough understanding of structure would necessarily incorporate knowledge of interactions among components and between components and their environment. In contrast to that view, ecological investigations of ecosystem functioning may be aimed at understanding effects of ecosystem processes, interactions, and components on their context, or on the larger system of which ecosystems are a part.

> "The loss of biodiversity should be of concern to everybody for three basic reasons . . . perhaps the most poorly evaluated to date, is the array of essential services provided by natural ecosystems, of which diverse species are the key working parts."

Ehrlich and Wilson (1991) use the term "ecosystem services" to describe the net output, or net effects of ecosystem function—or more generally, what an ecosystem *does* in a landscape or biosphere context.

These two quotations exemplify the contrast between questions basic to evolutionary/population/community ecology and ecosystem/landscape ecology. To answer questions about distribution and abundance, we need to "build up" from components by understanding their interactions. These questions often require historical or evolutionary answers [e.g., Hutchinson's (1959) famous query, Why are there so many kinds of animals?]. To answer questions about processes, it is possible to attempt mechanistic explanations at the lower level of communities or populations, or to seek consequences of those processes for higher-level phenomena or for systems at larger scales.

The aim of this introductory chapter is to identify key issues that are important to answering the question posed in its title: Why link species and ecosystems? The perspective is an empirical ecosystem ecologist's, who studies ecosystems (streams) with relatively easily defined boundaries. I will not do justice to species/ecosystem-linking models, but acknowledge that development of theory (models) and empirical work must go hand in hand. First, I will consider the subdisciplinary distinctions in ecology, then ask what evidence and logic support individual species' importance in ecosystem dynamics, then consider the role of assemblages and their interactions in ecosystems. Finally, what themes now integrate these subdisciplines? Is there an obvious conceptual basis for linkage? These last questions are more thoroughly considered in later chapters, but I hope to provide some insights that at least will stimulate discussion.

SUBDISCIPLINARY DISTINCTIONS

A good place to start in identifying distinctions among subdisciplines is in ecology textbooks. Common definitions of the science of ecology (with their different foci italicized) are these:

> The experimental analysis of *distribution and abundance* (e.g., Krebs, 1985).
>
> The scientific study of the *interactions* between organisms and their environment (e.g., Begon et al., 1990; Ricklefs 1990).

Begon et al. assert that "the ultimate subject matter of ecology is the distribution and abundance of organisms." This and the first definition reflect the unidirectional view—that ecology is oriented toward understanding structure—given in the Hagen (1992) quotation. Likens (1992) has proposed a definition that is the most inclusive of those listed, and allows activities of all subdisciplines to be called ecology: "[Ecology is] the scientific study of the *processes* influencing the distribution and abundance of organisms, the interaction among organisms, and the interaction between organisms and the transformation and flux of energy and matter."

Begon et al. (1990) and Krebs (1985) place ecology firmly among the biological sciences by stressing its overlap with behavior, genetics, physiology, and evolution. To many ecosystem ecologists, this list is not complete: ecology also overlaps with hydrology, geochemistry, geology and geomorphology, and perhaps even social sciences. Schimel (1993) has maintained that ecology is properly identified as an earth science. This possibility should be given some attention, if we are to explore effectively the links between the subdisciplines of ecology. We should also give attention to the relevance of evolution, genetics, and behavior to ecosystem ecology. These fields are now more closely allied with population and community ecology than is ecosystem ecology. The rich body of evolutionary theory may be irrelevant over the short term to ecosystem processes, but in light of rapid environmental change and assaults on natural ecosystems, it will become increasingly germane (Holt, Ch. 26). This seems a compelling reason to begin linking the ecological subdisciplines.

What theories underlie ecosystem ecology and population/community ecology? A strong theoretical basis for work on ecosystem energetics and biogeochemistry is the physical laws. Any system involving biota is also subject to evolutionary laws, and in this regard communication problems still exist. Aside from criticisms that the terms ecosystem and community are fuzzy and vague, common usage of "function" and "role," in particular, has fueled miscommunication, at least between evolutionary ecologists and others. The problem is that a term such as "function" implies to some an evolutionary origin, which must therefore imply operation of natural selection at supraorganismic levels, yet it remains a useful term for describing what a system does in the context of its surroundings. (In this chapter, I use the term function with no intended implication of an evolutionary origin.) On the other hand, communication will be facilitated by generalized interest in scale, disturbance, stability, and patchiness. These unifying themes are the areas in which development of general, multiple theories seems most likely.

Reiners' (1986) thoughtful commentary on conceptual models for ecosystem studies presents a view of how laws and theories may be organized to generate predictable hypotheses. He outlined the logical structure of two models that underlie most research in ecosystem ecology: one dealing with biogeochemistry and based on elemental stoichiometry and one dealing with energy flow and based on thermodynamics. Reiners argued that development of these models and a third complementary model (dealing with population or community structure) *and their ultimate linkage* is needed to facilitate progress in ecology. Reiners' model, based on the axiom that groups of organisms have regular chemical stoichiometries, has driven some excellent research on links between species and ecosystems, notably the work of Wedin and Tilman (1990; see also Wedin, Ch. 24) and Sterner and colleagues (Sterner et al., 1992; see also Sterner, Ch. 23).

Finally, in practice, many data necessary for evaluating the role of species in ecosystems are acquired in the course of typical ecosystem investigations. For example, Smock et al. (1992) determined rates of macroinvertebrate secondary production in a southeastern stream by dividing the stream ecosystem into patches and calculating secondary production for the whole system by weighted summation. But what information went into this calculation? Secondary production of macroinvertebrate consumers at the ecosystem level is most often calculated from an extremely detailed accounting of individual species, by microhabitat and by size class. The point is that aggregation of this fine-grained information, first to functional groups, then to subsystems or patches, and ultimately to the whole ecosystem, is the best way to arrive at the measure of energy flow through consumers. Incidentally, information on species composition, life histories, habitat preference, population size structure, and functional redundancy is obtained. How does this differ from a description of macroinvertebrate community structure? It differs only in the question originally asked.

Overall, ecologists aim to understand patterns in nature, be they community structure or gas flux across ecosystem boundaries, by resolving mechanisms explaining the patterns at lower scales and consequences of the pattern at higher scales. Ecosystem functioning is simply one higher-level consequence, or net effect, of organismal interactions with other organisms and the abiotic environment.

SHOULD WE CARE WHO'S INSIDE THE BLACK BOX?

What is the evidence that individual species can influence ecosystem functioning? The basic organizational question here is "Does the identity of species inside the black box matter at the ecosystem level?" The key to prediction is an understanding of what features of species and/or ecosystems are conducive to functional redundancy (Lawton and Brown, 1993) (or in contrast, keystone species). For ecosystems, we might consider such characteristics as the number of effective trophic levels, nutrient availability and limitation, types of functional groups, disturbance regimen, and degree of spatial and/or temporal heterogeneity; for species, competitive ability, predation efficiency, life history traits, and unique biochemical or metabolic abilities [e.g., nitrogen (N) fixation] may be important.

Since Paine (1969) first defined a keystone predator, the term "keystone" has been extended to species whose removal results in significant change in community structure or ecosystem functioning. When applied to ecosystems, keystone species are those species whose contribution to ecosystem functioning is unique. If we could divide an ecosystem into essential functions or processes ("ecosystem services" in Ehrlich and Wilson's terminology), then each process might be characterized by a complement of species that perform it (i.e., a functional group). Essential functions may be grossly classified as trophic, biogeochemical, and structural, and keystone species are the sole members of a trophic, biogeochemical, or structural functional group. It is obvious that the more detailed or specialized our definition of functions, the lower the likelihood of redundancy.

A few examples from streams serve to illustrate the interaction between species and ecosystem traits that determine whether functional redundancy or keystone species (trophic, biogeochemical, or structural) occur.

(1) Trophic keystone. In sunlit midcontinental North American rivers, grazing by minnows *(Campostoma anomalum)* is effective in reducing macroalgal abundance and productivity (Power et al., 1985; Power, Ch. 6). Predatory bass *(Micropterus* spp.) feed on the minnows in pools where both occur, either as a result of redistribution during river spates, movement between pools, or experimental introduction. When bass are added to a *Campostoma* pool, the algae are released from grazing pressure and grow rapidly. In small, species-poor prairie streams where biota are redistributed by floods (ecosystem characteristic), the single-species occupant of the large-grazer functional group *(Campostoma)* is a keystone species (possessing a unique ability to crop macroalgae). The cascading trophic effect of increased algal production due to effective bass predation is essentially a removal experiment that demonstrates this keystone effect. In larger rivers, the bass–*Campostoma*–algae interaction is less dramatic because of greater spatial heterogeneity (ecosystem characteristic).

(2) Biogeochemical redundancy? Desert streams in the southwestern United States are strongly affected by disturbance (flash floods). Because these events occur frequently (albeit unpredictably) and organismal life spans are short relative to the timeframe of ecological investigation, my colleagues and I have been able to describe repeatedly the process of succession (for summaries, see Fisher, 1990; Grimm, 1994). Ecosystem-level attributes (at the scale of 100 to 200-m reaches) such as primary producer biomass, organic matter storage, and energy flow variables of gross primary production and ecosystem respiration increase rapidly, often to an asymptote. Periphyton communities continue to change thereafter, however, often toward cyanobacterial dominance. In terms of the energy flow these community-level changes are irrelevant, but because cyanobacteria are N-fixers, there is no redundancy in terms of N cycling.

It is a frequent practice of stream ecologists to measure algal biomass using chlorophyll a, as this allows one to separate primary producer biomass from mass of detritus and microconsumers. Is this measure adequate to describe changes that might be of consequence to ecosystem functioning? In other words, can we lump species as functionally equivalent and summarize their effects at the ecosystem level using chlorophyll *a*? A study of succession at the scale of single rocks (Peterson and Grimm, 1992) provides an answer to these questions. As is typical of postflood succession at the reach scale, primary producer biomass and net primary production rose rapidly, indicating that assemblages on rocks are resilient. At the functional group level, however, biovolume of N-fixing species increased rapidly whereas there was little successional change in biomass of nonfixers. Chlorophyll *a* alone was inadequate to describe an important feature of this succession: the N fixer functional group was important throughout the sequence and would obviously affect N dynamics in a different way than would nonfixers. But what can be said about redundancy at the species level? The early dominant in the N-fixer functional group (*Epithemia sorex*, a diatom containing endosymbiotic cyanobacteria) was replaced within about 1 month of the 3-month sequence by *Calothrix* sp., which ultimately accounted for >95% of algal biovolume. Although we lose information about N dynamics using "chlorophyll *a*," the consequence of shifts from one species to another species of N-fixer may be less important; that is, there may be redundancy within that functional group.

(3) Structural effects. Keystone species also affect physical structure of ecosystems (see the detailed classification of "ecosystem engineers" offered by Lawton and Jones, Ch. 14). Biotically mediated change in hydrologic linkage between subsystems is an example from streams; macrophytes and beaver dams can alter patterns of water flow between streams and hyporheic zones. Earthworm and sediment-dwelling invertebrate burrowing, or trapping of fine sediment by macrophytes in rivers change the distribution of oxygen, which in turn alters biogeochemical cycles. In terrestrial ecosystems, plants and even snails (Shachak et al., 1987) affect physical structure by contributing to weathering. Finally, species can modify disturbance regimens. Introduced grasses can fuel fires (D'Antonio and Vitousek, 1992). In desert streams, introduced bermuda grass may anchor sandy sediments and prevent sediment movement during flash floods. Beaver dams may also ameliorate effects of spates. In all of these cases, the species involved possess some unique attribute that alters structure.

A final point is that the keystone concept has often been applied to groups of species or even functional groups, which diminishes its usefulness. It seems more sensible to assign an interaction strength to species, as Paine (1980) has advocated. In an ecosystem sense, interaction strength represents the potential of species to affect processes rather than species' distribution and abundance. Experiments should be conducted at realistic (whole ecosystem) scales to assess responses of ecosystem processes (functions) to manipulations of strong interactors (see also Lawton and Brown, 1993).

BIOTIC INTERACTIONS AND ECOSYSTEM FUNCTIONING

I turn now to the question of linkage between biotic interactions and ecosystem functioning. A brief description of biotic interactions in a desert stream ecosystem illustrates the kinds of questions we must ask to address this issue. The major players in this story are: *nitrogen*—the limiting nutrient, which is highly variable in space and time; *primary producers*—abundant and productive, dominated by diatoms, *Cladophora glomerata*, and *Anabaena* sp.; *microbial community*—associated with algae-derived particulate organic matter and invertebrate and fish feces; macroinvertebrates—collector-gatherers with high rates of secondary production and short life spans; *"megagrazers"*— snails and omnivorous fish; *flash floods and drying*—disturbances that dictate the successional and spatial context. Interactions between processes or phenomena associated with these players are indicated in Fig. 1-1 as having positive or negative effects.

A semistatic view of mid–late successional time is this: primary producers consume N as they grow, and in both time and space, their activities result in a decline in N availability. As they become N-limited, they may be more susceptible to being eaten or to losing in competition with N fixers. Fine particulate organic matter (FPOM) may be derived largely from megagrazer feeding, but whatever its origin, this material supports a rapid accumulation of collector-gatherer biomass. These small macroinvertebrate consumers feed on

Figure 1-1.   Effects of biotic and abiotic interactions in a desert stream ecosystem. Processes and interactions are linked by positive arrows when the net effect is to enhance or increase the likelihood of the recipient process, and by negative arrows when the net effect is to decrease likelihood. Disturbances (flood, drying) interact with this static system in midsuccessional time in complex ways (±). POM = particulate organic matter.

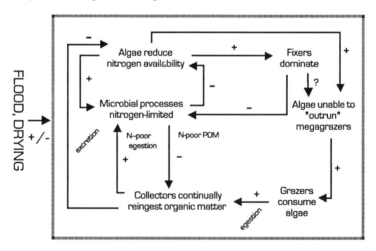

small particles collected from the stream bottom and sediment interstices. They feed at very high rates—up to one third of the total organic matter standing crop and seven times gross primary production each day. (This is most significant to the FPOM pool, because they do not feed directly on macroalgae.) These organisms must be continually reingesting fecal material because assimilation efficiency is extremely low (10%). Fisher and Gray (1983) hypothesized that this represents a reasonable tactic if residence time in the environment improves food quality (e.g., N content) through bacterial colonization. Collectors' continual reingestion of organic matter, however, makes the FPOM poor in N, potentially leading to N-limitation of microbial growth. In turn, invertebrate populations "crash" in sequences where N is especially low, perhaps owing to low food quality.

My purpose in delving in some detail into this system is to point out that such complicated sets of interactions can have strong effects on ecosystem-level patterns of nutrient retention and energy flow. Biotic interactions such as the replacement of nonfixers by fixers, competition for N between producers and decomposers, and grazing are at the heart of changes in system state. But this static view does not consider the effects of disturbances (floods and drying), which are ever-present in these systems. How does this scenario fit into that more inclusive scheme? In other words, *when* and *where* are biotic interactions such as these important in controlling ecosystem structure and functioning? Fisher and Grimm (1991) presented a conceptual model in which "control" of ecosystem processes varies temporally as a function of disturbance regimen. Spatial variation in control of ecosystem processes is equally likely, given that susceptibility to drying and flood varies spatially (Fisher, 1994). The point of both temporal and spatial conceptual models is that the question we should be asking is not *whether* biotic interactions affect ecosystem processes, but *when* and *where*—in the context of dynamic systems and disturbance.

Are these conceptual models applicable to other systems subject to disturbance? Or must they be restricted to streams? Bearing in mind that time frames over which disturbance regimens must be quantified vary tremendously among ecosystems, the answer is that such models do apply to all types of ecosystems. Other lessons from aquatic research might also be brought to bear on the question of linking biotic interactions and ecosystem processes. Research in aquatic ecosystems has already accomplished substantial integration of community and ecosystem ecology. Why is this so? At the most basic level, boundaries are relatively easy to define and populations are confined within them. As it is difficult to separate biotic from abiotic components, perhaps there has been more of a tolerance of "black boxing" in aquatic ecology. Finally, in streams and intertidal zones where biota grow attached to the substratum and are easily seen, research has emphasized disturbance and spatial heterogeneity. These are of consequence both to biotic interactions and community structure and to ecosystem processes.

INTEGRATING THEMES

The themes of disturbance, stability, patchiness, and scale are potentially integrating ones, and population, community, and ecosystem ecologists share a common interest in them. But researchers are not using the same language and therefore must carefully define terms *and variables of interest* for any discourse to be meaningful. Operational definitions of stability and the included concepts of resistance and resilience for the stream ecosystem study discussed earlier are taken from Webster et al. (1983). But even using consistent definitions, the benthic invertebrate assemblage in this system can be seen as resistant or not, depending on the variable measured (community structure or total biomass; Boulton et al., 1992). This underscores the need to identify common response variables in comparisons of stability across levels of organization.

Landscapes and ecosystems are spatially heterogeneous, and this key feature must be incorporated into any conceptual scheme for linking species and ecosystems. Lessons learned from studies of spatially heterogeneous land–water units would be well applied to all ecological studies. A particularly good example from a biogeochemical study is the description of patches along an arctic toposequence as nutrient sources or sinks linked by down-slope water movement (Giblin et al., 1991). For streams, my colleagues and I are attempting to determine how patch dynamics affect trophic structure and thereby the extent to which different patches (each with a characteristic trophic structure) are sources or sinks for N. Such approaches might provide a means for integrating biotic interactions and ecosystem processes—in the latter example, effects of trophic–nutrient interactions on nitrogen retention.

In population and community ecology, models involving sources and sinks explicitly consider different spatial units within landscapes or ecosystems. For example, sink populations (mortality rate $[d]$ > birth rate $[b]$) may persist when linked by active dispersal to source ($b > d$) populations (Pulliam, 1988). Stream reaches with few prey and many predators may be supported by prey immigration from upstream prey-rich reaches (e.g., Cooper et al., 1990).

Interactions between population/community and ecosystem ecologists would be facilitated by adopting, *as a starting point*, a spatially based conception of the units of study. I disagree with Allen and Hoekstra (1992), who contend that ecosystems are not geographical entities (although it is certainly true that their size and boundaries may be ill defined). The advantage of this spatial basis is the ability to scale up or down, and in so doing, to include all relevant, potentially interacting components (both biotic and abiotic) of ecosystems. Whatever the scale of the investigation, a spatially based perspective places species interactions (the traditional focus of community ecology) into a context in which their effects on ecosystem processes may be assessed. Interactions *between* patches may be critical to larger-scale processes, and include biotic interactions that occur within component subsystems.

CONCLUSIONS

Population/community and ecosystem ecologists ask different questions about nature. Why then should we be concerned with linking species and ecosystems? For ecosystem ecologists, the answer is clear: the activities of certain strongly interacting single species and certainly the interactions among species can have strong effects on functioning at ecosystem and landscape levels. Resolution of key issues will be needed to begin forging such a link. First, how can theory underlying these two distinct approaches be integrated? How important is evolution to ecosystems and geochemistry to populations? Second, can keystone species be identified based on traits of the ecosystem or of the species themselves? If we adopt the terminology of interaction strength, can it be applied to specific ecosystem functions? Third, biotic interactions clearly affect ecosystem processes. Under what conditions, in space and time, is this effect strong? Finally, what conceptual basis for linkage exists? Treatment of traditional population, community, and ecosystem processes in a spatial context, with the patch as a fundamental unit, may provide a useful tool for integrating species and ecosystems.

ACKNOWLEDGMENTS

I thank Stuart Fisher and James Collins for helpful comments on the manuscript, and James Collins and students in our graduate seminar, "Linking Species and Ecosystems," for stimulating discussions. Support for the author by the National Science Foundation (Grants BSR-8818612 and DEB-9108362) is acknowledged.

2

Organisms and Species as Complex Adaptive Systems: Linking the Biology of Populations with the Physics of Ecosystems

James H. Brown

SUMMARY

Historical constraints of paradigms, approaches, training, and methodology currently prevent the conceptual unification of ecology. They inhibit communication and collaboration between the biologists who study the structure and dynamics of populations and communities and the earth scientists, physicists, chemists, and biologists who study the distribution and fluxes of energy and materials in ecosystems. To link the attributes of individual organisms and species with the biogeochemical processes that occur in their environments requires an integration of biology and the physical sciences.

To make these linkages, I suggest that individuals and species be considered a special class of complex adaptive systems (CASs) with some uniquely biological attributes. This view encourages a focus on the physical relationships of the biological CAS: on developing an energetic/thermodynamic currency that can be used to characterize population growth and fitness in terms of basic physical processes; on characterizing the niche in terms of the environmental requirements of the CAS for survival and reproduction; and on understanding the impacts of species on ecosystems in terms of work performed in meeting these requirements.

INTRODUCTION

I was asked to introduce this volume by examining "why a knowledge of ecosystem functioning can contribute to understanding species activities, dynamics, and assemblages." I have found it surprisingly difficult to address this topic.

On the one hand, the answer is very simple and general: because all species live in ecosystems, they are part of and dependent on ecosystem processes. It is impossible to understand the abundance and distribution of populations and the species diversity and composition of communities without a knowledge of their abiotic and biotic environments and of the fluxes of energy and matter through the ecosystems of which they are a part. But everyone knows this. It is what ecology is all about (e.g., Likens, 1992). It is why the discipline has retained its integrity and thrived, despite a sometimes distressing degree of bickering and chauvinism among its various subdisciplines: physiological, behavioral, population, community, and ecosystem ecology.

On the other hand, I find our present ability to answer the question unsatisfactory. I could make a list of the many ways that ecosystem processes affect species, and vice versa, and then try to make some kind of functional classification of these linkages. There are excellent case studies, many of which are summarized later in this volume. There are organisms that move soil and alter the flow of water; that transport other organisms; that build large complex structures; that alter the physical and chemical composition of air, soil, and water; and that feed selectively on other organisms, altering the flow of energy and materials through food webs. These activities can perhaps be classified into groups or syndromes. Many of my ecologist colleagues like to make such classifications (e.g., of vegetation types, trophic levels, and pollination syndromes), but I find them unsatisfying. I am afraid that they often divide an inherently continuously distributed natural world into a number of discrete, arbitrary categories.

What seems to be needed to make linkages between species and ecosystems is not so much a better classification of the kinds of species according to their functional roles and ecosystem-level impacts, but a unifying conceptual framework to explain both the variety and the common, emergent features of these relationships.

Do all of these linkages between species and ecosystems have a common theme and a set of common properties? If there is such a unifying framework, what is it and why is it not more apparent in the studies that have been done and in the chapters of this book?

I suggest that there is a common conceptual basis for linking activities of species with biogeochemical processes. It has not yet been well developed, however, because it represents a new kind of integration between the biological and physical sciences.

In this chapter, I try to sketch the outline that this conceptual unification might take. I focus on four themes: (1) viewing individual organisms and species as a special class of CAS, (2) adopting a common energetic/thermodynamic currency, (3) operationalizing Hutchinson's concept of multidimensional niche to characterize the ecological uniqueness of species, and (4) viewing the effects of species on ecosystems in terms of physical work.

A DIGRESSION: THE DICHOTOMY AND ITS HISTORY

To link species and ecosystems will require integration of population and community ecology with ecosystem ecology. These subdisciplines have a long history of limited communication, as scientists with different backgrounds and training have pursued different questions, using different paradigms and techniques. Population/community ecologists are interested in questions about the abundance, distribution, and diversity of species, most of them have training in biology that includes genetics and evolution; and their research is based largely on uniquely biological theories of population growth and adaptation by natural selection. In contrast, ecosystem ecologists are interested in the biological and physical processes that affect the distribution and fluxes of energy and materials, many of them have training in the physical and earth sciences, and their research draws heavily on the fundamental laws of physics and chemistry. Elsewhere (Brown, 1981) I have attributed this dichotomy in large part to the enormous influence of Robert MacArthur and Eugene Odum on the development of modern population/community and ecosystem ecology, respectively.

Here, however, I note that this division between physical and biological subdisciplines of ecology reflects a more fundamental division between the physical and biological sciences. Biologists have often emphasized the unique features of living systems, rather than trying to explain these characteristics in terms of physical laws. Although a great deal of physics and chemistry has been incorporated into some areas of biology (e.g., physiology and cellular and molecular biology), many of the connections remain to be forged. They have the potential to contribute to several areas—such as the origin of life, developmental biology, and neurobiology—in addition to ecology.

ORGANISMS AND SPECIES AS COMPLEX ADAPTIVE SYSTEMS

A group of natural and social scientists, many of them associated with the Santa Fe Institute (e.g., see Lewin, 1992; Waldrop, 1992; Kauffman, 1993; Cowan and Pines, 1994), are taking a synthetic approach to the study of complexity. They are trying to identify and explain the common, emergent features of a wide variety of complex systems. These include physical systems such as spin glasses, Bernard cells, and thunderstorms; biological systems such as human brains, developmental programs, individual organisms, and ecological communities; and human systems such as machines, adaptive computer programs, economies, and languages. Increasingly, these are being referred to as complex adaptive systems (Cowan and Pines, 1994) (I use CAS as shorthand for one such system, and CASs for more than one). CASs share several properties: (1) they are comprised of many components of different kinds; (2) they have the capacity for self-organization, using relatively simple rules and materials to produce complex structures and dynamics; (3) they are open systems that exchange energy and materials with their surroundings; (4) they maintain a state far from thermodynamic equilibrium by the uptake and transformation of energy; (5) they have a capacity for adaptive change, changing their components and altering their performance as a consequence of interactions with their environment; and (6) they develop a unique structure and dynamics that reflect a history of cumulative changes.

Individual organisms, populations, species, communities, and ecosystems all possess the above characteristics. I do not want to argue here whether it is productive to consider all of them as CASs; I do want to make the point that both individuals and species represent special classes of CASs. They possess some additional features that are not shared either with most kinds of nonbiological complex systems or with populations, communities, and ecosystems (Brown, 1994a). They are discrete units with an inside and an outside, a beginning and an end. They have relatively well-defined boundaries that regulate the exchange of energy and materials between the system and its environment. They have finite life spans but they also reproduce themselves, resulting in historical ancestor/descendent lineages. They have a codified system of information storage and retrieval that serves as the basis for building, maintaining, and modifying the CAS. Finally, as a consequence of the last two attributes, both individual organisms and species obey a Malthusian/ Darwinian dynamic. This has four key elements: (1) both birth and speciation are inherently multiplicative processes, producing more offspring than parents and more than can survive; (2) the offspring differ in their characteristics from each other and from their parents; (3) some of this variation is heritable because of the information system; and (4) the environment acts as a filter, so that individuals or species possessing certain traits are more likely to survive and reproduce. The result is descent with adaptive modification of the CAS.

The CAS concept potentially provides a way to reconcile the uniquely biological features of organisms and species with emerging ideas in nonlinear mathematics, complex systems theory, and open-system, nonequilibrium thermodynamics (e.g., Kauffman, 1993; Cowan and Pines, 1994; Schneider and Kay, 1994; see also Odum, 1988). Both individual organisms and species are complex, unique entities that maintain their thermodynamically unlikely organization through the uptake and transformation of energy. They have evolutionary histories of adaptive changes, both in the structure and function of the CASs themselves and in their genetic information systems, that can be traced back to the common origin of all life on earth. During this history, they have become more complex and moved farther from equilibrium. In the process they have transformed not only their local environments (e.g., Grimm, Ch. 1; Power, Ch. 6; Pollock et al., Ch. 12; Lawton and Jones, Ch. 14), but the entire earth, affecting geological processes (e.g., Shachak and Jones, Ch. 27) and changing the concentration of gasses in the atmosphere (e.g., Holland, Ch. 13). The remainder of this chapter expands on this concept of organisms and species as complex adaptive systems, by focusing more explicitly on the relationships between the biological CAS and its environment.

A COMMON ENERGETIC/ THERMODYNAMIC CURRENCY

At present, perhaps the most severe barrier to linking species and ecosystems is the use of two fundamentally different currencies. Most population and community ecologists and evolutionary biologists use a dN/dt currency. The unifying and uniquely biological concept is fitness, which is measured in terms of rates of change in numbers of individuals (N), species, heritable traits, or units of genetic information— rates that ultimately depend on the effect of the environment on the differential survival and reproduction of these entities. In contrast, many ecosystem ecologists and also many behavioral and physiological ecologists use a dE/dt currency. The unifying concepts are the fluxes and transformations of energy (E) and matter as they are governed by physical laws, such as the laws of thermodynamics, chemical stoichiometry, and the periodic table.

The resolution of this problem is to recognize that ultimately dN/dt or fitness can be expressed as an equivalent dE/dt energetic/thermodynamic currency. This would have several advantages. It would avoid problems of near circularity or tautology that have plagued the concept of fitness since Darwin (e.g., Peters, 1976). It would encourage population biologists to explore the relationships between the biological components of fitness and their physical bases. Biological CASs maintain and reproduce their highly organized states by transforming energy to do work. The work is applied to concentrate essential resources from the environment, to build and maintain complex structures, and to avoid physical stresses and biological enemies. Elsewhere (Brown et al., 1993; see also Lotka, 1922, 1925; Odum and Pinkerton, 1955; Odum, 1983; Mansson and McGlade, 1993; Patten, 1993), it has been suggested that fitness can be defined as reproductive power, the rate of transformation of energy into offspring, and that such a redefinition offers new insights into the physiological constraints and evolutionary processes that are associated with body size.

To claim that there must be some energetic/thermodynamic currency of fitness does not imply that energy *per se* is always the proximately limiting resource. For example, water limits desert plants, and nutrients (usually nitrogen or phosphorus) often limit plants and herbivores in terrestrial ecosystems as well as phytoplankton in freshwater and marine ecosystems. But these organisms must expend energy to do the work of acquiring these materials from environments where they are scarce. Because survival and reproduction of all organisms requires expenditure of energy to attain states far from thermodynamic equilibrium, there must ultimately be some common energetic/thermodynamic currency (see also DeAngelis, Ch. 25; for an alternative way to deal with multiple currency, multiflow systems, see Shachak and Jones, Ch. 27). I do not want to underestimate, however, the thorny problem of how to measure empirically the energetic equivalence of other currencies. Often it will not be necessary; it will be sufficient to study the effects of proximally limiting resources on the structure and dynamics of ecological systems. Similarly, it will be adequate to use components of fitness, such as survival, the number and quality of offspring, and rates of phenotypic or genetic change, to study evolutionary phenomena.

However, a common energetic/thermodynamic currency is needed to go beyond the limited framework of traditional ecological and evolutionary studies and to make more explicit the interrelationships between physical and biological processes. This is what is required to develop a general conceptual framework for linking species and ecosystems: to understand how fitness and population dynamics affect and are affected by fluxes and transformations of energy and materials—how the CAS interacts with its biogeochemical environment.

THE MULTIDIMENSIONAL NICHE

Two features of individual organisms and species as CASs are embodied in the classic concept of ecological niche. The first is that each CAS has certain environmental requirements in order to maintain its thermodynamically unlikely state. These requirements can be divided for convenience into two categories: resources and conditions. Resources are particular forms of energy and materials that must be concentrated from the environment and usually altered in some way in order for the CAS to maintain itself and reproduce. Examples include services of essential mutualists, appropriate prey for heterotrophs; and sunlight, nutrients, and sometimes water for autotrophic plants. Conditions are ranges of abiotic and biotic variables that are necessary for the CAS to survive and reproduce. Examples include tolerances for certain ranges of temperature, concentrations of otherwise toxic substances, and levels of biological enemies (competitors, predators, parasites, and diseases).

The multidimensional Hutchinsonian niche (Hutchinson, 1957) characterizes the environmental requirements of each species, and hence also of the individuals of that species. Only a modest number of resources are likely to be in short supply and a small number of conditions are likely to be near the limits of tolerance so that they affect survival and reproduction. These variables are the dimensions of the niche. They limit the abundance and distribution of the species.

Thus, if the concept of the niche is made operational, niches of species can be measured, providing a quantitative representation of the demands that each one places on the ecosystem. The niche can be used to understand the environmental factors that affect abundance and distribution, and therefore to predict changes in the CAS in response to spatial and temporal variation in the environment.

The second feature of both individual organisms and species that is embodied in the concept of niche is the uniqueness of the CAS. By definition, every individual is unique; it is also a tautology that each species is different from all other species. Nevertheless, there has been much discussion recently—in this volume and elsewhere—about "functionally redundant" species. The suggestion that some species are functionally redundant implies that they are ecologically identical to other species, and that they can be lost from a local ecosystem, and even go globally extinct, without any ecological consequences. This has important consequences for basic questions about biological diversity as well as applied issues of conservation.

The distinctive morphology and genetic composition that enables each species to be identified and classified is also reflected in distinctive functional attributes. Each species has a unique niche, and this special set of requirements is reflected in its distinctive pattern of abundance and distribution in both space and time and in its interactions with other species and its abiotic environment. Just like individuals, some species are more similar than others. For certain purposes it may be useful to aggregate similar species into functional groups or other such classes. But to suggest that species are ecologically identical or functionally redundant is incorrect and unwarranted.

WORK, THE IMPACT OF SPECIES ON ECOSYSTEMS

The concept of niche characterizes the effect of the ecosystem on the species: the extent to which the environment meets the requirements for survival and reproduction and thus limits abundance and distribution. In meeting its niche requirements, the CAS does physical work on the ecosystem. In acquiring resources and avoiding intolerable conditions or creating tolerable ones, the species alters the distribution of energy and matter and the composition of materials and other organisms in its environment.

It is through this work that species have their impact on the structure and function of ecosystems. The work may take many forms: mechanical—transporting inanimate materials or other organisms; chemical—active uptake of nutrients or photosynthesis; and biological—selective predation on certain species or protection of other species from physical stress or predation. Ultimately, however, all of these effects represent physical work, because they are accomplished by the transformation of energy. To assess the effects of species on ecosystems requires understanding the nature, magnitude, and consequences of this work.

Unfortunately, I do not see any easy way to make general predictive statements about the impacts of individual species on ecosystems. As much as I would like to hope that ecologists could produce a "field guide to the strong interactors" (S. Carpenter, pers. comm.), I am pessimistic. What traits of kangaroo rats would predict that they would have a greater impact on one shrubland–grassland ecotone than grazing livestock (Brown and Heske, 1990)? And would these same field marks also predict that in other ecosystems the effect of kangaroo rats is much smaller than that of livestock?

The reason for such skepticism is straightforward. The work performed on the ecosystem by each species in meeting its unique niche requirements will also be unique—unique in its type, magnitude, distribution in space and time, and impact both on the abiotic environment and on other organisms. The impact will depend on the particular abiotic conditions and biotic composition of the ecosystem in which the species is embedded—and this environmental setting will almost always be somewhat different. Furthermore, the nonlinearities that are inherent in any complex system of interactions will cause some small effects to be amplified and other large ones to be diminished, and this in turn will make it very difficult to predict outcomes. Within small, highly specified systems—Canham and Pacala's (Ch. 9) forests may be one example—it may be possible to make such predictions, but at the sacrifice of the ability to generalize to other, even superficially similar systems. The impacts of species will be as unique and as dependent on the local environment (including the other organisms as well as the abiotic factors) as the effect of the local environment on the abundance and distribution of the species.

CONCLUSIONS

Both individual organisms and entire species are special biological forms of CASs (Brown, 1994a). The linkages between a species and an ecosystem can be viewed as the interactions between a CAS and its biogeochemical environment. The CAS framework encourages the search for a general physical and mathematical characterization of this interaction. Lotka (1922, 1925) attempted such a synthetic approach to ecology, but it was not pursued by his contemporaries or successors. Probably this was because the physics and mathematics of Lotka's time were able to offer only limited insights into the complexity of ecological systems.

Have things changed? Have recent advances in physics (e.g., nonequilibrium thermodynamics, spin glasses, and self-organized criticality), mathematics (e.g., nonlinear systems), and computer science (e.g., cellular automata, genetic algorithms, neural nets, and artificial life), and in biology (e.g., origin of life, evolution, and ecology) made possible a new, more productive look at the interface of physics and biology? If so, can this also contribute to a synthetic understanding of the relationships between the physical features of ecosystems and their biological components?

On the one hand, I think that there is reason for optimism. We have new conceptual, methodological, and technological tools to study complexity. And I believe that there is a large class of complex systems—physical, biological, social, and computational—that share common emergent features (see Lewin, 1992; Waldrop, 1992; Kauffman, 1993; Cowan and Pines, 1994). Biologists have much to learn by asking how living things both resemble and differ from comparably complex nonliving systems. Biologists also have much to contribute, because the systems that they study—a brain, a rain forest, an ant colony, and a language—represent the epitome of complexity and have special, uniquely biological attributes. Ecologists, in particular, have much to contribute, because the physical and biological processes that enable living things to transform energy to maintain and reproduce complex organizational states far from thermodynamic equilibrium are ecological processes. They involve interactions between organisms and their biogeochemical environments. They should be reflected in general, emergent structural and functional properties of ecological systems. They are central to understanding the origin of life, the development and reproduction of the individual organism, and the relationship between species and ecosystems.

On the other hand, I do not want to create unreasonable expectations. One message from CAS theory is that a few simple rules specifying interactions among a few basic kinds of components can produce an essentially infinite variety of complex structures and dynamics. This helps to explain the uniqueness of individuals and species, and also the uniqueness of the environmental requirements and impacts of these species. But it also makes clear the magnitude of the task that ecologists face—to develop a conceptual framework that incorporates this uniqueness and complexity, but also enables us to identify and understand the common, emergent features. Only then will we be able to make general, predictive statements about the organization of ecological systems.

ACKNOWLEDGMENTS

Discussions with many students and colleagues helped me to develop the ideas in this chapter. The comments of A. Kodric-Brown and two anonymous reviewers helped me to express these thoughts more clearly. The U.S. National Science Foundation supported my research with Grants BSR-8718139 and BSR-8807792.

SCOPE

It is obvious that organisms can affect the functioning of ecosystems, and that the functioning of ecosystems in which species and communities are embedded can affect their dynamics. But how, when, where, and why do species and ecosystems affect each other? The next 11 chapters present studies from marine, freshwater, and terrestrial environments, and the entire biosphere, that illustrate the diversity of these influences and help define the scope for integrating species and ecosystem perspectives. They provide examples of where ecosystems affect species, where individual species or communities affect ecosystems, and some cases where species have no detectable influence on particular ecosystem processes.

Aquatic habitats provide some of the best examples of integrated studies on population and ecosystem processes, as the first five chapters in this section make plain. In Chapter 3, Levinton summarizes the ways in which benthic bioturbating invertebrates largely determine the physical structure and chemical characteristics of their environment. Deposit feeding populations often change sediment water content, stability and particle-size distribution; and via their impacts on oxygenation, alter sediment chemistry and the environment for microbes. On the other hand, Giblin et al. (Ch. 4) argue that overall rates of decomposition in sediments appear to be only weakly affected by benthic animals, but that the converse is not true; the structure of benthic communities is influenced by a key ecosystem process, namely decomposition rates.

Away from the bottom, in the open waters of pelagic marine and freshwater ecosystems, very different processes operate (Silver et al., Ch. 5). The waste products and effluvia (e.g., body parts and carapaces) that constitute marine and freshwater "snow" make a distinct physical and chemical microenvironment that provides a habitat for entire communities, with far-reaching affects on the way the rest of the ecosystem may function, for example by transporting materials and organisms into deeper waters as the snow sinks.

Power (Ch. 6) focuses on disturbances in rivers, and shows how floods alter community structure, particularly trophic networks. The impacts of flood disturbance regimes on trophic structure vary among communities, depending on the attributes of constituent species. In the absence of regular flooding and scouring, patterns of biomass, energy flow, and nitrogen fluxes change markedly. Pace et al. (Ch. 7) provide evidence (from food-web manipulation experiments in entire lakes) for ecosystem-level effects on population dynamics. The variability of zooplankton populations was significantly higher in manipulated lakes. They point out that not only do ecosystem-level changes affect population dynamics, but also that shifts in population dynamics may have strong impacts on ecosystem processes, for example, primary production.

Huntly (Ch. 8) moves us from water to land, focusing on major effects of herbivores on ecosystem processes in space and time. Ecosystem-level phenomena constrain the populations and behavior of consumers, and the effects of consumers often appear to be mediated by changes in ecosystem functions such as productivity, nutrient cycling, and nutrient flows across system boundaries. Patchily distributed herbivores may be particularly important for the heterogeneity they create within ecosystems.

Patchiness of ecosystem processes is also dealt with in Chapters 9 and 10. Canham and Pacala (Ch. 9) use individual based models of plant population dynamics in forests to show how the dynamics of tree populations can largely determine patterns of nutrient cycling; idiosyncratic differences among tree species in traits such as litter quality predispose forests to strong linkages between relative abundance of tree species and spatial and temporal variation in forest ecosystem processes. These authors also show how the presence of a single understorey species, in this case a fern, can determine the entire structure (and, by implication, the functioning) of the forest. Anderson (Ch. 10) deals with the scale dependency of the effects of species, using examples from the soil environment. As the chapter that follows his demonstrates, it is relatively straightforward to quantify nutrient fluxes through soil invertebrate populations (Parmelee, Ch. 11), but as Anderson points out, it has proven to be extremely difficult to show that soil invertebrates regulate carbon and nutrient mineralization, for example, by showing that the activities of particular species determine rates of biogeochemical fluxes at the ecosystem level. He argues that the problem is one of scale in a patchy world, and that effects can appear at one scale but disappear at other, larger scales, because heterogeneities "cancel out."

The last two chapters in this section have been juxtaposed to show a real contrast in the magnitude of species' effects on ecosystems and vice versa. The beaver described by Pollock et al. in Chapter 12 illustrates how a single species can change virtually everything in the environment—energy and material flows, as well as physical and biological structure on the scale of entire drainage basins. Holland (Ch. 13) paints an even bigger canvass, reviewing the way in which the oxygen content of the atmosphere, a direct product of organisms (plants), is actually controlled at the planetary scale by a nonbiological process, the burial of sediments in the oceans. Although organisms produce the oxygen, Holland argues that they do not regulate its equilibrium concentration (currently 0.2 atm). Sometimes, the role of species in ecosystems is surprising and counterintuitive.

3

Bioturbators as Ecosystem Engineers: Control of the Sediment Fabric, Inter-Individual Interactions, and Material Fluxes

Jeffrey Levinton

SUMMARY

Deposit feeders are often strong habitat fabric interactors, altering the structural habitat of marine soft sediments. Through their feeding and burrowing activities, deposit feeding populations often change the sediment water content, near-surface sediment stability, grain size, and spatial distribution of grain sizes. As a result of deposit feeders ingesting sediment, the decomposition of particulate organic matter and microbial activity is also accelerated. Deposit feeders can rapidly turn over the sediment, and derive their nutrition from microbial sources, which are assimilated efficiently, and from particulate organic matter, which is far more refractory. Through alterations of the physical characteristics of the sediment, deposit feeders change sediment chemistry, principally by oxygenating pore waters, which in turn changes the environment for microbes. Nearly all aspects of life in muddy sediment are competitions between the activities of microbes, which usually consume oxygen, and the activities of burrowing sediment consumers, which usually bring oxygen to the sediment by mixing of surface sediment with the overlying water or by irrigating burrows. Alterations of the sediment plus direct interference often combine to cause exclusion of either competing deposit feeding species, or exclusion of functional groups that are ill-equipped to survive in the unstable, watery sediment conditions that are caused by deposit feeding dominance.

At one and the same time, deposit-feeding bioturbators perform the functions of major physical and biological forces in other communities. The vertical fluxes generated are analogous to the major vertical transport of nutrients seen in oceanic water column systems. But the processes seen within guts are analogous to the major contributions that ungulates and other herbivores make in the decomposition and growth of terrestrial plant communities. Deposit feeding bioturbators transport and transform nutrients over the entire physical and perhaps nearly the chemical scale of the soft bottom benthic ecosystem.

INTRODUCTION: SPECIES AND ECOSYSTEMS

Strongly interacting species may affect an ecosystem through two distinct routes. First, species might be *biological interactors* and affect other species by being predators or competitors. In marine ecosystems, predators often exert strong effects on prey species, to the point of devastation. In some cases, predators increase in abundance, change their foraging behavior, and direct the ecosystem into distinct and stable states (e.g., Harrold and Reed, 1985). Both predators and competitors change considerably the relative abundance of species and the nature of food webs.

Alternatively, species may be *habitat fabric interactors*, and alter the physical habitat itself (Lawton and Jones, Ch. 14). Beavers, for example, change the hydrology of streams and radically alter the immediate watershed, nutrient flux, and the structural habitat of aquatic and terrestrial species living in the vicinity. Although beavers cut down trees, of course, their effects on an ecosystem do not involve mainly competitive interactions, or predation on other animal species (see Pollock et al., Ch. 12).

Deposit feeding marine invertebrates burrow in soft sediments and eat some part of the sediment, digesting and assimilating some of the nonliving and living organic matter in the process (see Lopez and Levinton, 1987). Their burrowing and feeding activities alter the fabric of the sediment, which, in turn, changes the environment for the deposit feeders themselves but also for microbial organisms and for other marine benthic species. In the sense of Lawton and Jones (Ch. 14), deposit feeders are therefore *engineers* and affect the habitat fabric of the ecosystem. Deposit feeders, however, interact directly with other species via competition for space and food, and therefore are also biological interactors.

Josiah Wedgewood inspired his nephew, Charles Darwin (1881), to examine carefully the consequences of the activities of earthworms. Darwin's studies over several decades demonstrated that earthworm populations turned over large amounts of earth, which appeared to be beneficial for plants. Worms in densities of hundreds per square meter changed completely the fabric of the soil and introduced large amounts of particulate organic matter by drawing down leaf litter beneath the surface. Marine deposit feeders also consume large amounts of sediment and draw down organic matter beneath the surface, but their activities occur in a sediment whose interstices are filled with water and whose interface is with water, rather than air. It is the purpose of this brief chapter to introduce the reader unfamiliar with deposit feeders to their strong effects on the soft bottom marine ecosystem, both from the point of view of strong biological interactions and habitat fabric interactions. Owing to the brevity of this article, I cannot do justice to the many excellent studies that have led to some of the generalizations made in the sections that follow.

DEPOSIT FEEDERS STRONGLY CONTROL THE PHYSICAL CHARACTERISTICS OF THE SEDIMENT

Deposit feeders (Fig. 3-1) either swallow sedimentary grains whole, or scrape microbial organisms or organic matter from the surfaces. In either case, deposit feeders process large amounts of sediments, often many animal body weights per day (Lopez and Levinton, 1987). The burrowing and feeding activities usually raise the sediment water content, and the formation of copious fecal pellets increases the grain size of the sediment from a fine mud to a fine sand, often oxygenating the sediment in the process (e.g., Rhoads, 1967; Levinton, 1977; Levinton and Lopez, 1977). The establishment of burrows beneath the sediment water interface creates three-dimensional structuring of the sediment ecosystem. The physical stirring of sediment, or *bioturbation*, increases the penetration of oxygen. This usually results in an approximately horizontal interface known as the *redox potential discontinuity* or RPD, which is the border between oxidative reactions above and reducing reactions below (Fig. 3-1). Burrowing bivalves or tube-dwelling polychaetes may extend the RPD as an irregular surface, deep into the sediment. The oxidative and reductive reactions across this interface are facilitated by a variety of bacteria that derive energy from oxidation and reduction of sulfur compounds. Metals involved in sulfur reactions (e.g., Fe, Cd) also exchange across this interface. As a result, bioturbating organisms strongly affect the chemistry of sediment pore waters. Below the RPD, the sediment is anoxic, which allows hydrogen sulfide to persist. By burrowing into the sediment, deposit feeders oxygenate the pore waters and increase the hospitable environment for burrowing organisms that cannot gather dissolved oxygen directly from the sediment surface by means of a siphon or irrigated tube (e.g., Levinton, 1977).

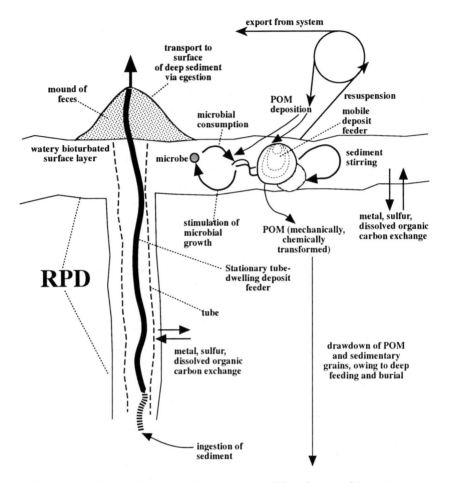

Figure 3-1.   General scheme showing cross section of the sediment and interactions generated by marine deposit feeding invertebrates. RPD = redox potential discontinuity, the interface between dominant oxidative processes above and reducing processes below; POM = particulate organic matter.

Other physical changes of the sediment generated by deposit feeders cause profound changes on the benthos. Deposit feeders increase the water content of the sediment and the interaction of near-bottom flow with watery sediment tends to destabilize the interface and increases turbidity in the boundary layer (Rhoads and Young, 1970). Near-bottom flow may also mix particles in the water column, resulting in an intimate coupling between the sea bed and the water column (Fig. 3-2). Suspension feeders have difficulty living in the watery sediment and may be excluded from muddy sediments dominated by deposit feeders (Rhoads and Young, 1970).

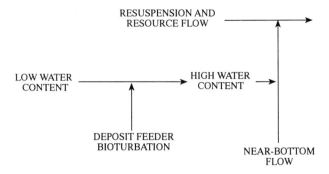

Figure 3-2.   The general effect of deposit feeder generated bioturbation on sediment water content and particle resuspension.

Many deep-feeding deposit feeders consume sediment, and transport it through the gut, to be defecated at the surface. On typical intertidal mud flats and the muddy sea floor, mounds of feces and pseudofeces (material that is collected by a feeding organ, but is rejected to the external environment before entering the gut) create a microtopographical surface that alters near-bottom flow (e.g., Eckman et al., 1981), and create microtopographical high points upon which smaller suspension feeding invertebrates can live (Rhoads and Young, 1971). Pits created by surface deposit feeders may slow flow and facilitate the deposition of particulate organic matter. Deep feeders may consume fine particles, transport them to the surface, and thereby create biogenically graded beds (Rhoads and Stanley, 1965). Whether in permanent burrows or free-burrowing, deposit feeders increase the flux of particulate organic matter downwards (Rice and Rhoads, 1989; Fig. 3-1), as well as dissolved pore water oxygen and a host of inorganic ions (Aller, 1982).

MICROBIAL STRIPPING AND FECAL PELLET FORMATION SUGGESTS RENEWABLE RESOURCE MODELS

Microbial organism abundance appears to be related to sediment particle surface area (e.g., attached bacteria) and the area of the sediment–water interface (benthic diatoms, other photosynthetic microorganisms). The stripping of microbial organisms, followed by recovery of microbial populations to an upper limit dictated by space, suggests a model of resource renewal (Fig. 3-3). As the sediment is pelletized, fecal pellets are often not available until the pellets break down to constituent particles, which also suggests a process involving renewable resources. Theoretically there is also a significant inter-

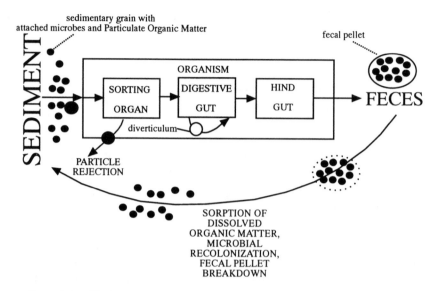

Figure 3-3.   The movement of sedimentary grains between the sediment and the feeding and alimentary systems of a deposit feeder.

action effect (Levinton, 1980). If pellets break down relatively slowly, particles "have time" for microbial recolonization. Thus, steady-state microbial abundance will be greater when pellet breakdown is slower, and when microbial recovery is more rapid. Using these models it is possible to predict the population sizes that might be supported, given known rates of microbial growth and fecal pellet breakdown. Pelletization–pellet breakdown models predict densities of the mud snail *Hydrobia* that fall within typical field densities (Levinton and Lopez, 1977)

Studies (e.g., Levinton and Bianchi, 1981) suggest that bacteria recover too rapidly to be grazed down by deposit feeders. Even at high standing stocks, however, bacteria appear to be insufficient to satisfy the carbon needs of deposit feeders (Cammen, 1980). Benthic diatoms are grazed down by natural population densities and appear to be a limiting food resource for some surface deposit feeders, such as the gastropod *Hydrobia*.

Sediment in intertidal and shallow subtidal muds also has considerable amounts of particulate organic matter (POM), which is probably derived mainly from deposition of decomposing sea grasses and seaweeds (see Marsh and Tenore, 1990). POM-derived material is probably the overwhelming majority of the carbon (C) and nitrogen (N) in intertidal soft muddy marine sediments (Levinton and Stewart, 1988). Several studies have demonstrated, however, that assimilation of this material is very poor, relative to living microbial organisms such as diatoms and bacteria (see Lopez and Levinton, 1987). This has led many to suggest that microbial organisms are the main food source for deposit feeders and the POM is a minor source, owing to its refractory nature. This is the basis for the microbial and particle renewable resource models mentioned earlier.

There is a major problem with this trophic hypothesis. As POM is the source of the overwhelming majority of C and N in muddy sediments, a relatively low assimilation efficiency can provide the same yield as a high assimilation efficiency on the relatively rare microbes (Cammen 1980; Levinton, 1980). An exception to this rule would be a species who could specialize on the diatoms living on the surface (species of the gastropod genus *Hydrobia* probably fit this description). This suggests that deposit feeders, particularly subsurface feeders, must rely upon relatively more particulate sources (see Levinton and Stewart, 1988).

POM DEPOSITS ON THE SEDIMENT–WATER INTERFACE AND APPEARS TO BE A MAJOR SOURCE OF FOOD FOR DEPOSIT-FEEDERS

In *Spartina* salt marsh mud flats, the most conspicuous source of POM would be from decaying *Spartina*, but seaweeds are less refractory and might be a major source of labile and digestible organic matter. Seaweeds such as the sea lettuce *Ulva* spp. grow prolifically on mud flats and are deposited on the sediment–water interface. Their movements can be traced by means of photosynthetic pigments, such as lutein (Levinton and McCartney, 1991). In False Bay, San Juan Island Washington, dense invertebrate populations are associated with dense *Ulva* beds and POM deposition (Levinton and McCartney, 1991). Deposition may be seasonal and some evidence suggests that spring detrital falls are nutritionally richer than POM that is deposited in the fall (Marsh and Tenore, 1990). By feeding an oligochaete ^{14}C-formaldehyde labeled POM, we found a progressive decline in absorption from spring to fall. In the spring, the most nutritionally rich organic matter appears to be near the sediment–water interface, presumably because it had recently been deposited on the sediment surface from the water column (Cheng et al., 1993).

"KEYSTONE DEPOSIT FEEDERS" EXERT STRONG CONTROLS ON OTHER SPECIES BY SEDIMENT MODIFICATION AND EXPLOITATIVE COMPETITION

As discussed above, deposit feeders may modify sediment water content, vertical chemical gradients, and food levels. Deposit feeders also take up space and may monopolize the environment, thus excluding other species. The simplest case is that of tube dwellers, which may inhibit the presence of mobile burrowers or other tube dwellers (Woodin, 1974). In subtidal New England muds the bivalve *Yoldia limatula* interferes with burrow establishment of the bivalve *Solemya velum* (Levinton, 1977). Species living at similar levels below the sediment–water interface may compete for space (Levinton, 1977, Peterson and Andre, 1980).

The mud snail *Ilyanassa obsoleta* dominates mud flats of the eastern United States and efficiently grazes the surface microbial layer of diatoms, while disrupting the sediment. It has strong negative effects on other species, often to the point of exclusion. It interferes directly with the smaller mud snail *Hydrobia truncata* and restricts its range in the intertidal (Levinton et al., 1985). *I. obsoleta* disrupts the sediment and causes emigration of tube-dwelling amphipods (DeWitt and Levinton, 1985). Perhaps by sediment disruption, it also inhibits population growth of the otherwise abundance free burrowing oligochaete *Paranais litoralis* (Levinton and Stewart, 1982).

CONCLUSIONS

The vertical and horizontal sediment stirring of bioturbators is the main driving force behind the transport of organic matter and chemical reactions in sediments. These effects are often mediated by microorganisms. Physical stirring also has a strong effect on species that might occupy the same space and bioturbators, often single species, regulate the species composition of sediment assemblages, both by modifying sediment chemistry and physical structure and by displacing other species. Marine bioturbators therefore are major controllers of sediment ecosystems.

4

Biogeochemical Processes and Marine Benthic Community Structure: Which Follows Which?

Anne E. Giblin, Kenneth H. Foreman,
and Gary T. Banta

SUMMARY

Overall rates of decomposition in sediments appear to be only weakly affected by benthic animals. Thus we predict that changes in benthic community structure would have only a small effect on decomposition although the pathways of decomposition (aerobic vs. anaerobic) might shift. The converse does not appear to be true, however; we expect the structure of benthic communities to change with a change in decomposition rates, with some communities being excluded from sediments with high rates of decomposition.

In contrast, microalgal primary production in shallow benthic sediments may be strongly controlled by the animal community. In many systems, grazers completely consume the benthic algae for at least a portion of the year. We predict that the absence of grazers would have a dramatic effect on rates of primary production but we know less about the effect of the addition or loss of individual species in the benthic community. Large epibenthic animals have the largest effect on primary production in shallow systems whereas smaller infauna seem less able to deplete stocks of benthic microalgae.

There is some evidence that the structure of the animal community alters sulfur cycling in the sediments. In laboratory experiments, the partitioning between oxic metabolism and sulfate reduction shifts with the type of animal community. Sulfur burial over geologic time has decreased, perhaps due to the evolution of deeper burrowing organisms.

Finally, it appears that changes in benthic community structure have a large potential to alter nitrogen dynamics, especially inorganic nitrogen release and denitrification. Because the nitrogen cycle is closely linked with primary production and carbon loading to sediments, it provides a feedback mechanism between the fauna and other biogeochemical processes.

INTRODUCTION

In marine sediments there is often a strong correlation between the rates of processes such as carbon (C) fixation, decomposition, and nitrogen (N) and sulfur (S) cycling and the composition of the animal community. This correlation reveals little about the underlying mechanisms linking biogeochemical processes to community structure. To what extent do differences in sediment biogeochemistry determine which animals are present, and to what extent does the nature of the animal community control the rates and types of biogeochemical processes? The question of "which follows which" becomes critical if we hope to understand and predict the effects of external changes on aquatic ecosystems.

Primary production, decomposition, and element cycling in sediments can be affected by animals through both direct and indirect mechanisms (summarized in Fig. 4-1), (Hargrave, 1970; Anderson and Macfadyen, 1976; Harrison, 1977; Aller, 1982; Levinton, Ch. 3). Animals may reduce primary production by overgrazing and depleting stocks of producers (Fig. 4-1, A and C), but may also stimulate production indirectly by increasing rates of nutrient regeneration and recycling (Fig. 4-1, H → T → V; I → J → T → V). Animals may alter decomposition directly by assimilating and breaking down organic matter in sediments (Fig. 4-1, H). Some species capture suspended matter from the water column, thereby increasing organic matter deposition and sediment metabolism (Fig. 4-1, D → G → P). Animals may affect biogeochemical processes indirectly by (1) irrigating sediments which alters the concentration of oxygen, other electron acceptors, and metabolites in porewaters (Fig. 4-1, K); (2) redistributing organic matter in sediments by bioturbation; (3) grazing microbes (Fig. 4-1, I); (4) physically breaking up particles which changes their surface-to-volume ratio and exposes new surfaces for colonization by microbes; and (5) excreting nutrients.

On the other hand, biogeochemical processes can affect benthic community structure. For example, benthic species composition is profoundly altered in

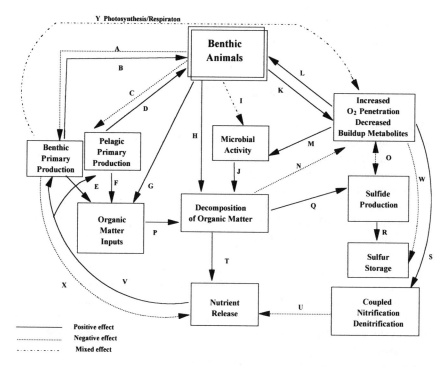

Figure 4-1. Some of the possible interactions between benthic animals and biogeochemical processes in sediments. Interactions may be positive (*solid lines*), negative (*dotted lines*), or mixed (*dashed and dotted lines*; see legend on Fig.). Letters are used to identify arrows which are discussed in the text.

sediments when decomposition rates are stimulated by high rates of organic matter loading. At high decomposition rates, oxygen penetration into the sediment is greatly reduced. Metabolic endproducts that can have direct negative effects on benthic organisms, such as toxic sulfides, accumulate in porewater (Fig. 4-1, P → N → L). Only a few surface-dwelling opportunistic species can colonize such organic-rich sediments. In contrast, sediments with low organic matter inputs typically support diverse communities of longer-lived, deep-burrowing forms (Pearson and Rosenberg, 1978).

Below we discuss several aspects of the interaction between community composition and C, N, and S cycles in unvegetated soft sediments. We ask which of these processes is most strongly regulated by the animal community and which most affect the taxonomic and functional character of the animal community. Although the fluxes we discuss are largely mediated by microbes, we have chosen to ignore the role of microbial community structure in altering these processes because microbes have rapid generation times, high dispersal rates, and a high degree of "functional redundancy" (Lawton and Brown,

1993; Rastetter and Shaver, Ch. 21; Frost et al., Ch. 22) which allows them to adapt quickly to changing conditions.

PRIMARY PRODUCTION

Although it is often assumed that sedimentary processes are driven by allochthonous sources of C, primary production by benthic microphytes can be locally important in shallow coastal sediments. In only a few instances have the controls on benthic microphyte production been well studied (e.g., Admiraal and Peletier, 1980). Nutrients, light, temperature, advection, and grazing have all been suggested as controls. After reviewing data from six seasonal cycles of benthic algal production and biomass, Foreman (1989) concluded that the importance of macrofauna in controlling benthic microphyte production has been underestimated. Foreman (1989) carried out caging and fertilization studies in temperate salt marsh creeks and determined that in this natural system, grazers appeared to limit microalgal production during the summer and fall (Fig. 4-1, A) whereas nutrients became limiting in the winter and spring when grazers are inactive (Fig. 4-1, V). In summer, when grazers were excluded by caging, benthic microalgal production was much higher inside cages than in uncaged areas, and production could be further stimulated within cages by nutrient additions (Fig. 4-2).

The majority of caging experiments have been carried out by excluding relatively large, mobile, "epibenthic" grazers such as fish, crabs, and shrimp. Smaller

Figure 4-2. Community respiration and gross primary production measured in intact sediment cores (mean ± SEM) from both inside and outside of cages. Data are from sites that had been amended with nutrients (fertilized) or left as controls. The caging effect was significant (ANOVA) but the fertilization effect was not. (From Foreman, 1989.)

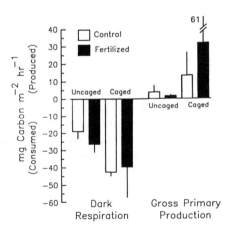

fauna living in the sediment (benthic infauna), such as polychaetes, also consume benthic microphytes (Reise, 1985), but grazing by infauna does not appear to depress benthic microphyte biomass and productivity as much as epifaunal grazing (Admiraal et al., 1985; Foreman, 1989). This suggests that disturbances that strongly affect benthic epifaunal abundance, such as severe winter icings and overfishing, could have large top-down impacts on benthic microphyte production whereas changes in the infauna may have less dramatic effects.

The presence or absence of benthic microphytes can affect other sediment biogeochemical processes. For example, benthic algae provide a source of labile fixed C which can stimulate decomposition (Fig. 4-1, E → P). On the other hand, nutrient flux from the sediment may be intercepted and taken up by benthic algae (Fig. 4-1, X). Benthic microphyte photosynthesis may shift the distribution of aerobic and anaerobic zones in the sediment on diurnal time scales (Nielsen et al., 1990). Alternating aerobic/anaerobic conditions enhances coupled nitrification/denitrification and decreases nutrient release (Fig. 4-1, Y → S → U). Decreased nutrient release from sediments due both to direct benthic algal uptake and enhanced denitrification could substantially alter the relative contributions of benthic vs. pelagic primary producers to overall ecosystem production.

ORGANIC CARBON DECOMPOSITION

Ultimately much of the organic matter produced in the water column, as well as organic matter produced by benthic microphytes, is deposited and decomposed in the benthos (Fig. 4-1, E + F → P). Many of the experiments examining the importance of benthic animals in altering rates of organic matter decomposition have been carried out under fairly artificial conditions using, for example, sieved homogenized sediments (e.g., Kristensen and Blackburn, 1987). When animals have been manipulated under more realistic conditions with intact communities in whole sediment cores (e.g., Banta, 1992), the effects of infauna on decomposition have been much less dramatic.

Banta (1992) suggested that the impact of macrofauna on decomposition depends on the quality and quantity of organic matter present in the sediment. If organic matter is in short supply, animals and microbes compete intensely for organic matter, and sediment respiration rates are similar regardless of whether or not macrofauna are present (Banta, 1992; Kristensen et al., 1992; Fig. 4-1, I negative, J competes with H). At the other end of the spectrum, when sediments are enriched with labile organic matter, animal activity may not further stimulate microbial decomposition because the availability of the substrate to microbes is high and not strongly affected by animal activities (Banta, 1992; Fig. 4-1, I zero). Thus, it may be intermediate levels of organic matter availability where animal activity has the greatest stimulatory effect on microbial respiration as has been shown, at least during short-term laboratory experiments (e.g., Kristensen and Blackburn,1987; Fig. 4-1, I positive).

Canfield (1989) and others have concluded that in the marine environment, the difference in the rate of preservation of organic matter, that is, the percentage of the C input which is permanently buried, is a direct function of the organic matter loading rate to the sediments. As sedimentation rates increase, less of the C reaching the sediments is decomposed and a greater percentage becomes permanently buried. This suggests that the eventual fate of organic matter reaching the sediment is not strongly regulated by the nature of the benthic animal community, but rather by sedimentation rate.

The relationships between organic matter deposition, decomposition, and benthic community structure have been generalized by benthic ecologists (Pearson and Rosenberg, 1978) into a successional scheme that posits that surface dwelling deposit feeders, such as capitellid polychaetes, process organic matter in enriched superficial sediments, increasing oxygen penetration, and making them habitable for larvae of deeper burrowing forms (Fig. 4-1, K → L). These opportunists over-exploit their resources and then crash, making vacant space available for a new wave of settlement by deeper burrowing invertebrates. The correlation between gradients of organic matter and benthic community structure exists in space (Anker, 1975) as well as time (Pearson, 1975). The persistence of opportunistic species in locations experiencing high rates of organic matter loading, and the inability of deeper-living forms to ever colonize these areas, demonstrates that benthic animals have only a limited ability to modify the sedimentary environment.

The large effect that the environment can have on the benthic community, and the limited effects that animals appear to have on organic matter decomposition, lead us to conclude that in this case, it is primarily biogeochemical processes that control animal community structure, not the other way around. Of course we recognize that factors such as competition, recruitment, and predation influence benthic community structure. But at high levels of organic matter loading, there is no doubt that biogeochemical processes such as sulfide production play a major role in determining benthic community structure.

SULFATE REDUCTION AND SULFUR BURIAL

In areas with high rates of organic matter deposition, such as coastal marine sediments, more than half of the C input to the sediments is decomposed using sulfate as the electron acceptor (Canfield, 1989). Sulfate reduction becomes less important as a mode of decomposition in deeper shelf sediments with lower rates of organic matter inputs and is insignificant in deep sea sediments. Some of the the sulfide produced during sulfate reduction is permanently buried as iron sulfide minerals, primarily pyrite (FeS_2) (Fig. 4-1, Q → R). The majority of sulfide, however, is eventually reoxidized back to sulfate and not stored in the sediments(Fig. 4-1,W).

Although benthic animals only weakly affect total sediment metabolism, they could affect the partitioning between oxic and anoxic metabolism by aerating the sediments (Fig. 4-1, K → O) or by mixing labile C into the sulfate reducing zone of the sediment (Fig. 4-1, G → P → Q). S storage is reduced by animals through increased porewater exchange and oxygen penetration which facilitates the oxidation of pyrite and other sulfides within sediments (Fig. 4-1, K → W).

The fossil record suggests that a shift in benthic community composition may have had a large effect on S storage in sediments (Berner, 1984). Early sediments have pyrite/C ratios greater than one, considerably higher than the average of 0.33 observed in modern sediments. This suggests that for the same C inputs, either sulfate reduction made up a higher percentage of sediment metabolism, or sulfide oxidation was lower in the past. One explanation for this shift in S cycling is that deep burrowing fauna capable of extensive sediment reworking were virtually absent in the Paleozoic, and have increased greatly in abundance and diversity over geological time.

Very few experiments have been carried out to determine explicitly the importance of benthic macrofaunal communities to S cycling. Banta (1992) found that small macrofauna in sediments did not affect either the partitioning between sulfate reduction and oxic metabolism or rates of overall metabolism. Kristensen and Blackburn (1987), however, found sulfate reduction was stimulated by the addition of the polychaete, *Nereis*. In contrast, large deep burrowing animals such as the lugworm, *Arenicola*, can greatly decrease sulfate reduction relative to aerobic decomposition (G.T. Banta, M. Holmer, M.H. Jensen, and E. Kristensen, unpublished). The mixed responses of sulfate reduction to different taxa suggest that species with different functional characteristics have quite a different effect on sulfate reduction in sediments and that benthic community structure may strongly influence the S cycle in coastal waters.

DENTRIFICATION AND NITROGEN FLUX FROM SEDIMENTS

Because regeneration of N to the overlying water is quite sensitive to redox conditions within sediments, benthic animals appear to have a greater effect on the form and amount of N released from sediments than on organic C decomposition. Although much of the N mineralized to ammonium during decomposition is released to the overlying water, a substantial amount may also be oxidized to nitrate. A portion of this nitrate is subsequently denitrified within the sediment, that is, converted into N gas which is largely biologically unavailable and leaves the ecosystem. In coastal marine sediments, denitrification is a major sink of N and may remove between 20 and 80% of the N mineralized by decomposition (Seitzinger, 1988).

Benthic animals affect both nitrification and denitrification rates by vertically mixing sediment particles and by building and ventilating burrows (Fig. 4-1, K → S). Laboratory and field data have shown that either the presence or absence of animals can alter denitrification rates and furthermore that the effects of animals may vary with the benthic community structure (Henriksen et al., 1980; Kristensen et al., 1991; Banta, 1992). The presence of small surface tube-dwelling polychaetes (*Mediomastus*) has a minimal effect on both nitrification and dentrification whereas larger animals such as *Nereis* and *Arenicola* can enhance denitrification.

Because benthic community structure shifts with the loading of organic matter to the sediments, there are potential feedbacks and interactions between community structure, N cycling, and ecosystem productivity (Fig. 4-1, K → S → U → V). Studies on the controls of denitrification at the ecosystem level have tended to focus on total N loading to the system and have largely ignored the role of benthic fauna. The feedback between benthic community structure and denitrification deserves further scrutiny.

LINKING BENTHIC ANIMAL SPECIES AND BIOGEOCHEMICAL PROCESSES

Many of the activities of animals that affect sediment biogeochemical processes, such as bioturbation and porewater irrigation, are carried out by a wide range of species, often with representatives in many phyla. The degree of "functional redundancy" (Lawton and Brown, 1993; Rastetter and Shaver, Ch. 21; Frost et al., Ch. 22) of the benthos in this respect is quite high. In the soft marine sediments discussed here, we may find we are able to understand the link between biogeochemistry and benthic animals by ignoring changes at the species level and defining guilds of functionally equivalent taxa, but there is much critical work to be done.

ACKNOWLEDGMENTS

The financial support of NSF OCE-86-15055 (for A.E.G. and G.T.B.) and NSF/EPA/NOAA OCE-89-14729 (for K.H.F) is gratefully acknowledged.

5

MARINE SNOW: WHAT IT IS AND HOW IT AFFECTS ECOSYSTEM FUNCTIONING

M. W. Silver, S. L. Coale, D. K. Steinberg,
and C. H. Pilskaln

SUMMARY

Particulate "snow" is a common feature of marine and fresh waters. Formed by both biological and physical processes, it is a short-lived home for pelagic and substrate-adapted organisms. Snow communities are often dense, serving as important foraging centers for small consumers. The particle matrix is a chemically distinctive microenvironment and a site of intense photosynthesis and mineralization. Sinking snow may also transport its associated populations to depth. The distinctive benthic-like nature of the snow habitat appears to have promoted the evolution of a fauna that adds species diversity to pelagic communities.

WHAT IT IS

Snow is the readily visible, nonliving particulate material of open waters, with individual particles ≥ 0.5 mm. Snow is often quite fragile, breaking apart when collected by standard sampling devices, and thus its study has required the development of new methodologies. Snow can be made directly by organisms or can result from the physical coalescence of smaller organic and inorganic particles. It is best known from oceans and estuaries, but similar material also occurs in fresh waters (Pilskaln and Johnson, 1991; Grossart and Simon, 1993). In the sea, abundances range from 10^{-3} to 10^2 particles/liter and particles occur at all depths (Alldredge and Silver, 1988). Snow received considerable attention from marine scientists starting in the late 1970s. However, much still needs to be learned, because individual particles are difficult to collect with their full consortium of associated organisms and because manipulation of the fragile particles disrupts their physical structure.

Bioengineered Snow

Some snow is produced directly by organisms and can be considered "bioengineered" (Lawton and Jones, Ch. 14). This bioengineered snow includes a variety of mucous structures, often used for food collecting. Well-known marine examples are the mucous webs of molluscan pteropods and the complex "houses" of tunicate larvaceans (Appendicularia). Larvacean houses are some of the most easily identifiable and abundant feeding structures, with multiple houses produced and abandoned each day. Feeding structures are normally highly enriched with food, including living organisms that often remain associated after the structure is abandoned. Mucous feeding structures of zooplankton in lacustrine environments, in contrast, appear relatively uncommon and rarely are sources of snow-like material there. Feces are another major class of bioengineered snow in both marine and fresh waters. Some are membrane-bound pellets whereas others are more amorphous and looser particles. Wastes often have enclosed populations of intact organisms and these include undigested food species and the remnants of gut flora. Disintegrating feces gradually transform into snow with their entrapped flora.

Physically Aggregated Snow

Physically produced "aggregates" are a second common type of snow and are formed by coalescence of smaller particles. The processes controlling aggregation and the physical characteristics of aggregates are presently one of the most active areas of research on snow. The components that aggregate include nonliving organic particles, lithogenic materials, and senescent and healthy cells. A variety of chemical and physical factors control the rates of aggregation. In marine environments, mass flocculation of algae is increasingly recognized and occurs in conditions of waning nutrient levels at the end of blooms (Hill, 1992). Changes in cell surface properties and sinking rates affect coalescence. In fresh water, flocculation can be greatly enhanced in some lakes by sediment inputs (Avnimelech et al., 1982). In lacustrine environments, cell surface properties of photoautotrophs (both algae and cyanobacteria), water chemistry, and chemical composition of suspended sediments, among other factors, control flocculation (Weilenmann et al., 1989).

Recently, aggregates have been recognized to be "fractals"—disorderly shaped structures with noninteger (fractal) dimensions found throughout the natural world. For natural snow from marine, estuarine, and freshwater environments, the measured fractal dimension indicates these particles are highly porous, open lattices, with proportionately less mass per volume as particle size increases (Alldredge and Gotschalk, 1988; Logan and Wilkinson, 1990). The significance of the fractal structure itself, for associated communities, has not yet received much attention. However, one aspect has been discussed: water flow through the porous snow is enhanced, resulting in increased advection past the surfaces of attached microorganisms and thus increased nutrient availability to them (Logan and Alldredge, 1989; Logan and Dettmer, 1990). Thus there is a physiological advantage for microorganisms to associate with snow, as compared with a life style in which movement occurs mainly through bulk fluid motion.

POPULATION AND COMMUNITY ECOLOGY OF SNOW

The presence of snow affects population and community processes. The unusual concentrations of microorganisms that occur here allow the survival of consumers that require dense prey aggregations. Furthermore, some species that are poorly adapted to life in the "plankton" can survive here because of the presence of a floating substrate.

The physical matrix of snow is usually heavily colonized with microorganisms and, occasionally, small invertebrates. These organisms become associated by various mechanisms. Bioengineered snow, as mentioned above, is often colonized when produced. Other organisms may actively immigrate to snow, as a favorable place to feed or live. Yet others become trapped on encountering the sticky surface. Whether the organisms are able to grow and reproduce on a snow particle depends on both the organisms' division rate and the particle's longevity. Some natural snow particles may survive for days (Riebesell, 1992), and these could support multiple divisions of microorganisms. In other cases, such as snow produced from feeding structures, the particles may survive only for hours (M.W. Silver, S.L. Coale, D.K. Steinberg, and C.H. Pilskaln, unpublished). The mass flocculation events observed in both marine and fresh water appear to produce aggregates that sink very rapidly; microorganisms may not be able to divide before being swept from the euphotic zone. The age of individual particles is poorly known and yet critical for understanding the types of adaptations required to successfully utilize these temporary structures.

Substrate-Adapted Species

The "pelagic" habitat is actually a "two-phase" system, at least for smaller organisms. One phase is aqueous, whereas the second is solid—a set of small, disconnected benthic-like habitats. Organisms obtained in water bottles and nets are a collection of both, and the existence of particle-associated populations helps to explain the presence of species adapted for life on substrates. These benthic-like organisms include those with locomotory mechanisms for moving over surfaces, such as the raphe-bearing pennate diatoms so common on marine snow (Silver et al., 1978). Some metazoans, such as copepods, have feeding appendages and locomotory morphology that appear adapted for life on surfaces (Ohtsuka and Kubo, 1991). Alternatively, snow may provide a site for important activities: one of us (D.K. Steinberg, unpublished) has found mating pairs of copepods on snow in mesopelagic waters. These examples suggest there are suites of organisms well designed to spend some stage or most of their lives on particles.

The chemical milieu inside snow differs from that of surrounding waters. Life in interstitial environments has been explored mostly through studies of sediments, and there are likely close parallels between fluid gradients in sedimentary and snow habitats. In sediments, the dissolved constituents in interstitial waters grade from those of concentrations typical of overlying waters to nutrient enriched, reducing, and anoxic conditions inside. Marine snow aggregates likewise have been shown to display similar chemical gradients (Alldredge and Cohen, 1987). A microbiota adapted to such benthic-like conditions probably is a part of the snow community. Comparisons of snow communities with benthic ones have not yet been made, to our knowledge, but may provide useful insights into the origin and adaptations of the particle-associated pelagic community.

Generalist Pelagic Species

Some species on snow appear to be accidental associates. An example includes centric diatoms on abandoned feeding structures in marine systems. Lacking motility, the diatoms may be entangled and unable to escape. Whether they are physiologically "healthy" and dividing may vary with the circumstances, but many species found here may not be adapted to such microenvironments. Other species, particularly consumers, actively immigrate to snow because of dense prey populations: some consumers may even be dependent on snow to obtain sufficient prey to balance their metabolic requirements (Caron et al., 1986).

Special Case: The Microbiota of Feces

Recently voided feces of many organisms house living microbial populations. In the plankton, some membrane-bound fecal pellets possess intact, autofluorescent cells and occasionally motile cells are seen to escape from broken pellets. Fecal pellets from field samples are frequently rich with ultrastructurally intact picoplankton, including *Chlorella*-like species and *Synechococcus*, cells that apparently are digestion resistant (Silver et al., 1986). Likewise, pellets from both surface and deep waters contain abundant intact bacteria: these could originate either as gut flora or be digestion-resistant species from consumed material.

ECOSYSTEM ECOLOGY OF SNOW

More is known about the ecosystem biology of snow particles than about the population or community ecology of organisms associated with snow. Bulk measurements have been performed on snow particles obtained from the field, and have provided insight into average particle characteristics and particle dynamics. These measurements indicate that the presence of snow critically affects the pelagic ecosystem by serving as a metabolic center for microorganisms and by conveying organisms and organic matter into the deep sea.

Biomass and Metabolic Centers

Snow contains highly concentrated populations of microorganisms, often one to three orders of magnitude or more above background levels (Alldredge and Silver, 1988; Grossart and Simon, 1993). Because of high population densities on flocs, processes conducted by associated organisms can be greatly enhanced. For instance, photosynthetic rates can be quite high, leading even to oxygen supersaturation, bubble formation, and, sometimes, rafting of particles to the surface (Riebesell, 1992). However, it is rare for snow particles to be major contributors to water column photosynthesis (Alldredge and Silver, 1988). Commonly marine snow particles are found to have elevated internal ammonia concentrations, likely resulting from intense protozoan grazing and excretion. Likewise, darkened snow particles can have low internal oxygen concentrations, some even anaerobic; very sharp concentration gradients can persist on the particle exterior, even for sinking particles (Alldredge and Cohen, 1987). Intense hydrolytic enzyme activity results from bacterial colonization, implying rapid particle breakup in the absence of further accretion (Smith et al., 1992). The relatively small numbers of snow particles, however, usually result in snow being a minor contributor to total water column photosynthesis or respiration.

Vertical Transport Role

The use of sediment traps for collecting sinking material has shown that snow is a major contributor to mass flux in both the sea and lakes (Pilskaln and Johnson, 1991; Silver and Gowing, 1991). The aggregate flux, in some cases, may be strongly seasonal (e.g., Pilskaln, 1991). Even fecal pellets, long speculated to be major agents of mass flux in the sea, may mostly be tangled in snow (Asper, 1987). In the sea, particulate carbon (C) is lost rapidly with depth; usually 75% of the C leaving the euphotic zone is mineralized by 500 m (Martin et al., 1987). Various mechanisms have been suggested to account for the depth-related losses in particles: these include microbial decomposition, consumption by the midwater fauna, and fragmentation.

Snow vertically transports not only nonliving material but also active populations. Forms adapted to settle out of the water during conditions unfavorable for growth, such as diatom resting spores, sink more quickly when they become associated with snow. It is possible that active populations of organisms also ride particles: this hypothesis is supported by the change in species composition of protozoa with depth (Silver et al., 1984). For some phytoplankton and bacteria, flocculation and the resultant increased sinking rates may enhance nutrient uptake (Logan and Alldredge, 1989; Logan and Dettmer, 1990). For other populations, sinking is detrimental and results in losses, against which mechanisms that allow escape after entrapment in particles may have evolved (Burkholder, 1992).

SUMMARY AND SPECULATION: DOES SNOW MATTER?

In some regions of the sea, especially the oligotrophic gyres, marine snow is thought to occur in very low abundance. In some lakes, flocculation events may be comparatively rare and waters are normally very clear. In the absence of flocs, are aquatic populations or ecosystems fundamentally different? Several decades of research suggest that particles are important.

As discussed above, the presence of snow has some immediate and relatively straightforward consequences. For example, it carries both particulate organic C and associated organisms to depth, supplying food to deep pelagic and benthic communities. This is the "rain of detritus" long discussed in oceanic biology. In some cases, snow in marine and fresh waters can act as a clarifying agent, removing organisms and setting the stage for regrowth by remaining suspended populations. Particularly impressive are mass flocculation events, most frequently described in the freshwater literature. Flocculation can be an episodic event that is catastrophic to the populations removed, but a renewal event for communities and individuals left behind.

On a less immediate scale, the existence of snow can be viewed as having much broader, evolutionary-scale implications. Basically, snow provides an alternative physical and chemical microenvironment, with suites of niches unavailable in the aqueous phase. The existence of alternative niches provides a partial answer to Hutchinson's "Paradox of the Plankton," or why there are so many species of "plankton" in an apparently uniform environment. Snow provides microhabitats that allow the coexistence of additional species adapted to substrates, yet living in the water column.

The abiotic aspects of the niches on particles are unique, including altered concentrations of oxygen, ammonia, and other biologically active substances. The presence of a physical substrate allows adsorption of dissolved materials and provides attachment sites for organisms. As a physical structure, some organisms may find refuges from predators too large to enter the matrix but unable to break it apart or possibly to detect their prey from outside. The morphological features required for organisms to prosper on substrates will also differ; for example, consumers of populations attached to snow require scraping or biting mouthparts, appendages not required for suspension feeders.

In some respects, the analogs to snow communities may best be found in aquatic sediments and terrestrial soils. Communities in these environments share the architecture of a grainy physical matrix with internal open space. Of the three environments, snow is clearly the most porous. The openness may select organisms that are not just creeping forms, but also ones that retain swimming ability. Compared to the sediments and soils, snow is both spatially patchy and short lived—with likely particle life expectancies of hours to weeks. Thus snow-adapted forms must have the ability to disperse to another particle, possibly using chemotaxis to locate the chemically distinctive environments. Adaptations to the snow environment may include rapid growth in primary producers, gorging in consumers, and other strategies for organisms that live in spatially patchy, short-lived, and greatly enriched environments. Such strategies are known for benthic protozoa (Fenchel, 1992) and some freshwater floc colonists (Burkholder, 1992).

Biotic aspects of the snow niches likewise differ from those of the aqueous environment. Organisms that require high food densities to survive may be obligate associates of snow. Likewise, parasites and infective agents may be favored in the dense communities, as shown by the elevated levels of viruses in prokaryotes from snow (Proctor and Fuhrman, 1991). Snow also provides dense populations of small organisms, which can be consumed en masse by larger predators if the entire matrix is ingested. Thus the existence of snow can support larger organisms that "short-circuit" typical, long pelagic food chains by feeding on snow. Overall, snow provides a unique suite of conditions that may allow the persistence of additional species and communities otherwise poorly adapted to life in open waters.

ACKNOWLEDGMENTS

The perspectives developed in this chapter have grown from research supported by NSF OCE grants over the last decade to M.W.S. and an NSF OCE 9015602 award to C.H.P.

6

Floods, Food Chains, and Ecosystem Processes in Rivers

Mary E. Power

SUMMARY

Disturbance regimens link species and ecosystems. For example, floods that scour channels in river drainage networks can also alter trophic networks that link river biota. The impacts of flood disturbance regimens on trophic structure vary among communities, depending on the attributes of constituent species. In the midwestern United States, where algivorous fish are the principal herbivores, floods may spatially rearrange predators and prey among pool habitats, but larger (reach) scale food chain patterns are not affected, or if altered, are rapidly restored. In California rivers, where fish faunas are relatively depauperate, invertebrates are the chief primary consumers. Here, hydrologic disturbance or its absence does affect food chain length at the reach scale. Scouring floods allow weedy invertebrate species to dominate early successional primary consumer guilds. These species are resilient following physical disturbance but subsequently vulnerable to predation. After prolonged low flow during drought, or in regulated channels with artificially stabilized hydrographs, lower trophic levels become dominated by armored or sessile taxa that are relatively invulnerable to predators. In these late successional communities, the biomass of primary producers is chronically suppressed, and energy flow to higher trophic levels appears to attenuate. In addition, other ecosystem functions may be changed, including nitrogen fixation and river–watershed exchange mediated by floating algal mats

INTRODUCTION

Until relatively recently, succession and food webs have been studied separately. Most studies of disturbance (events that remove large portions of the biota from habitats) and subsequent community recovery, or succession, have focused on species at lower trophic levels that are dominant space holders, typically plants or sessile animals. Succession was assumed to be driven by processes acting within trophic levels, such as competition or facilitation. Studies of multitrophic level interactions, on the other hand, generally ignored the temporal context (e.g., the time since the last environmental disturbance) of the food webs in question.

Only recently has the influence of consumer–resource interactions on the rate of succession gained experimental attention (e.g., Wootton, 1990; Farrell, 1991; and references therein Huntly, 1991 and Ch. 8). The impact of disturbance on trophic structure and dynamics has also received relatively little attention, although two distinct lines of reasoning suggest that disturbance should shorten food chains. Menge (1976; Menge and Sutherland, 1976) observed that mobile intertidal predators, such as whelks and starfish, forage more effectively in the absence of wave shock; disturbance frees sessile prey from mobile predators and shortens functional food chains. Pimm and Lawton (1977) and other theorists have found that model food webs with longer food chains are dynamically more fragile (slower to recover from disturbance). They therefore reason that short food chains should predominate in nature, particularly where environments are frequently disturbed.

In ecosystems where attributes of dominant species change over the course of succession, however, disturbance might lengthen food chains. Early successional species are often more palatable, or susceptible, to consumers than late successional species (Cates and Orians, 1975; Porter, 1977; Lubchenco, 1986). If life history tradeoffs cause species (or vegetative regrowth) that first colonize or recover after disturbance to be more vulnerable to predators, and if mobile predators also arrive during these early stages, the impacts of consumers on resources should be strongest during early phases of succession. Afterwards, the energy flow from prey to predators should wane. Hence, food chain length (in both the functional, top-down, population regulation sense, and the descriptive, bottom-up, energy flow sense) could decrease with time since disturbance.

I will illustrate the interplay of disturbance regimen, species traits, and trophic dynamics with case studies from rivers. In California rivers with seasonal, Mediterranean hydrologic regimens, disturbance susceptible grazers (insects) dominate primary consumer guilds. In a contrasting system, Oklahoma rivers under continental climates, floods can occur during any month, and grazers are herbivorous fish that are resistant to floods, but vulnerable to predators throughout their lives.

CASE STUDIES FROM RIVERS

Algal Food Webs with Insect Herbivores: Northern California

Rivers in Mediterranean climates, such as the Eel of northern California, have winter-flood, summer-drought hydrographs. Under these conditions, dramatic seasonal bloom–detachment–senescence cycles of macroalgae occur. Following scouring winter floods in the South Fork Eel River, macroalgae, dominated by the filamentous green alga, *Cladophora glomerata*, recover before animal densities build up. During this window of time, the food chain has only one functionally significant trophic level, and attached *Cladophora* turfs grow up to 8 m long. Toward mid-summer, these turfs detach to form floating mats that cover large portions of the river surface. Mats disintegrate in late summer, and remnants of turfs and mats take on a knotted, webbed architecture produced by heavy infestations of midge larvae (primarily *Pseudochironomus richardsoni*) that live in the algae and weave it into retreats, or "tufts." Midge populations explode shortly after floating algal mats form, and midges and oligochaetes associated with their tufts become the most numerous macroarthropods in the river (Power, 1990a). They crash as the algae disappear, but whether they contribute to this disintegration or simply track it is not apparent from observation.

Field experiments in the summer of 1989 revealed that higher trophic levels could strongly affect the maintenance and the taxonomic composition of attached algal turfs, as well as the production of floating algal mats. In early June, during the early summer bloom, 6 m² pens ("enclosures") built around bedrock or boulders that supported turfs of *Cladophora* were stocked with fish [California roach fry (*Hesperoleucas symmetricus*) and juvenile steelhead (*Oncorhynchus mykiss*)]. In other pens ("exclosures"), all fish were removed. After 5 weeks, algae in enclosures looked very different from algae in exclosures. *Cladophora* turfs in enclosures had collapsed to form a low prostrate mat 1–2 cm high, infested with tuft-weaving midges. In unstocked exclosures, *Cladophora* biomass remained higher, and turfs remained erect and became overgrown with nitrogen (N)-fixing *Nostoc* and an epiphytic diatom (*Epi-*

themia) that contains N-fixing endosymbiotic cyanobacteria. Kilograms of algae floated to the water surface in exclosures; virtually no floating algae were detected in enclosures (Power, 1990b).

Roach and stickleback fry, and large invertebrate predators (primarily damselfly nymphs) colonized unstocked exclosures, but were virtually absent in stocked enclosures with fish. Separate experiments showed that roach fry, stickleback fry, and lestid damselfly nymphs all had strong effects on midges, producing densities approximately one fourth of the midge densities observed in predator-free controls (Power, 1990b). These experiments, by revealing the importance of small predators in the Eel River food chain, showed that fish exerted indirect negative effects on algae mediated through four trophic levels. By suppressing small predators, fish released algivorous midges, with the predicted impact of greatly reducing standing crops of algae. In exclosures without large fish, three-level food chains maintained higher algal standing crops. The overgrowth of *Cladophora* by N-fixing algal taxa indicated increased N limitation, as predicted by theory (Hairston et al., 1960; Fretwell, 1977).

Rampant omnivory was a feature of this food web. For example, large roach fry consume algae and algivorous insects, as well as predatory insects. In 1991, experimental introductions of midge-free *Cladophora* into enclosures showed that the macroalga persisted equally well with roach, with steelhead (pure carnivores), with both, or with neither, supporting the hypothesis that the two-level effects of roach fry as herbivores were less important than their four-level effects as predators of predators. Perhaps more surprising, given the outcome of the experiments, is the observation that roach and steelhead consume many algivores such as mayflies; in fact, algivores make up >60% (by number) of the macroinvertebrates in guts sampled from both fishes. How can trophic cascades showing clear four-level effects be generated in food webs in which omnivory should blur distinctions between trophic levels? A key to this enigma in the Eel River food web is the predator-specific defense of the most abundant algivore, the tuft-weaving midge. Although algal tufts appear to be a completely effective defense for midges against fish, they are only partially effective against predatory invertebrates (Power et al., 1992). Odonates (lestid and aeshnid nymphs) can detect midges within algal tufts, and extract them with "surgical strikes" of their mouthparts. Naucorid bugs detect midges by probing tufts with their beaks. Therefore, when predatory fish eliminate these small predators, they release one guild of algivores that is capable of suppressing algae.

Multi-basin surveys and year-to-year contrasts in food web assembly. Observations of six rivers during 1 year, and of one river over 6 years, provided the opportunity to observe food web changes under contrasting hydrologic regimes. To study the effects of seasonal and hydrologic factors on northern California food webs and algal phenology, I surveyed river biota from 1988 to 1989 in six rivers. Four were unregulated, with a natural winter-flood, summer-drought hydrograph. In these channels, *Cladophora* (the dominant macroalgae in all six rivers) showed its typical bloom–detachment–senescence cycle. In two regulated channels with artificially stable low flow, short viable standing crops of attached *Cladophora* persisted throughout the year. The contrast in *Cladophora* phenology in regulated and unregulated rivers showed that *Cladophora* cycles were extrinsically driven by factors related to the hydrograph.

Survey data on river fauna suggested that flood effects on algae might be mediated through grazers. In regulated rivers, high densities of sessile, cased grazers (e.g., the aquatic moth larva *Petrophila*, the caddisfly *Tinodes*), or mobile grazers with heavy armored cases (e.g., the caddisfly *Glossosoma*), persisted year round. Few predators were observed in these regulated rivers, although the occurrence of isolated individuals indicated that physical-chemical conditions did not preclude them. In unregulated rivers, mobile, nonarmored grazers (e.g., baetid mayflies) initially dominated the fauna in the spring, but their numbers dropped and those of sessile or armored grazers increased later during the low flow season, as predators became more numerous. In a more productive (sunny) river, this transition occurred earlier than in an unproductive (dark) stream (Power 1992).

Further corroboration came during the drought of 1990–1992, when our study site at the South Fork Eel did not experience flooding. Unusually high densities of sessile aquatic moth larvae (*Petrophila confusalis*) and heavy-cased caddis larvae (e.g., *Dicosmoecus gilvipes*) survived through these three winters, and curtailed *Cladophora* blooms the following springs. The summer production of *Cladophora* in 1990, 1991, and 1992 was insufficient to produce the extensive floating mats observed in the South Fork during 1987, 1988, and 1989, when winter floods did occur.

High densities of late-successional grazers and low-standing crops of algae suggest that functionally significant food chains shorten from four to two trophic levels during the prolonged absence of scouring flood disturbance. This interpretation was supported by an experiment conducted during drought in the South Fork Eel, in which exclusion of the large, armored grazer *Dicosmoecus*, a caddisfly, strongly released algae so that floating mats were formed. The impact of steelhead in this experiment was statistically significant (suggesting that steelhead still exerted effects at a fourth trophic level), but was much weaker than the two-level effect of *Dicosmoecus* (J.J. Wootton and M.E. Power, unpublished).

Hydrologic effects on food chains may influence a number of ecosystem-level properties of rivers. If prolonged absence of flood scour leads to a truncated food chain of two functional trophic levels, energy flow to predators should attenuate, and secondary production of fish and other higher trophic levels should diminish. Macroalgae experiencing bloom–detachment–senescence cycles in rivers that periodically scour also provide huge areas of substrate for N-fixing epiphytes, which could significantly enhance fertility in these N-limited waters. Finally, extensive algal mats are food-rich, sun-warmed floating incubators for invertebrates. They increase both the production and the emergence rates of aquatic insects, and are likely to enhance the amount of this production exported from rivers to their watersheds. In the watershed of the South Fork Eel, where the old growth conifer forest vegetation is relatively inedible, this export could be important to terrestrial consumers such as spiders, birds, lizards, and bats.

Floating algal mats, insects, fish, and terrestrial consumers. Algal dynamics in river channels may affect terrestrial consumers by mediating both the rate of aquatic insect production and the amount that is exported from rivers to their watersheds. Algal mat formation coincides with order-of-magnitude increases in the densities of aquatic insects in the Eel River. Rates of emergence of adult insects were three to six times higher from floating mats than from benthic algal turfs (Power, 1990b). Floating mats may serve as refuges for insects from fish predation. In short-term experiments, rates of fish predation were 16 times higher in benthic algal turfs than in floating algal mats (Power, 1990b). As mentioned above, insect growth and development are probably accelerated in floating mats.

Floating algal mats not only increase production of certain insect taxa, but may also route this production from the channel to the land. The function of floating mats as potential valves diverting secondary production between adjacent ecosystem compartments depends on when mats form and how long they last. If algal mats disintegrate and sink before insect larvae emerge, they may serve as time-release capsules of food for fish. If mats last until insects complete their life cycles and emerge as winged adults, they can divert insect production from aquatic to terrestrial consumers. The second function is probably important, because the generation times of mat-dwelling insects such as chironomids are short. We are monitoring lateral penetration by aquatic insects from the river bank to sites up to 200 m back into the forest (M.S. Parker and M.E. Power, unpublished). The ratio of aquatic insects to total insect biomass on both sticky traps and spider webs is predicted to increase in "big algae years" relative to drought years during which floating mat formation is curtailed. The effects of aquatic and terrestrial insect availability on the growth and reproductive success of the spiders (M.S. Parker and M.E. Power 1993 and unpublished) will reveal how algal dynamics in the river affect the fitness of numerous consumers in its watershed. Future collaborations with bat biologists will examine ecological linkages of rivers and their watersheds over larger spatial scales.

Algal Food Webs with Herbivorous Fish: Midwestern Rivers

Grazing minnows, piscivorous bass, and stream algae. Prairie streams of Oklahoma can flood during any month of the year. The dominant herbivores in many of these streams are grazing minnows: *Campostoma anomalum*. In one such stream, Brier Creek of south-central Oklahoma, enormous pool-to-pool variation in algal standing crops occurs. Some pools are filled with filamentous green algae (*Rhizoclonium*, closely related to and to some phycologists, synonymous with *Cladophora*). Other pools appear nearly barren. The barren pools contained schools of *Campostoma* and the green pools lacked these grazers, and contained their bass predators (*Micropterus salmoides*, *M. punctulatus*). During periods of low flow, pools in Brier Creek were well isolated by long shallow riffles, but during floods, these riffles became corridors rather than barriers. Floods redistributed minnows or bass among stream pools, but complementary distributions of bass with minnows, and of minnows with algae, were maintained or reestablished within weeks in new locations. Similarly, when bass were experimentally removed from a green (three-level) pool and minnows were added, pool substrates were grazed to a barren (two-level) state within weeks. Where bass were added to a naturally barren minnow pool, minnows emigrated or were eaten, and algae built up to "green" levels, again within weeks (Power et al., 1985).

Campostoma are thin, soft fish, vulnerable to swimming predators throughout their lives. In contrast, neither adult *Campostoma* nor bass are vulnerable to floods (although their fry can be obliterated), and there is no disturbance-related switch in taxa dominating primary consumer guilds. The relative resilience of food chains following floods in these systems, in contrast to the changes seen in northern California rivers, illustrates that species attributes strongly affect the consequence of disturbance regimes for food chain length.

Fish, snails, and cyanobacteria. In Ozark rivers, *Campostoma* in the deeper channel and snails along shallow river margins both have strong effects on taxonomic composition of producer assemblages. Rock substrates are covered with dense cyanobacterial black felts <1 mm high, which, as our field experiments demonstrated, were maintained by intense grazing. When felt-covered rocks in the Baron Fork of the Illinois River were protected from grazers in in-stream flowing channels, they were overgrown within weeks by turfs of diatoms 8–10 cm high. When transferred to the open stream, these turfs were stripped off by grazing minnows within minutes, and after 11 days of chronic grazing, black felts reappeared. Cyanobacterial felts also developed under chronic grazing on unglazed tiles that had not previously been colonized by river flora. *Calothrix*, a dominant cyanobacterium in the natural stream flora, resists grazing by virtue of its basal growth. Cell division in *Calothrix* is restricted to five to six cells above the basal heterocyst. Grazers therefore remove the distal portions of trichomes and seston, including diatoms capable of overgrowing colonies, but leave behind both the basal regenerative portion of the cyanobacteria and the N-fixing heterocyst. This interaction, which appears analogous with maintenance by grazers of grasslands (e.g., McNaughton, 1984) in some terrestrial ecosystems, has ecosystem-level implications for N loading and fertility of rivers in this region (Power et al., 1988). During floods, *Calothrix* is scoured from rocks. It is not clear that it could reestablish as a dominant unless chronic herbivory favored it over attached green algae and diatoms. Because of the different species traits of both the N fixers and the grazers, flood disturbance may reduce N fixation in Ozark rivers, and enhance it in rivers of northern California. These ideas remain to be tested.

DISCUSSION

These and other experimental studies of river food chains suggest that trophic interactions make a profound difference to the function of river ecosystems. Simple theory (Hairston et al., 1960; Fretwell, 1977) can guide us, via "postdiction" from field experiments, in assessing how many functionally important trophic levels underly the patterns of distribution and abundance of river biota that recur between floods. We still need the much-sought "Field Guide to Strong Interactors " (S. Carpenter, pers. comm.) to predict, from the properties of species making up communities, which will be the strong threads through food webs that link predators or consumers to plants.

Such a guide may prove elusive, because interaction strengths, or impacts in general, of particular species are contingent on their temporal and spatial contexts (Mills et al., 1993). As a consequence, the length of functional food chains can vary in changing or heterogeneous environments, sometimes abruptly. For example, in Panamanian rivers, one can put one's finger on the point at which a two-level food chain dominated by grazing fishes becomes a three-level chain topped by fishing birds. This point occurs as one moves up from deeper heavily grazed substrates to shallow river margins rimmed with "bathtub rings" of algae (Power, 1984; Power et al., 1989). In rivers where lower trophic levels exhibit tradeoffs between resistance to scour and resistance to predation, interaction strengths and food chain lengths change more gradually, as months pass after scouring floods. As argued in the preceding sections, this apparent succession of grazers in northern California rivers allows functional food chains to shorten from four to two levels during drought or in regulated channels. After years of diverted and regulated flow, heavy growths of willow and alder tend to encroach into channels. This could be interpreted as further truncation of the food chain toward one trophic level (woody plants, whose spread and accrual are not suppressed by herbivores). The longer the period of artificial low flow, the larger are the plants in the deactivated channel, and the larger are the flood releases that would be required to restore the natural river system, along with favorable spawning and rearing habitats for top predators such as salmonids.

Better understanding of the consequences of flushing flows for river food chains and river–watershed linkages is crucial for river management and restoration, particularly in massively rearranged water systems such as those of California. In periodically scoured channels, food webs are reset to trophic configurations that divert more energy to higher trophic levels. Flushing flows are a key to geomorphic and food web restoration in rivers, when used with knowledge of how local key species interact with each other and with their physical environment.

ACKNOWLEDGMENTS

I would like to thank Michael Parker, Tim Wootton, Bill Dietrich, Frank Ligon, and Bill Trush for their many contributions to my thinking about river floods and food webs, and for their friendship which makes the work so much fun. This research has been supported by the National Science Foundation (BSR-9100123) and the California State Water Resources Center (W-825).

7

POPULATION VARIABILITY IN EXPERIMENTAL ECOSYSTEMS

Michael L. Pace, Stephen R. Carpenter, and Patricia A. Soranno

SUMMARY

A widespread change in ecosystems has been the alteration of food webs through human impacts on large-bodied predators. These changes should have strong effects on the abundance and the temporal variability of other components of the food web, but there have been few experimental tests at large scales. We evaluated population variability in response to whole-lake food web manipulations using a 7-year time series of zooplankton abundances in three lakes. Food webs were altered by manipulating fish populations in two lakes, and one lake served as a reference system. The variability of zooplankton populations was higher in the manipulated lakes relative to the reference system. The food web manipulations in the two experimental lakes differed in frequency and strength. Population variability was similar, however, in the two systems and not related to the experimental perturbations imposed. The most variable zooplankton populations were those most susceptible to fish predation. Ecosystem level changes clearly affect population dynamics, and shifts in populations can have strong impacts on ecosystem processes such as primary production. Integration of population and ecosystem studies should lead to improved understanding and prediction of the dynamics of ecological systems.

INTRODUCTION

Populations are sensitive to changes in ecosystems. Changes in the abundance and dynamics of species often occur even when there are no shifts in processes such as primary production or nutrient cycling (Schindler, 1987, 1990a). For example, experimental acidification of a lake ecosystem did not result in large differences in primary production or decomposition, but there were major shifts in community composition and in the ability of particular populations to persist in the lake with lowered pH (Schindler et al., 1985) (see also Frost et al., Ch. 22). The sensitivity of populations to ecosystem change is one of the most compelling reasons for a more unified approach to the study of ecological systems (Howarth, 1991; Carpenter et al., 1993). A key question at the interface between population and ecosystem studies concerns the relationship between changes in ecosystems and population variability. If ecosystem change induces increased population variability, there might be increased probabilities of local extinction (Pimm, 1991; but see Schoener and Spiller, 1992). More variable populations may also alter the dynamics of productivity, biogeochemical cycling, and the strength of trophic linkages in ecosystems.

One of the most pervasive changes in ecosystems is the alteration of food webs through human impacts on large-bodied predators. In lakes, shifts in these predator populations may cascade through food webs, altering community structure and ecosystem processes at lower trophic levels (Carpenter et al., 1985). Theory suggests that there should be both strong direct and indirect effects of such manipulations on population abundance, although the direction of these changes may be difficult to predict (Yodzis, 1988). Less attention has been given to how population variability might change as a consequence of food web manipulation (Pimm, 1991). Here, we ask if population variability in zooplankton communities is affected by alterations in food web structure.

We present this analysis as an example of how manipulations at the scale of an entire ecosystem affect populations. We argue that ecosystem studies benefit from considering population level responses and that certain species often have a large effect on ecosystem processes. A challenge for research is to identify which populations are likely to have important impacts on ecosystem processes and then to integrate study of these populations with research designed to address ecosystem level questions.

The Experiments

Ecosystem manipulations were conducted from 1985 through 1990 in Peter and Tuesday Lakes and compared to an unmanipulated reference system, Paul Lake. All systems were monitored in 1984 prior to the manipulations. These lakes and the experimental studies have been extensively described (Carpenter and Kitchell, 1993) and are only briefly sketched here. The reference system,

Paul Lake, contained a largemouth bass population (*Micropterus salmoides*) and a zooplankton community dominated by cladocerans (primarily *Daphnia pulex*). This system remained unaltered throughout the 7-year study.

Two manipulations were conducted in Tuesday Lake. In May 1985, piscivorous largemouth bass were added to Tuesday Lake, a system that had previously contained only planktivorous minnows (several species) and no piscivores. Bass substantially reduced the abundance of minnows, thereby decreasing planktivory. Reduced planktivory was followed by a shift in the zooplankton community to dominance by large cladocerans. At the end of the, 1986 field season, all the bass were removed from Tuesday, ending the first manipulation. The second manipulation returned Tuesday Lake to a minnow-dominated system. The minnow populations increased exponentially from 1987 through 1989. The increase in minnows led to increased planktivory and the eventual return of the zooplankton community to an assemblage dominated by small species (Soranno et al., 1993).

The Peter Lake experiment was a series of planktivore additions made in May of several different years. An assortment of minnows from Tuesday Lake was added in 1985; rainbow trout (*Oncorhynchus mykiss*) were added in 1988 and 1989; and golden shiners (*Notemigonus crysoleucas*) were added in 1990. Each of these additions increased planktivory and resulted in shifts in the zooplankton community. Because of the ontogeny of the introduced fish populations (e.g., rainbow trout shift from planktivory to benthivory and piscivory with development) and the presence of a remnant piscivore population (largemouth bass), increases in planktivory were episodic in Peter Lake. These manipulations, therefore, created a series of pulses of planktivory that influenced the population dynamics and community structure of the zooplankton (Soranno et al., 1993).

These experiments were not replicated. We sacrificed the security of traditional statistical analysis for the advantage of experimentation at the space and time scales actually of interest to ecologists and managers. Interpretations of responses rely on comparisons between ecosystems (outlined below). These interpretations must be considered cautiously given the limited number of systems we studied.

The Questions

We use these experiments and the 7-year time series of zooplankton to address three questions about population variability.

1. Were populations in the experimental systems more variable?
2. Was variation greater in the ecosystem that was perturbed repeatedly (Peter) versus the system where the perturbations were sustained (Tuesday) over a longer period of time?

3. Were the populations targeted by the fish manipulations (i.e., cladocerans) more variable than other populations in the community?

We expected that populations in the experimental systems would be more variable (question 1). It is well known that shifts in planktivory have a strong effect on zooplankton (O'Brien, 1979; Zaret, 1980). In addition, food web manipulations often lead to surprising changes in populations (e.g., Brown et al., 1986; Brown and Heske, 1990). Transitions in community structure should promote increased variability in constituent populations. The expectation with regard to question 2 was less clear. Repeated manipulations as were done in Peter Lake might promote increased variability over sustained manipulations as in Tuesday Lake, given repeated strong direct effects of the manipulations on the target populations. Alternatively, variability could be greater in sustained manipulations, because a variety of new competitive and predatory interactions might develop leading to greater shifts in community structure. Finally, we expected changes in planktivory to lead to greater variability of cladocerans relative to two other groups important in zooplankton communities, rotifers and copepods (question 3). Large cladocerans are particularly sensitive to increases in planktivory and are often replaced by smaller species (Zaret, 1980; Northcote, 1988).

SAMPLING AND ANALYTICAL METHODS

Zooplankton were sampled in each lake weekly from late May to early September and enumerated as described in Soranno et al. (1993). The zooplankton effectively sampled by our methods were rotifers, cladocerans, and copepods. We excluded protozoans and the dipteran larvae *Chaoborus*, because these groups were sampled separately with different methods and over different time periods. Juvenile copepods (nauplii and copepodites) were an abundant and consistent component of our samples, but these stages were not resolved to species. We recognize these are mixtures of species and that our lumping of species probably reduces the overall variability observed for these groups. We conducted analyses with and without juvenile copepods to test the effect of including them as "species" on our conclusions.

Data consisted of weekly estimates of the areal abundance (no. m^{-2}) for each species over 7 years. Time series were log_{10} transformed. The mean and standard deviation (s) for each series was then calculated excluding time points where abundances were zero. Our analysis, therefore, considers data only for species when they were present at detectable abundances (approximately 0.01 individuals L^{-1}). The actual number of observations (N) varied among species, and we analyze the effect of differences in N on our analysis and conclusions below.

RESULTS

Pitfalls Avoided

The measurement of population variability is subject to a number of technical and statistical problems. Measures of variability are often biased. They can be correlated with abundance, body size, and the number of times a population is sampled (Downing, 1986; McArdle et al., 1990; Pimm, 1991). Variability measures may not be comparable when populations are sampled over different spatial and temporal scales (McArdle et al., 1990; Pimm, 1991). In addition, time series of abundances usually require statistical transformation. These transformations can lead to biased estimates of variability and analytical problems (McArdle et al., 1990). Another problem lies in the estimation of population abundance. Many measures are based on indices that are potentially density dependent (e.g., rodents captured in a fixed number of traps). These indices may underestimate abundance at higher densities and consequently underestimate variability (Xia and Boonstra, 1992).

Our study avoided most of these problems (Fig. 7-1). First, our estimates of abundance were not indices, but closer to absolute estimates based on the number of animals captured in a given volume of water sampled. Second, we sampled each lake weekly from May to September over a 7-year period so that comparisons of variability were not affected by temporal differences in sampling. Third, the log transformation we applied successfully normalized the data when expressed as the standard deviation (s) of the transformed series (test of normality for each lake, all $P > .05$). Fourth, variability was not related to body size (Fig. 7-1), and so comparisons across communities with different compositions were not confounded by differences in size-related population dynamics. Finally, our estimates of variability were generally independent of mean abundance and N (Fig. 7-1). The only exceptions were when the data were partitioned into groups (i.e., cladocerans, copepods, rotifers). Significant correlations were observed between N and s for cladocerans ($r = -.57$, $P = .03$) and between rotifer mean abundance and s ($r = .33$, $P = .02$). The negative correlation between N and s for cladocerans is opposite to the expected sign for time series (Pimm, 1991) and is a consequence of species shifts within the cladoceran community, especially in Tuesday Lake (see Soranno et al., 1993). We would expect higher variability in the reference lake relative to the experimental systems if differences in N were determining differences in variability. This is the opposite of what we observed in comparisons of s for cladocerans (see section "Variability of Target Species Compared to Other Species"). In the case of rotifers, mean abundances were not significantly different (t-tests: $P > .05$) between the reference lake Paul and the two experimental lakes. The differences in vari-

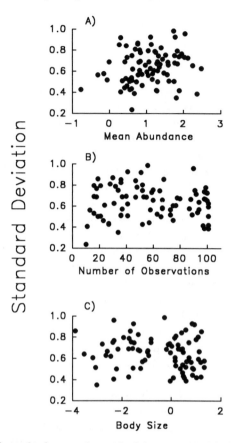

Figure 7-1. Relationship between the standard deviations (s) of the $\log_{10}$ transformed time series for each species in Paul, Peter, and Tuesday Lakes from, 1985 to 1990 and (A) log mean abundances (data are the logs of numbers $\times$ 10^4 m^{-2}), (B) the number of observations (N), and (C) log body size (μg dry weight). Each point represents a species except in the case of copepods, where juveniles were aggregated as nauplii and copepodites.

ability of rotifers we tested for below (see section "Variability of Target Species Compared to Other Species") were not a consequence of differences in mean abundance among the lakes.

Premanipulation Comparisons

We tested if zooplankton population variability in the premanipulation year of 1984 was similar by comparing s among lakes. There was no difference in variability prior to the manipulations (ANOVA: $F_{2,43} = 1.2$, $P = .31$). The reference system Paul actually had the highest mean standard deviation (Tab. 7-1).

Table 7-1. Means of the standard deviations ($\bar{s}$) for the log transformed time series of zooplankton in each lake.

	1984 $\bar{s} \pm 95\%$ CI	1985–1990 $\bar{s} \pm 95\%$ CI
Paul	0.479 ± 0.074	0.543 ± 0.050
Peter	0.436 ± 0.081	0.669 ± 0.062
Tuesday	0.396 ± 0.085	0.709 ± 0.053

Only time series with $N > 10$ were used. 95% confidence intervals (95% CI) of s are also presented.

Comparison of Variability Among Lakes

After, 1984, population variability was higher in Peter and Tuesday Lakes relative to Paul (Fig. 7-2). While mean abundances across lakes were not significantly different (ANOVA: $F_{2,84} = 0.2$, $P = .80$), s was significantly different (ANOVA: $F_{2,84} = 9.5$, $P = .002$). This result was not affected by considering only the more common species ($N > 50$) or by considering only species that were present in nearly all samples ($N > 90$). In both cases mean abundances were not different among lakes while s was again significantly different (ANOVA: for $N > 50$, $F_{2,48} = 8.4$, $P = .0008$; for $N > 90$, $F_{2,19} = 5.6$, $P = .01$). Exclusion of nauplii and copepodites from the data also did not affect the results of the analysis. We conclude that the food web manipulation led to increased population variability in Peter and Tuesday Lakes.

Comparison of Population Variability in the Manipulated Lakes

The distributions of s values in Tuesday and Peter Lakes were quite similar (Fig. 7-2). We used t-tests to analyze for differences in mean abundances and s between the two systems. Mean abundances and s were not different (all $P > .05$) between the two lakes when all data were considered or in the case of more restricted subsets of the data ($N > 50$, $N > 100$, exclusion of juvenile copepods).

Figure 7-2. Box plots of the distribution of standard deviations (s) for Paul, Peter, and Tuesday Lakes (1985–1990). *Open circles* represent the 5th and 95th percentiles. *Solid horizontal lines* represent the 10th, 25th, 50th, 75th, and 90th percentiles. The *dotted line* is the mean.

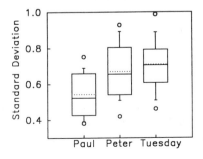

Zooplankton were particularly dynamic in Peter Lake during the years of annual planktivore additions from 1988 to 1990 (see Soranno et al., 1993). We, therefore, further restricted the data to only these last 3 years of the experiment and repeated the comparisons of means and variability (s) between Peter and Tuesday. Once again, however, there were no differences (both t-tests: $P = .9$).

The means of the standard deviations ($\bar{s}$) of all time series in the two lakes were essentially the same (Tab. 7-1). The differences in the experiments between Tuesday and Peter Lakes did not result in differences in the overall variability of populations in these systems. This result is dependent on the time scale of our analysis, which in this case encompasses the entire 6-year experimental period. It is likely that differences in zooplankton variability would be observed at shorter time scales because changes in populations and ecosystem processes (e.g., primary production) in Peter were apparent over shorter periods than in Tuesday (see Carpenter and Kitchell, 1993).

Variability of Target Species Compared to Other Species

One of the most general results of aquatic food web studies is that enhanced planktivory leads to the replacement of large cladocerans by smaller, more cryptic species including small bodied cladocerans (Zaret, 1980; Northcote, 1988). These effects were observed in both Peter and Tuesday Lakes, and we predicted that the relative change in variability would be greatest for the target group— cladocerans relative to copepods and rotifers. We compared the mean values of the standard deviations of time series (s) calculated in this case for each group. The variability of cladocerans was higher in the manipulated systems relative to the reference system (Fig. 7-3). Variability was also higher for rotifers, all copepods, and adult copepods, but the magnitude of these differences between experimental and reference systems was lower than for cladocerans (Fig. 7-3). We tested for differences in s using t-tests to compare reference and experimental systems (Tab. 7-2). Variability as measured by s was significantly higher in the experimental systems ($P < .1$) except for adult copepods (Tab. 7-2).

Our results support the prediction that increases in variability would be greatest for cladocerans, the group most directly affected by the manipulations. The higher variability observed in the nontarget groups indicates that in addition to the direct changes invoked by planktivory, indirect effects arising from the food web manipulations were important in determining population dynamics.

DISCUSSION

It may be argued that the food web manipulations reported here are not really ecosystem experiments but studies of the effects of community perturbations on population variability, a traditional area of community ecology.

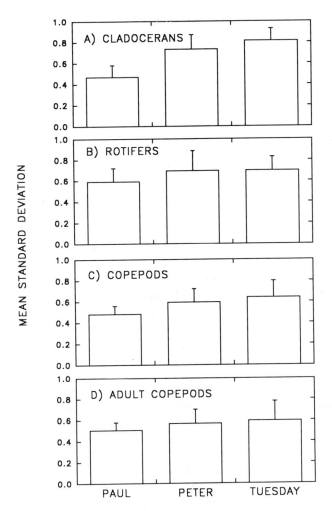

Figure 7-3. Means of the standard deviations for each major taxonomic group of zooplankton in Paul, Peter, and Tuesday Lakes from 1985 to 1990. Error bars represent 1 SD. Note the largest difference between reference and experimental lakes was observed for cladocerans, the group targeted by the manipulations. Smaller increases were observed for rotifers, all copepods, and adult copepods.

This view is incorrect for two reasons. First, an explicit focus of our studies was on how food web structure determines the fluxes of energy and nutrients within lakes. More importantly, our experiments were conducted at the scale of entire lakes where shifts in community structure lead to changes in interactions among populations that ramify through the ecosystem. This type of ecosystem change is important and occurring globally as a consequence of

Table 7-2.   Results of t-tests for differences in the standard deviations (s) of time series for the major groups of zooplankton (1985–1990).

	Paul – Peter			Paul – Tuesday		
	t	df	P	t	df	P
Cladocerans	2.81	7	.026	4.23	7	.004
Rotifers	1.74	28	.094	2.34	31	.026
All Copepods	2.18	16	.045	2.50	12	.028
Adult Copepods	1.05	12	.315	1.10	8	.304

The null hypothesis is that s is similar in the reference (Paul) and experimental (Peter or Tuesday) Lakes.

species invasions (e.g., Alpine and Cloern, 1992). It strikes us as unproductive to restrict the purview of ecosystem science to digging in the dirt, chasing molecules, and slinging radioisotopes around. Rather ecosystem studies must necessarily incorporate populations and communities.

Our analysis demonstrates that increased variability of populations is one consequence of ecosystem manipulations. Cladocerans and rotifers were the most variable species. These zooplankton share the life history traits of parthenogenetic reproduction and short generation times relative to copepods which reproduce sexually and have longer generation times (Allan and Goulden, 1980). As a consequence of these life history traits, rotifers and cladocerans respond rapidly to changes in food web conditions.

Species with high growth rates such as cladocerans and rotifers are considered resilient in that they can recover rapidly from perturbations (Pimm, 1991). Some past studies have argued that high resilience might lead to low variability in populations whereas other studies have suggested resilient populations should be more variable (summarized by Pimm, 1991). Our results indicate that in complex communities resilient species are the most variable. In the two experimental lakes, shifts in food web conditions led to large changes in the predatory and competitive interactions with strong impacts on the population dynamics of cladoceran and rotifers relative to copepods.

The increased variability of rotifers and cladocerans in the experimental lakes were related to shifts in the dominant species within each group (Soranno et al., 1993). For example, during the recovery phase of the Tuesday Lake experiment (1987–1990), a large species, *Daphnia pulex*, remained the dominant cladoceran in 1987. A medium-sized species, *Daphnia rosea* was most abundant in 1988 and then in 1989 and 1990 several smaller species including *Daphnia parvula*, *Bosmina longirostris*, and *Diaphanosoma birgei* were most abundant. These changes were consistent with expectations based on the increased rates of size-selective planktivory from the recovering minnow populations in Tuesday (He et al., 1993; Soranno et al., 1993).

It is generally considered that ecosystem processes such as nutrient recycling or primary productivity will be most sensitive to population variability when those processes are carried out by only a few species (Schindler, 1990a). The zooplankton communities in the lakes we studied, however, were highly redundant in the sense that many species feed on the same or similar resources, and all species regenerate nutrients that limit the growth of phytoplankton. This redundancy of functional roles within the zooplankton might be expected to blunt changes in ecosystem processes related to shifts in the community (see also Frost et al., Ch. 22; Rastetter and Shaver, Ch. 21). Nevertheless, we observed important ecosystem level changes mediated by the zooplankton including both long-term and transitory shifts in algal biomass, primary production, limiting nutrients, and sedimentation (Carpenter et al., 1988, 1993; Sterner et al., 1992). Variability in the populations of large cladocerans was the key to understanding these ecosystem level changes (Carpenter et al., 1993). These species are able to consume a broad array of phytoplankton at high rates and thus play a prominent role in determining fluxes and interactions within the ecosystem. Time series analysis revealed that the size of cladoceran herbivores was a key variable determining the dynamics of phytoplankton in the experimental systems (Carpenter et al., 1993). Much of the time-dependent behavior of ecosystems may be embedded in the dynamics of key populations as exemplified in our study by the importance of large cladocerans.

Population variability is the norm in ecological systems. This variability becomes interesting at an ecosystem level when population fluctuations have strong impacts on biotic and abiotic components of the system. Identifying these key interactions remains an important challenge. One approach that we have found useful is to incorporate the study of population dynamics into ecosystem experimentation. In the past, ecosystem studies often focused on fluxes and budgets of vital elements. Contemporary studies seek to understand how natural and human-induced changes will affect the structure and function of ecosystems. This more dynamic view requires better integration of population studies within the framework of ecosystem questions. A lesson from our experiments has been that the study of populations is a necessary component of understanding variability and change in ecosystems.

ACKNOWLEDGMENTS

Our work was supported by grants DEB 9019873 and DEB 9007196 from the National Science Foundation. This study is a contribution to the program of the Institute of Ecosystem Studies.

8

How Important Are Consumer Species to Ecosystem Functioning?

Nancy Huntly

SUMMARY

Species-level and ecosystem-level perspectives can often be combined to better understand nature. To date relatively few studies have explicitly combined these two perspectives; however, the insights these studies offer have been impressive. Considering species as embedded in ecosystems and as potentially having feedback effects that can alter ecosystem function is a powerful way of conceptualizing and analyzing ecological systems. Ecosystem-level phenomena constrain the populations and behavior of consumers, and the effects of consumers seem often to be mediated by changes in ecosystem functions such as productivity, nutrient cycling, and nutrient flows across ecosystem boundaries. These changes often are caused by nontrophic as well as trophic activities, and they may involve changes in behavior of plants. Just how commonly particular herbivore species will prove to control ecosystem functioning remains to be tested in many systems, using techniques with adequate power to discriminate ecologically meaningful effects.

HOW IMPORTANT ARE CONSUMERS
TO ECOSYSTEM FUNCTIONING?

It is relatively easy to find examples where a particular herbivore or higher level consumer has caused significant shifts in species composition, diversity, productivity, decomposition, nutrient cycling, nutrient loss or input, or geomorphic or hydrologic processes (Huntly, 1991). Nevertheless, the generality of consumer effects on ecosystem dynamics is hotly debated. Do most consumers affect the dynamics of their ecosystems and in species-specific ways, or do relatively few consumers have key effects? Are freshwater ecosystems more commonly influenced by herbivores or trophic cascades than are terrestrial or marine? Are grasslands more strongly influenced by herbivores than are forested ecosystems? Do planktonic systems function differently than benthic systems with respect to consumer effects? What proportion of consumer species significantly mold ecosystem function, at what scale, and by what mechanisms?

Here, I review ways in which consumers influence ecosystem processes and ask how significant consumer species are to ecosystem functioning. My focus is on primary consumers; by inference, higher level consumers that alter the behavior and population dynamics of herbivores may have like effects. These "cascading trophic effects" of higher consumers are discussed elsewhere in this volume (e.g., Grimm, Ch. 1; Power, Ch. 6; Estes, Ch. 15; see also Matson and Hunter, 1992). I use the term "consumer" interchangeably with "herbivore" and "primary consumer."

There have been far more studies of effects of consumers at the ecosystem level in aquatic systems than in terrestrial (for instance, Carpenter, 1988, Carpenter et al., 1991). It is not yet clear that the strong effects demonstrated on species composition, productivity, and nutrient dynamics in these systems are not also present or have not historically been present in terrestrial systems as well. Studies of the soil food web have also often focussed on effects of consumers at the ecosystem level (for instance, Ingham et al., 1989 and references therein; Anderson, Ch. 10; Parmelee, Ch. 11; Bengtsson et al., Ch. 16), but the dominant consumers in these food webs are detritivores rather than herbivores.

The most unequivocal evidence that herbivores affect ecosystem processes comes from experimental manipulations of herbivore populations, either directly excluding herbivores or manipulating higher-level consumers that control size or composition of the herbivore trophic level. These experiments have produced some clear demonstrations of large herbivore effects on ecosystem dynamics (for instance, primary productivity and rates of nutrient cycling) in both terrestrial (Cargill and Jefferies, 1984; McNaughton, 1985; Pollock et al.,

Ch. 12) and aquatic (Carpenter, 1986, 1988; Power, 1990c) ecosystems. In many other cases, alteration of ecosystem function was not directly measured but is implied by extensive changes in community structure (Huntly, 1991).

Another approach has been to use statistical techniques to separate effects of consumers from those of other attributes controlling ecosystem function using observational data. Such studies have shown comparably large and independent effects of herbivores and resources (phosphorus in lakes, Carpenter et al., 1991; nitrogen or rainfall in grasslands, McNaughton, 1985) on primary production and producer standing crop.

There are not enough direct tests to infer how often consumers are important, how important they tend to be, or if there are differences among types of ecosystems or herbivores in kind or strength of effects; however, the studies cited previously do indicate that consumers can have a strong influence on ecosystem dynamics. One can argue that biotic and abiotic components of ecosystems should be viewed as an integrative system (Shackak and Jones, Ch. 27), rather than as driven from below by abiotic factors or primary productivity or as driven from above by consumers. McNaughton (1985) has argued this point particularly well; he suggests that "the physical setting controls the bounds of ecosystem development, and biotic evolution and interactions control ecosystem function within these."

EFFECTS OF ECOSYSTEM STRUCTURE/FUNCTION ON CONSUMERS

The effects of consumers on ecosystem processes commonly are feedback effects. That is, the kinds and numbers of organisms in an ecosystem are fundamentally constrained by various ecosystem-level traits. These limit the number of animals within an environment, and influence their location and activities within it. Both the population density and the behavior of consumers, the way in which they are distributed across and use their habitat, are strongly controlled by ecosystem-level traits.

Most simply, the primary productivity of an ecosystem sets a maximum limit on the population density of consumers that can be supported. This is the basis of Oksanen's and Fretwell's arguments that consumers will strongly limit plant biomass at levels of ecosystem productivity that support high herbivore biomass, but that are not sufficient to support enough carnivores to limit the herbivores (Oksanen et al., 1981).

Variation in primary productivity on a smaller scale also affects the behavior of many herbivores, and thus determines the pattern of their impacts within an area. The degree to which animals concentrate their activities in particular microhabitats is widely appreciated. Work with pocket gophers in

Minnesota has shown that populations, burrow systems, and foraging tunnels all appear to be located in areas of higher than average soil nitrogen, primary productivity, and food availability (Inouye et al., 1987a; Huntly and Inouye, 1988). Price (1992) gives many examples of herbivores, particularly insects, that selectively exploit high-quality resource patches, choosing both among and within individual plants, plant species, and habitat patches. He also poses the interesting hypothesis that eruptive or outbreak species result from population biology that precludes effective resource assessment. If this is the case, it provides a clear example of a species-level trait that translates into a particular pattern of ecosystem-level effects.

Nutritional factors other than productivity or nitrogen availability may similarly affect animal population density and foraging. McNaughton (1988, 1990) has shown that both the spatial location of concentrations of nonmigratory ungulates and the movement patterns of migratory ungulates in the Serengeti are strongly correlated with availability of Na, Mg, and Ca in forage. The historical density and migration patterns of North American ungulates also appear related to mineral availability, particularly Na, Ca, and Mg (Jones and Hanson, 1985). A number of small terrestrial consumers also are reported to be sensitive to differences in mineral availability. In one case, the foraging responses of voles to an experimentally created difference in Na content of vegetation was sufficient to cause local nitrogen enrichment (Inouye et al., 1987b). It is thought that mineral nutrients are not important to diet selection of marine or aquatic herbivores (Lodge, 1991); however, nutrient fractionation and regeneration by consumers can alter which nutrients limit algal productivity (Sterner, Ch. 23).

Other aspects of the environment also constrain both population density and behavior of herbivores. Water sources are foci of animal activities in many terrestrial ecosystems, resulting in greatly altered vegetation and nutrient dynamics. Refugia from predation appear to be a major factor generating pattern in the activities of animals in all sorts of ecosystems (Jefferies and Lawton, 1984). Examples of this include riverine and lacustrine fishes, zooplankton, caterpillars, reef animals such as crabs and fishes, sea urchins, pikas, moose, and numerous other animals that use dens or burrows (Jefferies and Lawton, 1984; Menge et al., 1985; Huntly, 1991).

Habitat fragmentation and isolation also can affect the kinds, numbers, and effects of consumers. Animal species may differ in their abilities to exploit habitat patches of differing size or isolation (Foster and Gaines, 1991). A recent study of tent caterpillar outbreaks in forests in Ontario, Canada, shows that outbreak duration is strongly correlated with forest fragmentation (percent edge) but not with abundance of the primary host plant (Roland, 1993).

The activities of consumers also may be patterned in ways that are largely independent of preexisting patterns of plant resources or the physico-chemical environment. These too may feed back to alter vegetation, productivity, and nutrient dynamics. For instance, the position of animal trails has been shown to vary predictably with animal size and topography (Reichman and Aitchison, 1981). Social interactions such as territoriality and mate location also affect the spatial distributions of consumers and their impacts.

HOW DO CONSUMERS AFFECT ECOSYSTEM FUNCTIONING?

Consumers are constrained by the ecosystems they inhabit, but they also are capable of modifying those ecosystems. Consumers eat plants or plant parts, but they also do many other things. Many herbivores clip or tear and waste plant tissues, providing inputs of green litter and detritus. Various herbivores also trample, scrape, or bulldoze plants. In addition consumers affect the physico-chemical environment in a variety of ways. They form trails, nests, mounds, burrows, and wallows; and they move, mix, and structure soils, sediments, and other materials.

Below I distinguish trophic effects, the effects that arise directly or indirectly from removal of biomass by herbivores, from nontrophic effects, those that are caused by other things that herbivores do. Trophic effects are reasonably well dealt with in theory and have been the focus of the majority of studies of herbivores' effects on ecosystems. Nontrophic effects are far less well studied and little theory has been developed for them. Many nontrophic effects are of the sort that Lawton and Jones (Ch. 14) term ecosystem engineering.

I also suggest that herbivores frequently are a cause of spatial and temporal structure in environmental conditions. Both trophic and nontrophic activities contribute to production of structure, which provides opportunities for different sorts of primary producers to persist in communities, often changing species composition, tending to increase diversity, and probably affecting spatial and temporal stability of ecosystem productivity and nutrient dynamics.

Trophic Effects

Herbivores clearly often affect the species composition of the plant communities they inhabit (Huntly, 1991). Plant species differ in their growth phenology, maximal growth rates, typical size and allocation patterns, nutrient and water uptake and retention, and litter quality and quantity (Pastor et al., 1993; Wedin, Ch. 24). Thus, herbivores that alter plant community composition should often alter ecosystem dynamics. Moderate grazing frequently produces more productive communities, and high grazing can result in conversion to a much less productive system.

In addition to removing plant biomass, herbivores process that biomass and return fractions of it as feces, frass, urine, or other excretions. These herbivore-mediated chemical conversions can have significant effects on productivity and on the nutrient supply to plants (McNaughton, 1985; Bianchi and Jones, 1991; Huntly, 1991; Sterner, Ch. 23).

Although less often considered, herbivores also may affect the behavior of a plant, that is, they may change the way in which a plant of a particular species functions and interacts with other species or abiotic components of its environment. By using the term "behavior," I intend to point out the conceptual similarities to processes well appreciated for animals with predators: a predator can change the abundance of a prey population and thereby change that species' quantitative effects on other species or ecosystem components but not change the rules describing those interactions. Alternatively, a predator can cause change in the behavior of its prey, change in the way in which it interacts with its environment (Kotler and Holt, 1989; Huntly, 1991). Examples of this include consumer-induced changes in plant chemistry or form that alter the rate or phenology of nutrient, water, or light use or the litter loss characteristics. These changes appear to be very common (Carpenter, 1988; Brown and Gange, 1990; Louda et al., 1990) but have rarely been studied explicitly in the context of consumer effects on ecosystems, particularly for terrestrial ecosystems. Holland et al. (1992) recently modelled this sort of plant behavioral change, focussing on changes in allocation in terrestrial plants, which they concluded may be more important in stimulating productivity of terrestrial systems than the more commonly studied herbivore inputs of dung and urine. Clearly more work is needed on this aspect of herbivore effects on ecosystems.

The effects of herbivores, even when caused by straightforward trophic interactions, may be manifested through landscape-level dynamics. An example may illustrate what I mean by plant behavior being a critical element of herbivore effects on ecosystems, as well as showing the value of landscape-level considerations in what might at first be considered a smaller-scale question. The northern range of Yellowstone National Park is grazed by elk and bison, which currently number roughly 20,000 and 500, respectively. The effects of these large native grazers on their range is of continuing interest, and Frank and McNaughton (1992) recently showed that grazing of grassland sites either increased or did not affect short-term primary productivity. The effects of these migratory grazers, however, have a larger scale context as well (see Turner and O'Neill, Ch. 19). My colleagues R. Inouye, D. Frank, W. Minshall, J. Anderson, and I hypothesize that the exchanges of nutrients between upland grasslands, riparian areas, and streams are affected by the large grazers, which not only graze uplands, but also alter the form and phenology of the dominant riparian vegetation, particularly willows and cottonwoods. The balance of nutrient flows among these system components should be very important in determining the net effects of consumers on the long-term productivity of the ecosystem they depend on and influence. We are investigating this scenario, which requires constructing nutrient budgets at several spatial scales and at various herbivore densities.

Other sorts of consumers also affect nutrient flows between terrestrial and aquatic ecosystems. Emergence of terrestrial adult forms of insects or amphibians that have aquatic larval forms can provide nutrient transfers (Seale, 1980; Power, Ch. 6), and terrestrial consumers can alter the quantity and quality of litter inputs to streams (e.g., Irons et al., 1991). Consumers that graze or otherwise disrupt aquatic macrophytes can affect fluxes between pelagic, littoral, and shore regions (Carpenter and Lodge, 1986; Carpenter, 1988).

Nontrophic Effects

Nontrophic effects result from direct effects of consumers on the physico-chemical environment (such as moving and structuring soil, sediment, or other materials; burrowing, forming trails) as well as from their nontrophic effects on plants (trampling; clipping, scraping, or cutting but not eating plants, or otherwise causing greenfall). Although experiments in which herbivores are removed from areas are often interpreted as showing effects of herbivory, these in fact show net effects of herbivores, including all of their activities.

These nontrophic effects, many of which fall into the category that Lawton and Jones (Ch. 14) call "allogenic engineering," are far less well studied than are trophic effects; a preliminary conceptual framework for understanding them is laid out by Jones et al. (1994). Certainly the nontrophic effects of beavers are responsible for a major portion of their influence on ecosystems, although they also forage selectively and alter plant species composition (Pollock et al., Ch. 12). Many consumers may have similarly far-reaching nontrophic effects. Bertness (1984a) suggests that snails such as *Littorina littorea* limit formation of shallow soft-bottom habitat and marshes, as a result of bulldozing sediment while grazing. Numerous consumers influence soils by mounding, mixing, and tunneling, and these affect aeration, water movement, topographic relief, and nutrient dynamics (Hole, 1981). Even such small invertebrates as scarab beetle larvae can move substantial amounts of soil (Kalisz and Stone, 1984). The geomorphology and sediments of marine and aquatic ecosystems are similarly affected by consumers (Huntly, 1991; Levinton, Ch. 3).

Engineering of the physical environment is a bit harder to visualize for planktonic systems, but effects of consumers on macrophytes clearly provide environmental structure that influences planktonic communities via its effects on fish and zooplankton and that may have additional effects via changes in lake chemical and physical properties (Carpenter and Lodge, 1986; Carpenter, 1988). Phytoplankton intercept light, influencing the light environment, water temperatures, and mixing patterns of marine and fresh waters; and zooplankton produce fecal pellets that may sink and transfer nutrients to deeper aquatic subsystems (Jones et al., 1994). Also, generation of chemical patches may be analogous structural engineering in planktonic systems (Sterner, Ch. 23).

Trophic and nontrophic effects may be intertwined, as in the case of fishes that scrape sediment from algal-covered rocks, thus retaining a more productive community to exploit (Power, 1990c) or of pocket gophers whose mounds maintain annual plants of relatively high productivity and nutrient content in fields and meadows (Huntly and Inouye, 1988). In these cases, the nontrophic effects of animals have potential fitness paybacks in the form of trophic rewards.

Environmental Structure

Herbivores often, through either trophic or nontrophic pathways, strongly modify the spatial and temporal structure of the environment, and this may be a major overlooked way in which ecosystem function is patterned by consumers (Grimm, Ch. 1). Environments can be viewed as having structure of various sorts: they have spatial patches that differ in their characteristics and a given area may change in its characteristics seasonally and among years. Habitat quality for organisms thus varies in space and time, and the feedback effects of these organisms on ecosystem function will also vary.

Variation in the suitability of habitat patches in space and in time has strong effects on populations and communities. Environmental variation provides opportunities for populations to respond in nonlinear or nonadditive ways that result in species abundance patterns that are not simple averages over habitat conditions (Chesson and Huntly, 1989, 1993). It seems that ecosystem dynamics should also be affected by the sorts of environmental structure that consumers create.

Disturbance—spatiotemporal structure. The effects of herbivores in ecosystems have sometimes been conceptualized as disturbance, as herbivores may physically disrupt patches or may consume plants in a patchy fashion. Many of these are among the most conspicuous effects of herbivores, for instance gopher mounds, or trampled areas of high animal activity, or patches scraped clear of algae or sediment. However, consumers produce many less conspicuous sorts of disturbances, and these tend to be overlooked. Vole runways provide a good example of relatively inconspicuous disturbed patches that animals create. Although these are small (typically a few centimeters wide) and disappear quickly in the absence of voles, their effects may be much more significant than their size and longevity would suggest. Ericson et al. (1992)

report that more than 90% of establishment and survival of willow seedlings in a successional field in Sweden occurred in vole runways. Furthermore, establishment was largely limited to years following a vole population peak.

Spatial structure.   Many environmental characteristics cause herbivore feedbacks resulting in a relatively long-term spatial structure to the environment. Refugia around which animals concentrate their actions provide a common example. Also, less frequently but intensively used areas such as buffalo wallows may provide relatively permanent patches of differing soil and plant characteristics and dynamics. Gopher mounds may differ in soil characteristics from surrounding undisturbed areas and therefore be persistently different patches (Inouye et al., 1987a).

The feeding activities of herbivores may in themselves cause feedback that results in the development of persistently grazed patches of higher productivity than adjacent ungrazed patches within a previously more homogeneous environment. This is reported for a wide variety of consumers, including domestic livestock, native free-ranging ungulates, zooplankton, crayfish, small mammal and insect species, snow geese, fishes, and a number of reef animals (McNaughton, 1985; Bianchi and Jones, 1991; Huntly, 1991).

Temporal structure.   Many herbivores, both vertebrate and invertebrate, and including marine, freshwater, and terrestrial animals, have large population fluctuations. Both their population lows and their population highs can have lasting effects on ecosystem functioning (see also Pace et al., Ch. 7). Ecologists have sometimes incidentally observed a consumer population high that resulted in a major shift of species composition and presumably ecosystem function (e.g., Berdowsky and Zeilinga, 1987; Noy-Meir, 1988). Forest insects with outbreak population dynamics are known to have these sorts of effects (Schowalter et al., 1986), but in many other cases, the effects were entirely unexpected and the fact that a consumer was responsible for the change in system state would be difficult to infer in retrospect.

Research on blacktailed jackrabbits in Idaho sage-steppe provides an interesting example (A. Porth, N. Huntly, and J. Anderson, unpublished). The population peaked in fall 1992, and, shortly after snowfall, the rabbits disappeared from most areas of our 2,300 km² study area. We subsequently located radio-collared animals in an area of several square kilometers in which several thousand rabbits were present. The jackrabbits browsed virtually all sagebrush in this area and attracted high concentrations of predators. Within 2 months, the population had collapsed, and the snow was covered with patches of blood, carcasses, fecal pellets, and urine spots. Animals that had fed and gained weight over a large area of sage-steppe had concentrated and died in a few small areas, resulting in large imports of nitrogen and other nutrients into the concentration areas. Clearly, this phenomenon can rarely be observed, but it may occur commonly during the roughly decadal population peaks and could have long-term effects on moderate-scale pattern of vegetation and soil fertility.

In a number of cases, plant species appear to have established only during periods of low herbivore density (Huntly, 1991). Presumably the herbivores are usually sufficient to prevent establishment of seedlings or germlings of these species, but once established, the plants can persist in the face of herbivory and their characteristics affect ecosystem dynamics.

Also, the effects of consumers on ecosystems may depend on particular weather conditions and therefore arise only when both the consumers and the physical conditions are appropriate. Hobbs and Mooney (1991) showed an effect of pocket gophers on vegetation that occurred only with sufficient rainfall, and Dayton (1985) notes effects of weather on both consumers and algal propagules that control establishment vs. demolition of kelp beds. For long-lived plants or plants with long-lived propagules, a small window of time is sufficient to produce lasting effects on their populations and any ecosystem-level effects they may have.

HOW CAN WE BETTER UNDERSTAND THE IMPORTANCE OF CONSUMERS?

I have discussed ways in which herbivores can significantly influence the structure and function of ecosystems and have suggested that it may be particularly interesting to consider (1) effects of consumers on plant behavior, (2) the influence of consumers at a landscape level, (3) nontrophic effects of consumers, and (4) effects of consumers on ecosystem structure and its consequences to ecosystem processes. The empirical data base is not yet sufficient to resolve when, where, why, and how much consumers matter to ecosystem functioning. The following are practical suggestions for how to obtain better data for answering these questions. These suggestions do not concern only methodology, rather they involve important conceptual issues as well.

A Long-Term Perspective

Both the physical environment and the population dynamics of species vary over time. Herbivores that are only occasionally abundant, because their dynamics are cyclic, because their abundance depends on certain weather conditions, or because their primary food is only occasionally abundant, may nevertheless have strong and lasting effects on ecosystem function. Agents of successful biological control certainly illustrate this, as do a number of examples cited earlier. Furthermore, the physical environment varies in time due to weather patterns and geomorphic processes, and the life spans of primary producers and higher-level consumers constrain the rates at which consumer effects can be seen. An appreciation of consumer effects on ecosystem functioning must take this long-term environmental framework into account.

Although all studies cannot be long-term, it is necessary to estimate the probable long-term dynamics of a system. A careful analysis and linkage of organismal biology, the physical environment, and various feedbacks among these may often allow reasonable extrapolation from the shorter-term work we must do to the longer-term context we seek to understand (e.g., Chesson and Huntly, 1989; Likens, 1989; Huntly, 1991). Perhaps the need to understand typical long-term dynamics, rather than simply current function, is an important reason for putting species biology into ecosystem models.

Comparing Relative Effects

Absolute rate of change is often confused with strength of effect. This is arguably at the root of many disagreements over the importance of consumers in different ecosystems. More productive systems can show larger absolute changes in biomass over a given time interval; they also can show faster responses to a given perturbation. Ecosystems in which the organisms have inherently faster growth rates, such as pelagic systems dominated by phytoplankton, as opposed to meadows or forest, in which the dominant primary producers have life spans of decades and centuries, also can show more rapid changes in species composition and nutrient dynamics. However, more rapid change should not be mistaken for evidence of larger effects. Rather, the relevant response for comparisons of importance of herbivores among ecosytems is proportional to relative change in productivity, nutrient flux, or whatever variable is being compared.

The Concepts of Power and Effect Size

Statistics are now widely used to guide interpretation of ecological data. Statistical significance levels (for instance, $P < .05$) are expected to be reported in ecological studies, but the arbitrary nature of this statistic is poorly appreciated. In particular, the interrelated concepts of power of a test and effect size are often ignored. The power of a test, the probability of rejecting a false null hypothesis (or detecting a real effect), is a measure of how strong a test is. A very powerful test, one with large sample size relative to the background variation, can detect very small effects. Thus, an effect may be highly statistically significant, but account for a tiny fraction of the variation in the data; it may very well not be ecologically significant. To find no effect is trivial when a test has too little power. Just what effect size is ecologically meaningful is open to differences of opinion, but to consider statistical significance alone, without consideration of power, is to abdicate responsibility for thinking about what size effects are ecologically meaningful. These issues apply to interpretations of surveys as well as of experiments and are discussed in detail elsewhere in the ecological literature (e.g., Toft and Shea, 1983; Yoccoz, 1991) and in many experimental design texts (e.g., Winer, 1971; Steele and Torrie, 1980), which present how to calculate such useful things as power of a test, sample size needed to detect an effect of a particular size, or degree to which an effect occurs.

Animal Behavior and Ecosystem Effects

I am continually impressed with the necessity to consider animal behavior when designing schemes for study of animals' influences on the ecosystems they inhabit. Random placement of samples or exclosures is probably rarely the best way to study these systems, because the effects of animals are likely to vary spatially. When animals have effects that vary systematically from place to place, a design in which relevant sorts of habitats or patches are treated as blocks will provide much more power to detect those effects. A good knowledge of species biology informs these sampling decisions.

A study of effects of pikas on vegetation illustrates this concern (Huntly, 1987). The feeding preferences and selectivity of pikas varied predictably along a habitat gradient, distance from talus. Pikas grazed more heavily but less selectively near talus. Exclosures that were blocked with respect to distance from talus demonstrated that pikas significantly affected vegetation in the area adjoining talus. However, because the animals had different effects at different locations, an analysis of all exclosures pooled would not detect those effects, which effectively cancelled each other out in the pooled data.

ACKNOWLEDGMENTS

I appreciate support of my research by the National Science Foundation, Idaho State University, and the U.S. Department of Energy. I thank Eric Burr, Dan Lawson, and Adam Porth for helpful comments on the manuscript.

9

Linking Tree Population Dynamics and Forest Ecosystem Processes

Charles D. Canham and Stephen W. Pacala

SUMMARY

A new, empirically based, spatially explicit model (SORTIE) predicts that tree population dynamics in forests of southern New England are (1) nonequilibrial, (2) strongly dependent on species composition immediately following disturbance, and (3) highly spatially aggregated. The population dynamics predicted by SORTIE are the result of a clear set of physiological, morphological, and life-history traits of the individual tree species. However, those traits generally do not depend on or determine the species effects on productivity, hydrology, or nutrient cycling. As a result, idiosyncratic differences among tree species in traits such as litter quality predispose these forests to strong linkages between the relative abundance of tree species and spatial and temporal variation in forest ecosystem processes. It is clear that energetics and biogeochemistry place constraints on forest ecosystem processes, and that the rates of many ecosystem processes are strongly influenced by physical, environmental factors. However, in the absence of evidence of stronger control of tree population dynamics by ecosystem processes and physical factors, we conclude that the spatial and temporal dynamics of productivity, nutrient cycling, and hydrology in these forests will be largely driven by the interactions between microbes, plants, and animals—often acting in extremely local neighborhoods—that determine tree population dynamics. Thus, ecosystem dynamics in these forests should be characterized by the same nonequilibrial, spatially aggregated dynamics exhibited by tree population dynamics.

INTRODUCTION

Models of forest ecosystem dynamics can be divided into two general classes: models that ignore the dynamics of component species and that seek the mechanisms regulating ecosystem processes in physiology, biogeochemistry, and energetics (e.g., Parton et al., 1988; Raich et al., 1991; Rastetter et al., 1991); and models of the population dynamics of major forest species, in which ecosystem dynamics are both driven by and potentially exert feedbacks on changes in species composition (e.g. Shugart, 1984; Pastor and Post, 1986). Inherent in this second approach is the assumption that ecosystem dynamics represent the aggregate outcome of the plant–plant and plant–animal interactions—often acting on extremely small spatial scales in local neighborhoods—that govern community composition. Our goal in this chapter is to outline a series of results from a new, spatially explicit model of tree population dynamics (SORTIE, Pacala et al., 1993). We will describe the basic structure of the model and the empirical data used to parameterize the model for the transition oak–northern hardwood forests of central New England, and then outline the model's basic predictions about tree population dynamics. We will then summarize our ongoing studies of the effects of different tree species on productivity, hydrology, and nitrogen (N) cycling within these forests, and suggest potential forms of feedbacks between these processes and tree population dynamics. Our basic theme is that many aspects of forest ecosystem dynamics can best be viewed as the aggregation of interspecific differences in individual plant and animal effects on ecosystem processes (also see Turner and O'Neill, Ch. 19; Schimel et al., Ch. 20), and that the mechanisms that underlie many of the spatial and temporal dynamics studied by ecosystem scientists are best sought at the population and community level.

THE SORTIE MODEL

The basic structure of SORTIE is conceptually very simple (Pacala et al., 1993): tree population dynamics are described by four submodels that predict (1) recruitment of new seedlings as a function of the spatial distribution of adults; (2) growth of seedlings, saplings, and mature trees as a function of local resource availability (i.e., light in the version of the model used for this chapter); (3) mortality as a function of growth rate; and (4) local resource availability as a function of the distribution of plants in the forest. The recruitment submodel predicts the spatial distribution and density of new seedlings of each tree species from empirical relationships between the densities of new seedlings and the distribution of adults (or adult females for dioecious species such as *Fraxinus americana*) (E. Ribbens, J.A. Silander, and S.W. Pacala, unpublished). The growth submodel predicts the diameter and height growth of individuals from empirical relationships between growth and local (individual plant) measurements of light availability (Pacala et al., 1993). The rationale for the mortality submodel is derived from our observations that mortality, particularly for seedlings and saplings, is a predictable function of recent growth history (R.K. Kobe, S.W. Pacala, J.A. Silander, and C.D. Canham, unpublished). The mortality submodel was parameterized using measurements of average diameter growth for the previous 5 years in both live and recently dead saplings. The resource (light) submodel predicts the seasonal average light level experienced by each individual seedling, sapling, and adult tree as a function of (1) the distribution of neighbors that are taller than the target plant (including their specific location, size, and crown geometry); (2) empirically derived, species-specific light extinction coefficients; and (3) solar geometry at the latitude of the study sites (Canham et al., 1994).

The spatially explicit nature of the model allows us to incorporate processeses such as recruitment limitation that are ignored in models such as FORET (Shugart, 1984) and JABOWA (Botkin, 1992). It also allows us to mechanistically model neighborhood competition for light by specifically incorporating the effects of both solar geometry and tree geometry on spatial variability in gap and understory light levels. Virtually all of the parameters required for the model have been estimated from field research at Great Mountain Forest in northwestern Connecticut. The forest occupies a broad plateau at elevations of 300–500 m, and represents a transition between the oak-dominated forests of southern New England, and the northern hardwood forests of northern New England. The six most abundant species in the forest (listed in order of increasing shade tolerance) are *Quercus rubra* (red oak), *Fraxinus americana* (white ash), *Acer rubrum* (red maple), *Acer saccharum* (sugar maple), *Tsuga canadensis* (eastern hemlock), and *Fagus grandifolia* (beech).

There are three distinctive features of the population dynamics predicted by SORTIE (Fig. 9-1). First, even long-term dynamics of shade-tolerant species such as beech and hemlock are nonequilibrial. The relative abundances of shade-tolerant species drift significantly over time, in part because of stochasticity associated with seedling colonization of gaps. Our field data clearly indicate that beech is the competitive dominant in the system; however, our data also reveal a fundamental tradeoff between competitive ability and dispersal in these

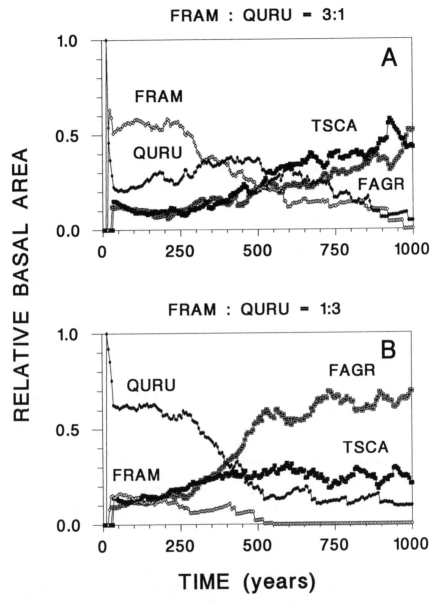

Figure 9-1.   SORTIE simulations of changes in relative basal area over time in a 1-ha stand, using only four species for simplification. FRAM (◇) = *Fraxinus americana*; QURU (◆) = *Quercus rubra*; FAGR (⊡) = *Fagus grandifolia*; and TSCA (■) = *Tsuga canadensis*. Species-specific parameter values used in the simulation are given in Pacala et al. (1993). The simulations begin by randomly distributing 100 seedlings each of TSCA and FAGR (the two most shade-tolerant species in the system) within the hectare. The initial seedling abundances of FRAM and QURU were varied from 300 FRAM : 100 QURU in Fig. 9-1A to 100 FRAM : 300 QURU in Fig. 9-1B.

species. Analytical models indicate that such tradeoffs can act as powerful mechanisms for species coexistence (S.W. Pacala and D. Tilman, unpublished).

Second, the spatial distributions of species within stands become highly aggregated during succession, again because of the relatively limited effective dispersal distances of all six of these species (i.e., mean dispersal distances of < 20 m; E. Ribbens, J.A. Silander, and S.W. Pacala, unpublished). Striking patchiness in the distribution of tree species develops even in the absence of any physical, environmental heterogeneity within the stand. Patches of a given late successional species tend to be only slowly invaded by other shade-tolerant species because locally produced seeds and seedlings are generally much more numerous than seedlings of the invading species, and the competitive differences among the shade-tolerant species are fairly small.

Third, long-term successional dynamics are strongly affected by the initial abundances of species following disturbance (Fig. 9-1). As expected from a large body of research on forest succession, the relative abundances of different early successional tree species during the first several hundred years following catastrophic disturbance are strongly influenced by the relative abundances of both advance regeneration (i.e., juveniles that survive the disturbance) and seedlings established immediately following the disturbance (whether from seed banks or seed dispersal) (Canham and Marks, 1985). SORTIE also predicts that differences in the abundances of early successional species have effects well beyond the time when the species are displaced from the stand. In the example in Fig. 9-1, a high initial abundance of northern red oak led to an approximately 250-year period of strong dominance by that species, with exclusion of the other early successional species (white ash) within approximately 500 years. However, as the abundance of red oak declined, the abundance of beech increased dramatically (Fig. 9-1). In contrast, the simulation with a comparable density of initial white ash seedlings did not result in comparable levels of dominance by white ash, and as white ash declined, there was a transient increase in red oak abundance, followed by eventual codominance by both beech and hemlock (Fig. 9-1). These long-term, persistent effects of initial conditions following disturbance appear to reflect differences in the competitive interactions between specific pairs of shade-tolerant and -intolerant species in the system.

The current version of SORTIE describes the dynamics of tree species, without explicit consideration of other plant growth forms. However, an understory plant species also appears to have a profound effect on stand structure and succession in these forests. Hayscented fern (*Dennstaedtia punctilobula*) often forms dense, vegetatively spreading layers of foliage 30–50 cm tall in the understory of northeastern forests. There has been widespread concern about its ability to inhibit regeneration of commercially important tree species in the Allegheny Plateau of Pennsylvania (Horsley, 1986). In stands at

Great Mountain Forest, hayscented fern cover is directly proportional to incident, understory light levels: in areas of the understory that receive relatively high light levels, hayscented fern cover is also high (Hill and Pacala, 1992). As a result, the fern effectively eliminates many of the light gaps that are normally exploited by newly emerging tree seedlings (Hill and Pacala, 1992). When this simple effect is incorporated in SORTIE, the model predicts two fundamental changes in forest dynamics: first, succession is accelerated, because only the most shade-tolerant species can successfully penetrate the dense fern layer; and second, total tree biomass is significantly reduced (by as much as 40%) because the sparse understory is much slower to fill gaps formed when canopy trees die (J. Hill, unpublished).

LINKAGES BETWEEN TREE POPULATION DYNAMICS AND ECOSYSTEM PROCESSES

Productivity and Carbon Cycling

We now turn to the issue of the linkages between these features of tree population dynamics and forest ecosystem processes. During the initial phases of secondary succession, the process of self-thinning results in predictable changes in stand structure and productivity that appear to be largely independent of the precise composition of a stand (e.g., Mohler et al., 1978; Bormann and Likens, 1981). However, there are well-known differences among tree species in (1) total biomass for trees of similar stem diameter or crown volume (e.g., Fig. 9-2A); (2) the relative allocation of that biomass to different structures (i.e., roots, stems, leaves, and seeds) (e.g., Tritton and Hornbeck, 1982); and (3) the chemical constituents of the biomass that determine the fate of the organic matter (including both the risks of death of the tissue, particularly due to herbivores, and the rate of eventual decomposition) (Melillo et al., 1982; McClaugherty et al., 1985; Melillo et al., 1989). Models that have linked forest stand dynamics with productivity and decomposition clearly indicate that the long-term dynamics of carbon allocation, storage, and decomposition in forests are sensitive to changes in forest species composition (e.g., Pastor and Post, 1986). We suggest that as the realism of the population dynamics of these models is improved, the dynamics of productivity, decomposition, and carbon storage in forests will be found to be even more sensitive to the kinds of nonequilibrial, spatially heterogeneous dynamics exhibited by SORTIE.

Hydrology

It has long been appreciated that streamflow can be strongly affected by the structure and composition of vegetation within the watershed (e.g., Kittredge,

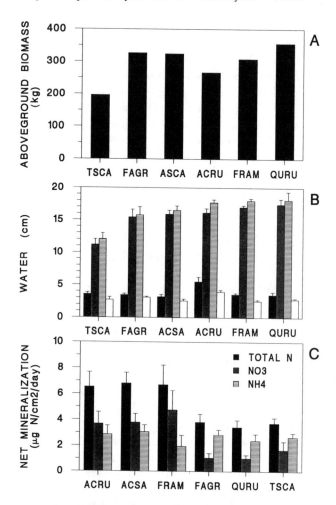

Figure 9-2.   (A) Aboveground biomass (kg) of 25-cm DBH trees of the six dominant species at Great Mountain Forest. Biomass estimates were calculated from dimension analysis equations given in Tritton and Hornbeck (1982). (B) Effects of the six dominant canopy tree species on growing season water budgets (1991). Data (means and standard errors) are from six replicate trees of each species within a 3-ha area. Solid bars are initial soil water content (in cm) in the top 15 cm of soil at the start of the growing season. Cross-hatched bars represent total growing season precipitation in rainfall collectors placed beneath the tree crowns (at 1 m above the soil surface). Clear bars represent the soil water content at the end of the growing season, and horizontally hatched bars represent the surface soil water losses estimated by mass balance. (C) Midsummer net N mineralization rates (means and standard errors) in the surface soil beneath the crowns of the six dominant tree species. Mineralization was estimated from 28-day incubations of *in situ* cores of the top 15 cm of forest floor plus mineral soil beneath six replicate trees of each of the species in each of two stands. Solid bars show total net N mineralization; cross-hatched bars give net nitrification rates, and horizontally hatched bars show net ammonification rates.

1948; Swank and Douglass, 1974; Swank et al., 1988). We have documented the effects of the major tree species at Great Mountain Forest on surface soil water budgets, primarily to test for interspecific differences in the effects of canopy trees on soil water availability for seedlings and saplings (Fig. 9-2B). While we did not find evidence of significant differences among the species in their ability to deplete surface soil moisture content during the growing season, there were significant differences among the species in interception losses due to evaporation from foliage (Fig. 9-2B). The relative differences in interception losses during the growing season from hemlock (TSCA) crowns vs. the five deciduous species would be amplified even more during the dormant season due to hemlock's evergreen foliage (Swank et al., 1988). Physiological differences among species in transpiration rates may further amplify the effects of tree population dynamics on hydrologic yield from watersheds (Swank et al., 1988). Hence, as forest composition changes (due to combinations of directional, successional trends and the more unpredictable drift and fluctuations predicted by the model), the hydrology of the watershed could change markedly.

Nitrogen Cycling

The dominant tree species at Great Mountain Forest also clearly differ in their effects on N cycling. Net N mineralization rates at any given location within a forest vary more than twofold as a function of the identity of the canopy tree directly overhead (Fig. 9-2C). These differences occur even within a spatial scale of <10 m between adjacent trees (Finzi et al., 1993). Net ammonification rates were uniform across all six species; however, the six species can be divided into two functional groups on the basis of differences in apparent nitrification (Fig. 9-2C). The three species with high net nitrification (sugar maple, red maple, and white ash) are all known for the production of high-quality leaf litter relative to the other three species (beech, hemlock, and red oak). These effects on rates of N mineralization persist even after the canopy tree dies and forms a gap (data not presented). Although there has been a great deal of concern over the potential for nitrate export from forests and contamination of water supplies because of excess deposition of N from anthropogenic sources (e.g., Aber et al., 1989), our results suggest that changes in the relative abundance of species within a stand may also play a major role in determining the potential for nitrate export. It is worth noting that of the three major late successional tree species in this system (beech, hemlock, and sugar maple), the two species with low rates of apparent nitrification (beech and hemlock) are both threatened with reductions in relative abundance due to the spread of introduced pests or pathogens [i.e., beech bark disease (*Nectria coccinea*) and both the hemlock wooly adelgid (*Adelges tsugae*) and the hemlock looper (*Lambdina athasaria*)]. Thus, these two introduced pests may dramatically increase the potential for nitrate export and contamination of groundwater, by indirectly causing an increase in sugar maple abundance, the remaining late successional species within the forests.

These differences in local rates of N mineralization within stands clearly have the potential to feedback upon the population dynamics of the component species by altering the relative growth of seedlings and saplings. For example, of the three late successional species in these forests, sugar maple saplings show stronger responses to soil N availability than do beech and hemlock saplings (unpublished data). This suggests that the high litter quality and N cycling rates of sugar maple trees may act as an autocatalytic feedback loop that promotes persistence of the species in these forests (see also Wedin, Ch. 24). We also have preliminary evidence that these differences in N cycling rates may have an indirect effect on tree regeneration that is mediated by herbivores. In particular, tissue N concentrations of saplings in the forest understory appear to vary not only as a function of the identity of the seedling (presumably due to inherent interspecific differences in N metabolism), but also as a function of the identity of the canopy trees overhead (presumably as a result of the effects of canopy trees on local N availability in the soil and luxury consumption of N by seedlings in deep shade) (C. Tripler and C. Canham, unpublished). These differences in tissue N concentration appear to affect the risk of browsing of the seedlings by white-tailed deer (*Odocoileus virginianus*) (C. Tripler and C. Canham, unpublished).

SYNTHESIS

In summary, SORTIE predicts that tree population dynamics in these forests are (1) nonequilibrial, (2) strongly dependent on initial conditions following disturbance; and (3) spatially aggregated, with relatively little linkage over distances greater than even 25 m. The population dynamics predicted by SORTIE are the result of a clear set of species-specific physiological, morphological, and life-history traits (Pacala et al., 1993). However, those traits generally do not depend on or determine patterns of carbon allocation and storage by a species, or the species' effects on hydrology and N cycling. As a result, idiosyncratic differences among species in traits such as litter quality predispose these forests to strong linkages between the relative abundance of tree species and spatial and temporal variation in forest ecosystem processes. Moreover, although the list of tree species present in a given environment may be generally predictable, SORTIE indicates that the absolute abundances of species at a site, and the changes in relative abundance during secondary succession, may vary immensely due to historical effects, particularly the distribution and abundance of colonists following disturbance. There is no question that energetics and biogeochemistry place constraints on forest ecosystem processes, and that the rates of many ecosystem processes are strongly affected by physical, environmental factors. However, in the absence of evidence of stronger control of tree population dynamics by ecosystem processes and physical factors, we conclude that the spatial and temporal dynamics of productivity, nutrient cycling, and hydrology in these forests will be largely driven by the interactions between microbes, plants, and animals—often acting in extremely local neighborhoods—that determine tree population dynamics. Thus, ecosystem dynamics in these forests should track the nonequilibrial, spatially aggregated dynamics exhibited by tree populations.

ACKNOWLEDGMENTS

We would like to thank the Childs family for making the forests and facilities at Great Mountain Forest available for our research, and the Bridgeport Hydraulic Company for allowing us access to their property for additional research sites. We would also like to acknowledge the colleagues, graduate students, and the research assistants who have worked on this project. This research was supported by the National Science Foundation (BSR 8918616 and DEB-9220620) and the Mary Flagler Cary Charitable Trust. This study is a contribution to the program of the Institute of Ecosystem Studies.

10

SOIL ORGANISMS AS ENGINEERS: MICROSITE MODULATION OF MACROSCALE PROCESSES

J. M. Anderson

SUMMARY

Soil invertebrates and microorganisms are proximate factors regulating carbon and nutrient mineralization. It has proved intractable, however, to relate the activities of species in the community to rates of biogeochemical fluxes at the ecosystem level. A major reason for this is that the metabolic contributions of individual species affect soil processes orders of magnitude below the scales at which the ecosystem is defined and process measurements are made in the field. In addition, however, soil organisms also have indirect effects of longer duration and extent where they act as modulators of carbon, nutrient, or water fluxes. By changing the physical controls over biogeochemical processes, invertebrates and microorganisms can regulate much larger fluxes than their direct effects on soil processes. The mechanisms often involve the formation of durable artifacts, such as stabilized aggregates, burrows, or buried organic matter, which continue to function in the absence of the organisms creating them. The effects of these artifacts, most notably those created by earthworms and termites, are cumulative and can therefore be manifested in macroscale measurements. It is shown, however, that the net effects of these activities may not be detectable on larger areas, or over longer time periods, because sink/source processes operate within patches at the scale, or domain, at which the species populations function. Perturbations to the system can also synchronize these activities in space or time so that invertebrate activities become apparent at higher levels of organization. It is concluded that the role of invertebrates should be more explicitly considered in ecosystem studies and that this is facilitated by a "top-down" approach to defining the scales at which their regulatory effects on biogeochemical processes are emergent.

INTRODUCTION

The spectacular variation in biomass, production, and biodiversity of terrestrial ecosystems ranging from tundra meadows to tropical rainforests has attracted the attention of ecologists for decades. Just as remarkable is the shallow depth and fragility of soil maintaining these systems. On a global scale the top meter of the soil contains twice the organic matter found above ground, with most of the annual flux of carbon (C) and nutrients occuring within the top 5–10 cm of the profile. Within this narrow zone, the activities of the soil biota regulating processes at the scale of soil pores and aggregates determine the chemistry of trace gas emissions, C and nutrient turnover, and leachate fluxes at the ecosystem level and above. The roles of these invertebrate and microbial communities are, however, rarely explicitly considered in ecosystem studies although their biomass, diversity, and community metabolism usually exceed those of herbivores above ground.

Soil fauna contribute directly and indirectly to soil processes (Anderson, 1988a,b). The direct effects can be quantified by budgeting trophic transfers through food webs. Studies in a wide range of systems show that the fauna mediate about 15% of the C and 30% of the nitrogen (N) turnover. These fluxes are mainly due to protozoa and nematodes (microfauna), and to a lesser extent collembola and mites (mesofauna). Populations of these groups turn over rapidly, consuming bacterial and fungal biomass with low C/N ratios. Macrofauna, such as earthworms and termites, that consume materials with high C/N ratios and have longer generation times make less of a direct contribution to community metabolism. The coupling of these population processes in food web models is attractive as a means of linking community structure and nutrient fluxes (see, for example, Bengtsson et al., Ch. 16). However, they have proved insensitive to community structure at levels below that of broad functional groups and appear to be unaffected by the interations of higher trophic levels (Beare et al., 1992). In addition, food web models also exclude the indirect effects of invertebrates on C, nutrient, and hydrologic fluxes considered here.

Energy transformations and water fluxes at the interface between air and soil are regulated by physical parameters such as litter cover, aggregate stability, macroporosity, and surface crusting to a depth of a few millimeters. The transfer of dead plant material across this interface from the surface to the edaphic environment has a major impact on decomposition and the stabilization of soil organic matter. Here too, soil fauna, and their interactions with microorganisms, affect all these soil physical and chemical properties and act as modulators of soil processes by engineering the environment (see Lawton and Jones, Ch. 14).

The indirect effects of soil fauna are difficult to quantify and relate to community structure for two reasons. First, they often operate through the creation of physical artifacts (soil aggregates bound by mucopolysaccharides, soil macropores and channels; litter incorporated into soil). These constructs continue to function in the absence of the living organism that created them and act cumulatively on soil processes. Second, the effects are often aggregative and nonlinear. Averaging process measurements for soil volumes or areas may result in no detectable net flux unless sampling is carried out at an appropriate scale to detect the activities of micro-, meso-, or macro-fauna (e.g., earthworms and termites). These organisms operate within spatial and time domains that may be four to six orders of magnitude below those at which ecosystems are often defined and are consequently not explicity considered at the catchment or plot scale.

This chapter initially adopts a "top-down" approach to illustrate the scales at which the engineering activities of soil fauna emerge as regulators of C, nutrient, and water/sediment fluxes alongside more distal environmental factors. The second half then describes a "bottom-up" approach where soil fauna affect the dynamics of soil processes within the zone of influence, or domains, of their populations. Finally, I consider how the synchronization of these activities can result in the signals from microsite processes being detectable at higher levels of organization.

A "TOP-DOWN" VIEW OF SOIL FAUNA ACTIVITIES

The operational principle of ecosystem ecology is the concept of a functional unit regulating energy and mass fluxes across the boundary. The components of the system can then be defined at progressively finer scales of organization to determine scales at which species populations, or guilds, can be identified as the primary rate determinants of ecosystem properties and processes. This concept is illustrated in Fig. 10-1, where the attributes of species are shown nested within the aggregative properties of different levels of ecosystem structure. The emergence of species effects from the "top down" is analogous to

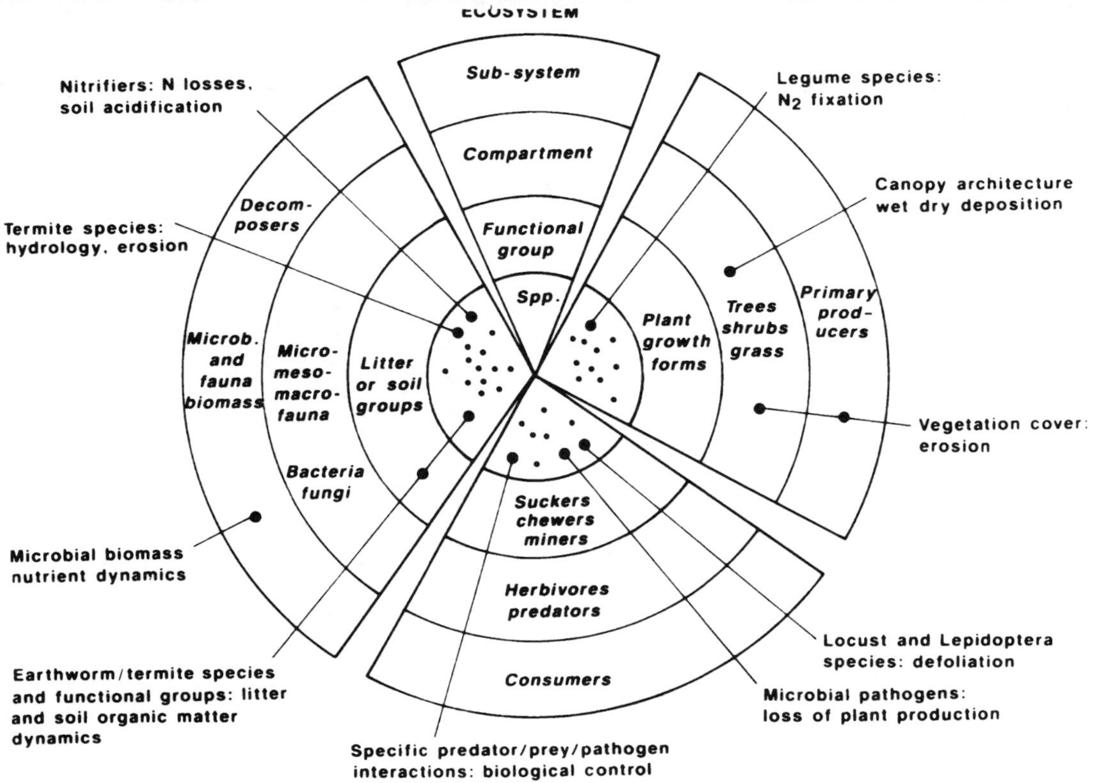

Figure 10-1. Hierarchical scales of organization in terrestrial ecosystems. The structure of an ecosystem is illustrated by concentric circles representing different functional scales from subsystems down to the activities of species. At present there is poor understanding of the relationships between community structure and the proximate controls that organisms have on process rates. Hence as processes are measured over larger areas and longer periods of time the processes are related to more distal levels of organization. The figure illustrates the point, however, that some species are sole representatives of functional groups whose effects are therefore manifested at the system level (shown by lines linking species to ecosystem functions on the outer circle) (From Anderson 1994.)

draining a lake. The first features that emerge are the plant subsystems; then components of subsystems (wood, leaves, roots; soil organic and mineral pools) processed by functional groups of organisms; and finally the effects of individual species that are located at the lowest ("benthic") level of organization. However, when a functional group, or an entire subsystem, is dominated by a single species, its specific attributes can be manifested at a much higher level of organization. In terms of the lake analogy, these are "water-lily" effects where a species rooted in the "benthos" (i.e., operating proximally) affects the functioning of the lake ecosystem in terms of energy or material transfers across the boundary (i.e., more distally).

Examples where soil fauna can be readily identified as having ecosystem-level impacts on the plant/soil system occur mainly where the activities of macrofauna regulate the direction of energy, water, or nutrient fluxes. Here it is useful to distinguish between epigeics (which process organic matter on or near to the soil surface), endogeics (living in mineral soil) which are often humivorous, and anecics which transfer materials between soil and litter habitats. The epigeic and endogeic groups can passively affect the transport of solutes and particulate matter by water, wind, or gravity, in a manner analogous to "shredders" in aquatic communities (Anderson, 1989). Anecics, however, not only bring about the active redistribution of organic matter but, in doing so, can modify the physical controls over biogeochemical processes.

Water and Sediment Transport

Thermal and water balances in soils are mediated by surface phenomena: litter dissipates the kinetic energy of incident precipitation or throughfall and affects reflectance and surface evaporation. The surface architecture of the mineral soil, aggregate stability, and porosity determine storage of water, runoff, infiltration, and water supply to the underlying horizons. Invertebrates can affect all these regulatory processes (Fig. 10-2) but their activities must be placed within the context of vegetation cover (and rooting) which not only has similar control functions at larger spatial scales, but also influences the patchiness and magnitude of invertebrate effects.

Rates of infiltration and splash erosion are usually low under undisturbed vegetation providing that the surface is protected by ground vegetation or litter cover. If the litter is removed, surface macropores may be blocked by rain drop impaction and processes of erosion are initiated. van Hooff (1982) showed that the spatial patterning of splash erosion and overland flow in a mixed deciduous woodland was determined by the palatability of tree leaf litters to earthworms. The seasonality and extent of patches of bare soil was determined by the removal of hawthorn leaves (*Crataegus laevigata*) from the surface by *Lumbricus terrestris*. Beech (*Fagus sylvatica*) leaves were a less preferred

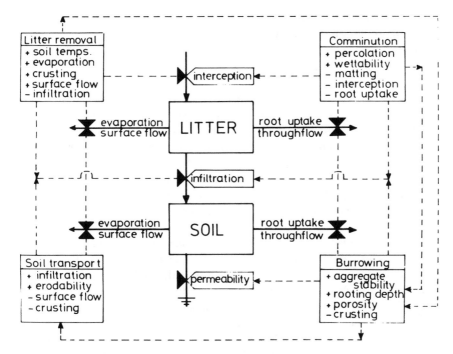

Figure 10-2. Effects of soil invertebrates on water flux pathways. The consequences of litter removal, litter breakdown (comminution), burrowing, and soil transport are shown in terms of positive (+) or negative (–) effects on rainwater transfers through the litter and soil horizons. (From Anderson, 1988b)

food source and the incidence of bare patches of soil was found to be an inverse function of beech contribution to litter cover. Infiltration rates will have been high under the beech litter cover and provided a sink for surface water and sediment. In such a case no net water and sediment yield from surface runoff will be detectable on scales greater than the spatial patterns of sinks and sources. A similar sink/source pattern has been shown for New Zealand pastures by Sharpley et al. (1979) where splash dispersal of earthworm casts increased the sediment load of surface runoff, and burrows acted as sinks for the suspended material. As a consequence net sediment losses (and phosphorus associated with suspended clay) were lower in runoff plots with earthworms than when these animals were eliminated by pesticides.

The stability of soil surface aggregates is an important factor preventing the sealing (capping) of macropores in exposed cultivated soils. It is well established that the casts of earthworms can show exceptional aggregate stability, and hence protection against erosion in some soils, especially after drying (Shipitalo and Protz, 1989). The kinetic energy of raindrops required to disperse earthworm casts in a sequence of Nigerian soils was 5–54 times higher than that for soil macro-aggregates not derived from earthworm activities (De Vleeschauwer and Lal, 1981).

In overgrazed rangeland in Ethiopia, the scavenging of the small amounts of grass and litter remaining in the dry season by *Macrotermes, Odontotermes,* and *Pseudacanthotermes* left the soil bare and susceptible to erosion (Wood, 1991). The bare and consolidated soil around termite mounds can also initiate erosion processes affecting a wider area. High densities of *Trinervitermes* mounds in grassland fallows of Cote d'Ivoire may lead to land degradation (Janeau and Valentin, 1987). Conversely the longer-term elimination of a single termite species, *Gnathitermes tubiformans,* in a Chiuhuahuan desert system by Elkins et al. (1986) resulted in changes in soil structure, and increased runoff and sediment loads in areas with low or no vegetation but insignificant effects under shrub cover. The reduced water supply to soils in the open areas resulted in a reduction in grasses and increased shrub cover. Another good example of a landscape comprising dynamic patches of water sources and sinks associated with termite activities has been described for mulga (*Acacia aneura*) woodlands in Western Queensland (Tongway et al., 1989). These mulga woodlands have a characteristic pattern of alternating groves and intergroves with a runoff zone of stony, severely sealed, surface soil with cryptogamic crusts, and a runon zone of mulga grove with infiltration rates 5–10 times higher than the runoff zone (Greene, 1992). The breakdown of dead wood in the mulga groves by termites produces mounds of fertile soil with high infiltration rates, imparted by a coarser soil texture and termite foraging galleries. Coupled with runoff interception this supports higher production by perennial herbs on the up-slope than down-slope aspects of the grove and creates a successional pattern of patches moving across the landscape on an unknown time scale.

Carbon and Nutrient Fluxes

Consumption of above-ground woody litter by termites in arid and semi-arid soils ranges from 17% to more than 90% and up to 75% of total grass and forb litter (Jones, 1990). Food collected by foraging (anecic) termites is used entirely within the nest system which may be constructed on or above the soil. Carbon and nutrients are thus concentrated in patches from the surrounding area over distances of centimeters to a few meters for diffuse nest systems (e.g., nonmound building humus feeders and the fungus gardens of *Microtermes* and *Ancistrotermes*) or up to 50 m for large *Macrotermes* mounds (Darlington, 1982). The effect of these foraging activities is a depletion of organic C in bulk soils under conditions of low plant production and high termite population (Jones, 1990) while the mounds contribute patterns of heterogeneity in soil nutrients and plant growth across the landscape (Lal, 1987).

Anecic earthworms, in contrast to mound- or nest-building termites, incorporate organic matter into soil and thereby initiate important indirect effects on C and nutrient cycling. The turnover of mineral soil, which can amount to 50,000–250,000 kg/ha/yr or more in temperate regions, produces a more even vertical distribution of organic matter and nutrients in the soil profile. This is nicely illustrated by the profile of ^{137}Cs (from nuclear fallout) in forest soils in Ontario (Tomlin et al., 1992). Sites where *Lumbricus terrestris* had been introduced had ^{137}Cs distributed throughout the surface horizons whereas sites with endogenous species had higher ^{137}Cs remaining at or near the soil surface. There are also a number of well-documented cases where the introduction of European lumbricids into pastures in New Zealand and the Netherlands have had dramatic initial effects on herbage production as a consequence of the accelerated decomposition rate and nutrient release when the turf mat has been incorporated into soil by anecic species (Lee, 1985). Here again, the energy expenditure by the worms in incorporating litter into soil, and the low fraction assimilated, is small in relation to the timing and location of energy and nutrient fluxes initiated by changing the environmental controls over microbial activity. The introduction of the peregrine earthworm *Pontoscolex corethrurus* into tropical cropping systems has also shown marked effects on crop production at the plot scale (Spain et al., 1992) and has interesting implications for the management of small-holder farming systems. Conversely, where earthworms are absent or excluded, soil physical and chemical properties can change markedly (Tab. 10-1).

At the microsite level the intimate mixing of organic matter with mineral particles may result in the physical or chemical stabilization of C depending on the mineralogy of the soil (Scheu and Wolters, 1991; Lavelle and Martin, 1992). The formation of casts can also optimize conditions for N transformations. Elliott et al. (1991) have shown that the quality of dung and herbage in pastures was reflected by high concentrations of available C and N in casts

Table 10-1. Physical and chemical parameters of topsoil in fertilized grassland plots with or without earthworms (eliminated by pesticides) from 1969 to 1989.

Parameter	With Earthworms	Without Earthworms
Surface litter (g/m^2)	13	1,088
Soil organic matter (%)	5.90	4.53
Shear strength (kPa)	79.3	99.7
Bulk density (g/cm^3)	1.39	1.65
Soil moisture (% w/w)	13.4	11.9

(From Clements et al., 1991).

that formed "hot spots" of microbial activity. Denitrification rates were up to five times higher from casts than from uningested soil. The total earthworm-mediated flux was estimated to be as high as 20% of total denitrification and increased proportionately as a function of fertilizer applications to pastures. This flux did not represent direct losses of fertilizer N but was a consequence of the linked effects of increased herbage quality and production, higher cattle stocking rates, dung deposition, earthworm biomass, and cast production.

SOIL FAUNA ACTIVITIES FROM THE "BOTTOM-UP"

The "top-down" examples of soil faunal activities impacting processes at the ecosystem level occur where a single abundant or active species dominates a particular functional role. Consequently the specific attributes of that species' niche are manifested at a higher level than in more complex communities where the functional group comprises many species with slightly different activities varying in space and time. Furthermore, in complex communities there may be rapid compensation for functional roles with fluctuations of species populations. For example, fumigants have been used in the laboratory (Brooks et al., 1991) and field (Ridge, 1976; Rovira, 1976; Klein et al., 1986) to virtually eliminate most bacterial and fungal species from soils. The lysis of microbial biomass causes a flush of C and N mineralization over a period of 24 h or so but rates then return to prefumigation levels, indicating that these processes are insensitive to the composition of microbial communities. Clearly there is a considerable degree of functional compensation in these communities by the few remaining species (see also Frost et al., Ch. 22). Nitrification, however, is far more sensitive to fumigation (Klein et al., 1986), as might be expected from the specific transformations and slow growth rates of these autotrophs in comparison with ammonification, which is carried out by most heterotrophs.

Similarly, the feeding activities of invertebrates have been shown to have a wide range of indirect effects on microbial populations and processes which have been extensively reviewed elsewhere (Edwards et al., 1988; Verhoef and Brussaard, 1990). These interactions include shifting the competitive balance between fungal species, switching from fungal to bacterial dominance in litter, affecting the activities of mycorrhizas, rhizobia, and pathogens in the rhizosphere; and dispersal of microbial inocula. Predation at higher tropic levels can also produce a cascade of feedback effects through microbivores to microbial processes. In general, however, it has proved difficult to relate understanding of these complex interactions to biogeochemical processes in the field. Beare et al. (1992) concluded that the impact of soil fauna on N dynamics in cultivated plots could be linked to broad trophic groups and that a species-specific focus was unnecessary for understanding these mechanisms (see Parmalee,

Ch. 11). Similar conclusions have been reached for grassland soil by Hunt et al. (1987) using food web simulations and by Ingham et al. (1986) using selective biocides.

These examples suggest that the main problem in linking community structure to ecosystem processes is the disparity between the scales at which different organisms operate and the scales at which measurements of their effects are made. Processes have no inherent dimensions; we impose operational scales of space and time as a matter of convenience. With increasing sample size, or duration of measurements, the "signals" of indirect species effects are difficult to detect against the "noise" of direct contributions of the total biota to C and nutrient fluxes; or the effects of vegetation and abiotic factors have an overriding influence on water and sediment fluxes. Process measurements made at any particular scale are therefore the net effects of organisms operating over many orders of magnitude in space or time. To isolate the signal of species populations or functional groups, sampling must be carried out at the ecological scale, or domain, at which the effects of those organisms are expressed (see Shachak and Jones, Ch. 27).

FUNCTIONAL DOMAINS OF FAUNA MEDIATING SOIL PROCESSES

The domain of a species is the zone of influence determined by the magnitude of local effects of the individual within the spatial and temporal context of the species population. The indirect impact on soil processes is therefore a function of body size and activity of individuals, and the density and aggregative characteristics of the population. Hence, collembola or mites feeding on fungi and bacteria at a scale of millimeters or less form aggregations affecting microbial processes within an area of a decimeter or more; earthworms forming burrows of a few millimeters in diameter have aggregative effects on hydrologic processes in patches of several meters. Poier and Richter (1992) show that the larger the earthworm species the more the spatial distribution of the biomass and abundance varied within a range of 20–50 m. These effects are also nested within the pattern of litter cover (providing food and surface protection) and patches of plant species cover operating at still larger spatial scales. The same concept can be extended to fungi, where the hyphae can constitute the mycelium of a single genetic individual with a biomass of tons, dominating decomposition processes over many hectares (Smith et al., 1992). The activities of other microorganisms or invertebrates operating in smaller domains are therefore expressed against the background of effects from the domain of a larger organism.

The functional properties of the domains of species or functional groups can therefore be considered as a hierarchy of nested interacting systems.

Moving up this hierarchy, successive levels may accommodate the same processes, but with slower dynamics covering a larger area. As these scales increase, the correlates with soil processes also shift from proximate factors that may be related to species activities, through functional groups to distal environmental factors. For soil faunal activities to be explicitly expressed at more distal scales, the signals from their effects on processes within their domains must be reinforced. It is easier to consider the spatial and temporal elements of these interactions separately.

Interactions of Organism Domains in Space

The spatial element of these faunal effects involves the dynamics of patches within and between the domains. If fauna are considered to have positive and negative effects on a process then the balance between sinks and sources of nutrients, trace gasses, water, or sediment will determine whether the net effect is detectable at a higher level. The domains of different organisms will also interact to amplify or dampen the signal (Fig. 10-3). This is a general phenomenon for soil processes and a few examples will illustrate the point. Collembola grazing on hyphae have been shown to have stimulatory, neutral, or inhibitory effects on fungal growth according to the population density of the animals and the nutritional regimen of the fungus (Leonard and Anderson, 1991a,b). Patches with high grazing intensity can therefore be sources of mobilized N which can be translocated to adjacent areas, with low collembola population densities, which act as N sinks. No net mineralization would be detected if the flux was measured above the domain level. Plant roots or mycorrhizal hyphae innervate soil or litter within this domain scale so that the N flux can be intercepted by the plant. Similarly, the examples described earlier for the sink/source effects of earthworms on surface water and sediment fluxes, or anaerobic aggregates acting as sinks for nitrate produced in adjacent aerobic microsites, illustrate the dynamic interactions between patches of activity at different scales. In these cases the specific faunal effects would not be detectable unless sampled at the domain level (Fig. 10-3). The examples for termites show plant domains acting as sinks for surface water which was a limiting factor in those systems. Here the extensive nature of the faunal effects and the legacy of their activities on soil surface hydrology combine to produce a synchronous effect in large domains which determines functioning at the subsystem level. At the ecosystem boundary, above the vegetation domains, no net fluxes would be detected.

Interactions Over Time

Temporal reinforcement of faunal activities occurs where the system is perturbed and the signs of patches (+/0/−) become synchronized. This results in a net effect of the microsite events that is detectable at a higher level. There are many examples that fit into this pattern: the invasion of new habitats by exotic earthworms; the initiation of faunal activities after drought; changes in land use; seasonal pulses of litter inputs into soils; etc. Many of the examples of mesofaunal effects on N dynamics have been demonstrated using litter bags where decomposition of an age class of litter creates cohort effects of organism activities. Beare et al. (1992) showed that the exclusion of microarthropods from crop residues resulted in increased fungal densities and a 25%

Figure 10-3.   Interactions of the functional domains of soil invertebrates. Micro-, meso-, and macro-fauna have functional domains of different sizes (represented by *circle diameter*). As discussed in the text the indirect or modulating effects of soil fauna can act positively (+) or negatively (–) on soil processes. The upper figure represents the "top-down" view of these nested domains. Flux measurements carried out at the scale of the hemicircle A, where the signs of the domains have been brought into synchrony by some event, would detect a net faunal effect but it would not be possible to assign the flux to species or functional groups. Conversely, no net flux would be detectable in hemicircle B because sources and sinks of water or nutrient fluxes are balanced. To identify the modulating effects of species or functional groups it is necessary to sample at an appropriate macro-, meso-, or micro-scale in time or space (lower figure) at which the signs of domains are apparent.

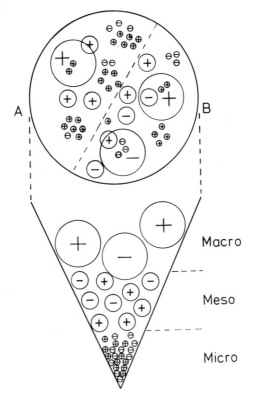

greater N retention in the litter. These effects are less readily detected in the horizons where the slower dynamics of humus formation and mineralization approximate to steady-state conditions.

Soil invertebrates have been shown to influence soil conditions affecting different phases in the growth of arable mono-crops such as seedling establishment, root development, nutrient availability, soil moisture, etc. Few of these processes are widely recognized as being agronomically significant. There are several reasons for this. First, biological processes are overridden by anthropogenic effects of tillage, fertilizer management, and irrigation. These biological effects can reemerge when external inputs to the system are reduced (see the comparison of conventional and no-till systems by Parmelee, Ch. 11). Second, unless parameters affected by soil fauna, such as macroporosity or the timing of nutrient release, are limiting plant growth their effects may not be measurable in terms of yield (Clements et al., 1991). Third, processes operating over longer time scales average out the effects of soil fauna. For example, earthworms may promote seedling emergence in capped soils (Kladivco et al., 1986) but these initial effects may be obscured by subsequent plant demographic processes such as self-thinning or tillering (Harper, 1977) which are more proximate determinants of yield for the plant population than seedling establishment. Finally, agronomic measures of crop performance at a plot scale of hectares, over a cropping season of several months, integrate across small-scale, short-term events such as those affected by specific activities of the soil biota.

CONCLUSIONS

The terms ecosystem and processes are inherently without dimensions of space or time. The scales we set are usually for operational convenience and are rarely defined by the domains of organisms that determine the functioning of the system (but see Shachak and Jones, Ch. 27). Where components of the systems are defined on a scale of meters, rather than hectares, or when processes are measured over shorter intervals, the modulating effects of plants, and then of invertebrates and microorganisms, emerge according to the size of the domain. These emergent effects are most readily identified in species-poor, perturbed systems where the activities of dominant representatives of functional groups are synchronized by climate or management practices. Linking species populations and ecosystem processes is therefore a matter of defining measurements at scales defined by the spatial and temporal domains of the organisms and the effects of their activities in litter and soil.

11

SOIL FAUNA: LINKING DIFFERENT LEVELS OF THE ECOLOGICAL HIERARCHY

Robert W. Parmelee

SUMMARY

Soil fauna play a significant role in ecosystem-level processes in terrestrial ecosystems through their interaction with the microbial community and alteration of the physical environment. Analyses of soil fauna populations and communities can provide important information on nutrient cycling processes. Four examples demonstrate the utility of this approach. First, analyses of nematode and microarthropod community trophic structure in agroecosystems help to identify the underlying mechanism of why nutrient cycling rates are different under different management practices. Second, by combining earthworm population growth rates and estimates of community secondary production, the amount of nitrogen flux through earthworm biomass in an agroecosystem is estimated. Third, analyses of soil fauna communities in the rhizosphere of pine trees lead to inferences about nitrogen turnover in the rooting zone. Finally, analyses of the trophic structure of soil nematode and microarthropod communities provide a sensitive indicator of disruption of soil food web structure and function following exposure to chemical pollutants. In terrestrial ecosystems, soil fauna are ideal for conducting studies that link population, community, and ecosystem levels in the ecological hierarchy.

INTRODUCTION

Ecologists have traditionally limited the scale of their research to a single level of the ecological hierarchy, either the population, community, or ecosystem. This approach has led to fragmentation of the discipline, and has arguably impeded development of the science of ecology in general (Grimm, Ch. 1). Regardless of their particular bias, many ecologists fail to recognize the hierarchical organization of ecological systems, and to realize that no level exists as a separate or isolated component. It would seem to be the obvious step to view any study in a more holistic perspective, and to attempt to consider the possible ramifications of the outcome of any particular study on levels lower or higher in the ecological hierarchy. The objective of this chapter is to use examples from soil ecology to demonstrate that analyses of populations and communities can provide important insight into ecosystem structure and function. To illustrate this point, I will first present a brief overview of why studies of soil fauna populations and communities can be of particular value in obtaining information on ecosystem structure and function. Then, I will present examples in which analyses of soil nematode, microarthropod, and earthworm populations and communities were used to investigate various aspects of ecosystem-level processes in terrestrial agricultural and forested systems.

ROLE OF SOIL FAUNA IN NUTRIENT CYCLING

In terrestrial ecosystems, soil fauna are major components of the decomposer food web and are key regulators of decomposition and nutrient mineralization/immobilization processes. Although in most systems the microbial community accounts for about 90% of carbon (C) and nitrogen (N) mineralization, its activity is regulated by soil fauna. Soil fauna regulate microbial activity by directly feeding on bacteria and fungi, and by the transport of microbial propagules to new substrates. After consumption of microbial biomass, soil fauna excrete inorganic N which is then available for further microbial or plant uptake. Soil fauna can increase the surface area for microbial attack by the fragmentation of organic matter and by the production of fecal pellets. Soil fauna also indirectly alter the microbial microenvironment by fecal and cast production, which, in turn, affect soil pore space and aggregate size and stability. In general, the net effect of the activities of soil fauna on ecosystem-level processes is to increase the decomposition rate of organic matter and turnover of nutrients.

Two groups that often dominate numerically in the soil fauna community are nematodes and microarthropods. Within these groups, species diversity is very high, and all trophic groups are represented (bacterivore, fungivore, herbivore, omnivore, predator). Because of high abundance and diversity, and the presence of all trophic groups, soil nematode and microarthropod communities can be sensitive detectors of change in many ecosystem parameters.

Nematodes are microscopic roundworms that inhabit the water film on soil particles, and often number over a million individuals per square meter. Two of the most important nematode groups affecting nutrient turnover by grazing on microbial biomass are the bacterivores and fungivores. Bacterivore nematode population and community dynamics are known to be positively correlated with bacterial densities and production (Golebiowska and Ryszkowski, 1977; Clarholm et al., 1981; Sohlenius and Bostrom, 1984), suggesting an increase in grazing activity and resultant increase in turnover of nutrients stored in microbial biomass. Likewise, fungivore nematodes are known to reduce densities of fungal hyphae (Mankau and Mankau, 1963; Wasilewska et al., 1975; Santos et al., 1981; Parmelee et al., 1989). In this case, however, nematodes remove fungal cytoplasm by sucking materials through a hollow stylet which leaves some nutrients bound in relatively recalcitrant hyphal walls.

The microarthropod community is composed largely of mites and collembola that inhabit the air-filled pore spaces in soil, and typically number hundreds of thousands per square meter. Fungivorous microarthropods may be particularly important in affecting litter decomposition and N dynamics in terrestrial ecosystems. When microarthropods were eliminated from litterbags by the use of selective insecticides, fungal hyphal densities increased, suggesting that grazing by microarthropods reduced fungal abundance (Parker et al., 1984; Beare et al., 1992). In addition, the increase in fungal densities following microarthropod elimination coincided with an increase in the amount of N immobilized in surface litter. Microarthropods can also affect nutrient cycling rates by predation on bacterivore nematodes (Santos et al., 1981).

Earthworms, when present, can dominate soil fauna biomass and have large effects on nutrient cycling processes. Earthworms consume and incorporate large amounts of coarse organic matter into the soil with resultant effects on microbial activity, C and N mineralization rates, and soil structure (Lee, 1985; Anderson, Ch. 10). Unlike nematodes and microarthropods, whose effects on nutrient turnover are largely indirect, earthworms may have a large direct effect on nutrient cycling rates, particularly for N. For example, estimates for the amount of N flux directly through earthworm tissue can exceed 100 kg ha^{-1} yr^{-1} (Lee, 1985).

NUTRIENT CYCLING IN AGROECOSYSTEMS

Recent investigations on soil fauna in agroecosystems suggest that different management practices have strong effects on soil fauna populations and community structure. Nutrient cycling rates are also affected by changes in management practices, and it has been suggested that these changes may occur, in part, because of alterations in the structure of belowground food webs (Hendrix et al., 1986). This hypothesis has been investigated in some detail in a comparison of conventional tillage and no-tillage agroecosystems on a flood-plain at the Horseshoe Bend Experimental Area near Athens, Georgia. With conventional tillage practices, the soil is plowed, organic matter is incorporated, and there is disruption of soil structure. In no-tillage agriculture, there is minimum soil disturbance and organic residues remain on the soil surface. This practice prevents soil erosion and, theoretically, takes advantage of biological processes to develop soil structure and increase biologically mediated nutrient cycling processes.

Because of the differences in placement of crop residues and amount of soil disturbance, nutrient cycling rates are faster in conventional tillage than in no-tillage systems. In conventional tillage, crop residues are incorporated and buried within the soil profile, and often decompose faster than surface residues in no-tillage (Parmelee et al., 1990; Beare et al., 1992). For example, after 300 days only 20% of rye litter remained in the conventional tillage system compared to 40% in no-tillage. To explain the differences in rates, Hendrix et al. (1986) proposed that ecosystem nutrient cycling rates were faster in conventional tillage because burial of the litter increased fragmentation and maintained more stable moisture levels, factors that would promote a bacterial-based food web with higher metabolic rates and more rapid turnover of nutrients. In the no-tillage system, litter is not as fragmented and is more subject to extremes of moisture fluctuations, factors that favor a fungal-based food web with a greater tendency for nutrient immobilization and slower turnover of nutrients incorporated in fungal biomass.

This food web hypothesis was based on microbial data and on analyses of the trophic structure of nematode and microarthropod communities. Holland and Coleman (1987) found that decomposition rates were slower and net N immobilization was greater in surface vs. incorporated litter, and concluded that the microclimate of surface litter favored fungal over bacterial abundance. Coupled with the microbial data was evidence that the fungivores were a more important component of nematode and microarthropod communities in no-tillage compared to conventional tillage agroecosystems. Parmelee and Alston (1986) reported that fungivore nematodes were more prevalent in no-tillage during the growing season, and House and Parmelee (1985) observed larger numbers of potentially fungivorous microarthropods in no-tillage compared to conventional tillage. While the importance of fungivores was seasonal in no-tillage, bacterivore nematode abundance was higher in conventional tillage than in no-tillage throughout the year.

The hypothesis of Hendrix et al. (1986), based on microbial data and trophic structure of soil fauna communities, has since been largely verified by field experiments with biocides targeted at bacteria, fungi, and microarthropods (Beare et al., 1992). We were able demonstrate that interactions between fungi and fungivore microarthropods were important in regulating litter N dynamics in the no-tillage system. Although the results were less clear for bacteria, it was apparent that bacteria–bacterivore interactions were relatively more important in conventional tillage than in no-tillage systems. For these agroecosystems, nematode and microarthropod community analyses provided important insights into ecosystem-level processes.

Another important biological consequence of no-tillage is the development of large earthworm populations. At the same research site in Georgia, earthworm densities and biomass were consistently much higher in no-tillage compared to the conventional tillage system (Parmelee et al., 1990). Because of the large biomass, earthworms in these no-tillage systems contributed significantly to total C efflux from soil and processed large amounts of coarse and fine organic matter (Hendrix et al., 1987; Parmelee et al., 1990; Anderson, Ch. 10). From these results, it was reasonable to conclude that earthworms were also important in the cycling of N in these systems.

To attempt to quantify the direct role of earthworms in the N cycle of no-tillage agroecosystems, we estimated annual secondary production for the earthworm community (Parmelee and Crossley, 1988). Secondary production estimates the C flux through earthworm tissue as the sum of production of new earthworm tissue and loss through mortality. By multiplying the annual secondary production by the N content of earthworm tissue, it is possible to estimate the annual N flux through the earthworm community. To estimate secondary production, both population (temperature-dependent species-specific growth rates) and community (change in earthworm communtiy biomass over an annual cycle) parameters are required.

Laboratory growth rates at different temperatures were determined for the two dominant species at our field site, *Lumbricus rubellus* Hoffmeister, and *Aporrectodea turgida* Eisen. Growth rates were then regressed against temperature. With this regression, field soil temperature was used to estimate earthworm growth rates in the field. To estimate earthworm community biomass (all earthworm species included), we determined earthworm biomass (ash-free dry weight m^{-2}) at approximately monthly intervals for 1 year.

Secondary production for earthworms during any given time interval is calculated as the growth rate times the change in biomass $\{P = IGR \times [(B_f + B_i)/2] \times t$; where P equals production, IGR is the Instantaneous Growth Rate, and B_f and B_i are the final and initial standing stock biomasses observed over a time interval (t) measured in days; Romanovsky and Polishchuk, 1982$\}$. The values for all time intervals are then summed to calculate annual secondary production. To obtain an estimate of the N flux through earthworm biomass, the annual community secondary production estimate is multiplied by the average percent N of earthworm tissue. For our study in Georgia, we estimated that there was a 40 kg N ha^{-1} yr^{-1} flux through the earthworms in the no-tillage system (47 g ash-free dry mass m^{-2} yr^{-1} × 8.4% N earthworm tissue). When estimates of N excreted in urine and mucus were included, the total N flux through earthworms increased to 63 kg N ha^{-1} yr^{-1}. We concluded that this N flux was significant for the no-tillage system, and that earthworms could process 50% of the N input from plant residues, and could account for 38% of the N uptake by plants. Furthermore, the N flux through earthworm biomass exceeded the losses from the system by denitrification and leaching. This example, again, demonstrates that both population and community analyses can provide valuable information on ecosystem-level processes in agroecosystems.

RHIZOSPHERE PROCESSES

The rhizosphere is defined as the soil in intimate contact with live roots, which, because of this close association, has distinct chemical and biological characteristics compared to bulk soil. Live roots actively excrete a variety of low molecular weight C compounds that are readily assimilatable by microorganisms. The secretion of labile C compounds can stimulate the microbial community, and can provide energy resources to attack the more recalcitrant C compounds in the soil. With the increase in microbial activity, CO_2 concentrations increase in the rhizosphere. This can attract soil fauna such as protozoa, nematodes, and microarthropods that graze on bacteria and fungi. In the process of grazing on microbes, soil fauna incorporate a proportion of the N into their own biomass, but the rest is excreted as inorganic N, largely NH_4, that then becomes available for uptake by either microbes or plants. This process is commonly refered to as the "rhizosphere effect" (Clarholm, 1985; and others). With such a mechanism, soil fauna may act as regulators of N release in the rhizosphere, and their abundance should provide some insight into the rate at which N becomes available.

We have been studying rhizosphere processes in the mineral horizon of the spodosols of the New Jersey Pinelands (Parmelee et al., 1993a). The mineral horizon soils of these forests contain very little C or N, and, therefore, root inputs should be important substrates to support biological activity. To investigate the effects of live root (+mycorrhizal) inputs on ecosystem nutrient cycling activity, we measured microbial growth rates and the nematode and microarthropod community response to different amounts of pine seedling root biomass in a series of laboratory microcosms. Pitch pine seedlings (*Pinus rigida* Mill.) were planted at densities of zero, one, two, and four plants in mesh tubes that isolated the rhizosphere soil from surrounding bulk soil. The plants were grown for 20 weeks, and then the rhizosphere soil was analyzed.

As the amount of root biomass increased, there were significant increases in the growth rate of the microbial community. The number of nematodes and microarthropods also increased in rhizosphere soil with greater amounts of root biomass (Fig. 11-1). The nematode community was dominated by bacterivores whereas in the microarthropod community fungivorous types were the most prevalent, indicating there were active populations of both bacteria and fungi in the rhizosphere. In soils without roots, faunal numbers were close to zero, providing further evidence for the importance of roots in these soils in maintaining biological activity. The increase in soil fauna, and associated accelerated microbial growth rates as root biomass increased, suggested we were observing the classic rhizosphere effect. Ammonium and nitrate levels were very low at high root biomass levels, and it is likely that soil fauna were important in the turnover of N by grazing on microbes in the rhizosphere. We conclude from these studies on microbial growth rates and nematode and

Figure 11-1. Nematode (A) and microarthropod (B) numerical responses (±1 SE) in rhizosphere soil to increased pine seedling root biomass (as seedling density) in the mineral horizon of a coniferous forest spodosol. Different shaded bars are two different experiments, and different letters indicate significant differences among seedling density treatments at $P = .05$, within each of the two experiments (Modified from Parmelee et al., 1993a.)

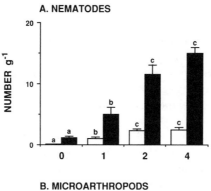

A. NEMATODES

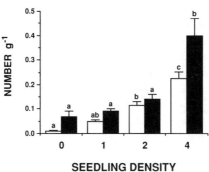

B. MICROARTHROPODS

SEEDLING DENSITY

microarthropod communities that live root inputs were important substrates for maintaining biological activity in the mineral soil horizons, and that soil fauna could be important regulators of ecosystem processes such as N turnover because of their abundance in the rhizosphere.

ECOTOXICOLOGY

Nematode and microarthropod population and community analyses can also be useful in addressing more applied ecological problems such as examining the effects of pollutants on terrestrial ecosystem function. The methodologies used in terrestrial toxicology have not been developed to the extent required to answer ecological questions concerning pesticide registration, Superfund site cleanup, the Toxic Substances Control Act, and other environmental problems. Many of the methods used to assess the toxicity of chemicals to soil fauna are single-species or a battery of single-species tests. Single-species tests are not adequate to assess effects of chemicals on exceedingly complex soil decomposer food webs consisting of many species. Clearly, more sophisticated tests are needed to enable scientists to accurately assess ecological damage to soils in terrestrial ecosystems. Tests that detect changes in community trophic structure or ecosystem-level processes will generate valuable information for ecological risk assessors and risk managers. Trophic structure and community analysis of soil nematodes and microarthropods in response to chemicals may prove to be particularly useful in achieving this goal.

In order to determine the utility of nematode and microarthropod community analyses in detecting ecological effects of pollutants, a laboratory microcosm technique was developed. Soil was collected from a mature oak-beech forest. Soil containing natural nematode and microarthropod communities was then treated with the heavy metal, copper, at a rate of 0, 100, 200, 400, and 600 μg g^{-1} (Parmelee et al., 1993b). After treated soils were incubated for 7 days, nematodes and microarthropods were extracted and enumerated. Total numbers of nematodes declined significantly at 200 μg g^{-1} copper, whereas total numbers of microarthropods did not decline until 400 μg g^{-1}. More detailed trophic and functional group analyses revealed even higher sensitivity to copper with certain groups showing negative effects at 100 μg g^{-1} (Fig. 11-2). The groups that showed the highest sensitivity to copper were potential predators of nematodes, omnivore–predator nematodes, and mesostigmatid and oribatid mites. With the decrease in these predators, there was a relative increase in abundance of herbivorous nematodes compared to controls where predation was unaffected. This suggested that we were able to detect a disruption of soil food web structure which, as discussed previously, has been shown to affect ecosystem function. For example, in forests exposed to copper pollution, our results suggest there could be increased levels of belowground herbivory with potential effects on forest productivity. The high sensitivity of nematode and

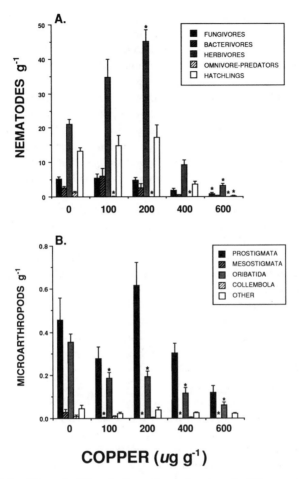

Figure 11-2.   Nematode (A) and microarthropod community (B) numerical responses
(±1 SE) to different levels of copper application. Asterisks indicate significant differences from
control treatment at $P = .05$ (From Parmelee et al., 1993b.)

microarthropod communities, and the ability to detect disruption of food web
structure and function indicates that analyses of the soil fauna community could
prove to be a useful tool for ecological risk assessment.

CONCLUSIONS

An important question that remains to be answered is: Which and how many
species are critical in affecting soil processes? The majority of the examples
presented in this chapter have relied on analyses of soil fauna communities to
make inferences about ecosystem structure and function. For nematodes and
microarthropods, the community is characterized by identifying trophic or

functional groups and not by identifying the species present in these groups. A weakness of our approach in attempting to link population and community parameters with ecosystem function is that we sacrifice fine-scale resolution. It is likely that within our trophic and functional groups dominant or keystone species are present that may largely explain our observed results. While the questions of which and how many species are critical in affecting soil processes need to be addressed, the success of our approach in answering a variety of questions in different ecosystems indicates that species-level research may not always be necessary to link levels of the ecological hierarchy.

The examples presented in this chapter demonstrate that population and community-level analyses of soil fauna can provide important insight into ecosystem-level processes. Soil nematode and microarthropod community trophic analysis contributed to the development of food web models for conventional and no-tillage agroecosystems. These conceptual models provided a framework for investigations on nutrient cycling processes in the agroecosystems. In the no-tillage system, population growth rates and community biomass estimates for earthworms were used to estimate the N flux through the earthworm community. The estimate indicated that earthworms were an important component of the N cycle in no-tillage agroecosystems. Soil nematode and microarthropod community analyses also provided insight into nutrient cycling processes in the rhizosphere of coniferous trees, and provided a sensitive indicator of disruption of food web structure and function following application of a pollutant.

The success of our approach to use soil fauna to link levels of the ecological hierarchy can be attributed to key aspects of soil faunal ecology. In the case of earthworms, we are dealing with communities that have a small number of species yet have large direct and indirect effects on nutrient cycling processes. Both of these aspects make it relatively easy to work at and link different levels in the ecological hierarchy. For nematodes and microarthropods, the usefulness of community analyses in predicting ecosystem structure and function is a result of feedback interactions with the microbial community, high diversity and abundance, and the presence of all trophic groups. If similar critieria could be identified in other nonsoil systems, then our approach could have broader applicability in promoting a more a holistic perspective in ecology. From the examples presented in this chapter, and an increasing number of similar studies in soil ecology, I conclude that studies of soil fauna are ideal for linking population, community, and ecosystem levels in the ecological hierarchy.

12

BEAVER AS ENGINEERS: INFLUENCES ON BIOTIC AND ABIOTIC CHARACTERISTICS OF DRAINAGE BASINS

Michael M. Pollock, Robert J. Naiman,
Heather E. Erickson, Carol A. Johnston,
John Pastor, and Gilles Pinay

SUMMARY

We review how beaver (*Castor canadensis*) affect ecosystem states and processes (e.g., biogeochemical cycles, nutrient retention, geomorphology, biodiversity, community dynamics, and structural complexity) by altering the physical properties of stream channels and riparian forests. Beaver influence the distribution, standing stocks, and availability of chemical elements by changing the hydrology; alter stream geomorphology as dams trap sediments; and change microclimates as water surface area increases and with the opening of previously closed forest canopies. Beaver also increase plant, vertebrate, and invertebrate diversity and biomass, and also alter the successional dynamics of riparian communities. Both abiotic and biotic influences are spatially extensive and long-lasting, affecting system level patterns and processes for decades to centuries. We conclude that ecosystem-level states and processes can be altered radically by the population dynamics of certain species such as beaver and that the magnitude and diversity of animal influence on ecosystems has been underestimated.

INTRODUCTION

Historically, beaver (*Castor canadensis*) have had an enormous impact on the streams and riparian forests of North America. Beaver modify drainage network morphology and hydrology by cutting wood and building dams. These activities retain sediment and organic matter in the channel, create and maintain physically diverse wetlands, modify biogeochemical cycles, alter the structure and dynamics of the riparian vegetation, influence the character of water and materials transported downstream, and broadly influence biotic composition and diversity (Naiman et al., 1994). The result of this ecosystem engineering (Lawton and Jones, Ch. 14) is a mosaic of temporally and spatially variable communities with long-term effects on landscape level features (Ives, 1942; Johnston and Naiman, 1990a,b).

The historical population of beaver is estimated to have been more than 60 million individuals (Seton, 1929) occupying an area of 15 million km², from the Canadian Arctic to the Sonoran Desert of Mexico (Jenkins and Busher, 1979). Trapping brought beaver close to extinction by 1900 A.D. (Jenkins and Busher, 1979), but today the population has rebounded to an estimated 6–12 million individuals and appears to be increasing (Naiman et al., 1988). Historical records refer to many low-order streams as a series of beaver impoundments characterized by slow meandering streams with extensive riparian zones (Morgan, 1868; Rudemann and Schoonmaker, 1938; Rea, 1983). Currently such streams are rare except in places where beaver receive some protection. While beaver existed historically at much higher densities than today, modern limnological research began in an era of low beaver density. Consequently, much of our understanding of how stream systems function is derived from stream systems that had been altered radically by the removal of beaver (Naiman et al., 1986, 1988).

By studying beaver influences on ecosystem processes, we gain an understanding of how drainage networks operated historically and how aquatic systems change as beaver populations fluctuate. The objectives of this chapter are to examine the abiotic and biotic influences of beaver on streams and forests and to identify key ideas and questions raised by contemporary research efforts.

ABIOTIC INFLUENCES

Beaver exert their influence on abiotic processes by altering the hydrological conditions of streams and adjacent forests. Changes in hydrology ultimately result in short- and long-term alterations to geomorphology, biogeochemistry, and microclimate. Widespread geomorphic alterations to stream and valley floors have occurred as a result of beaver dams. As beaver dams reduce stream velocities, sedimentation rates increase. For example, approximately 10,000 m^3 sediment km^{-1} of stream are retained by beaver dams in second- to fourth-order streams in the Matamek River and Moise River watersheds, Quebec, Canada (Naiman et al., 1986). Paleoecological evidence suggests that over millennia, entire valley floors have been raised by the continual building of ponds (Rudemann and Schoonmaker, 1938; Ives, 1942). Organic matter also accumulates in pond systems because anaerobic conditions there decrease decay rates (Naiman et al., 1994).

It has been well established that beaver impoundments increase the standing stock of carbon (C) in aquatic systems and that the export rate of C in the water column can be higher in drainages containing beaver, as compared to drainages where beaver are absent . In Beaver Creek, Quebec, standing stocks of C were approximately three times that of a nearby riffle (12×10^3 g m^{-2} vs. $4.3 \ 10^3$ g m^{-2}) (Naiman et al., 1986). In the same stream, export rates of dissolved organic C and fine particulate organic C were also higher from ponds than from riffles (121 g m^{-2} yr^{-1} vs. 51 g m^{-2} yr^{-1}). Ponds also export a small but significant portion of sediment C as methane gas. Annual methane fluxes ranged from 8–11 g C m^{-2} in permanently flooded areas to 0.2–0.4 g C m^{-2} in seasonally flooded meadows and adjacent upland forests (Naiman et al., 1991). Combining these data with information on beaver population size and areal extent of habitat alteration suggests that beaver may be responsible for about 1% of the recent rise in global atmospheric methane.

Biogeochemical changes associated with the formation of beaver ponds are quite variable, both in time and in space. Morphological variation in ponds creates a spatial array of patches, each with a distinct hydrologic regimen and resultant biogeochemical characteristics. Seasonal and storm-related fluctuations in water levels shift the redox status of soil patches within a beaver pond. These shifts in soil redox conditions have important implications for biogeochemical processes such as nitrogen (N), C, and phosphorus (P) cycling. For example, under anaerobic conditions the conversion of ammonium to nitrate (nitrification) is inhibited. This results in an accumulation of total organic N and ammonium. Preexisting nitrate is removed from the sediments by denitrification. Under aerobic conditions during drawdown, organic N and ammonium are converted to nitrate, and vegetative uptake increases rapidly.

Phosphorus availability also changes. With sediment reduction, adsorbed and chemically bonded P, mainly by ferrous ion, is released into solution making P potentially more available to plants . Moreover, the timing and frequency of water table fluctuations can affect the ability of the system to retain nutrients. For example, if water tables decline during periods when there is little vegetative growth, ammonium is nitrified and the soluble nitrate can be leached from the system.

Beaver impoundments also retain and accumulate other biologically important nutrients and elements such as potassium, calcium, magnesium, iron, and sulfate as well as nitrate and ammonium. When an area is flooded due to beaver activity, most of the antecedent soil nutrient pools are retained (Naiman et al., 1994). Nutrients released from the decay of vegetation killed by flooding, and upstream erosion further increase the standing stocks of nutrients, as well as increase downstream export rates. On Kabetogama Peninsula, Minnesota, over a period of 61 years, nutrient stocks in ponds and meadows increased from 39% for potassium to 400% for ammonium (Tab. 12-1). Such accumulations have important implications for aquatic system stability. Systems able to retain nutrients can more rapidly recover from disturbances than systems where nutrients are easily lost (Webster et al., 1975).

Beaver often abandon ponds due to predation or a reduction in food supplies. Without regular maintenance, most dams will eventually collapse. Once ponds are drained, N and P mineralization rates vary with hydrology and may determine successional pathways. Phosphorus and N availabilities were greater in the wetter, more hydrologically dynamic, sedge-dominated communities as compared with the drier, primarily grass-dominated communities (Fig. 12-1, H.E. Erickson, unpublished data). The greater nutrient availabilities correspond to higher net primary productivities in the plant communities.

Beaver can alter nutrient availability across the landscape as well. Naiman et al. (1994) determined the changes in total N and plant-available N within various communities affected by beaver during a 46-year period over an area of 294 km². Although beaver did not significantly change the amount of total N, they did increase the amount of N available to plants in moist meadows,

Table 12-1. Absolute amounts of ions and nutrients in 1927 and 1988 associated with habitat influenced by beaver activities*

| Parameter | Absolute Amounts (kg) | | Percentage Increase (%) |
	1927	1988	
Total Nitrogen	5.8×10^5	1.0×10^6	72
Nitrate-N	3.9×10^2	1.2×10^3	208
Ammonium-N	2.1×10^4	8.3×10^4	295
Total phosphorus	7.0×10^4	1.0×10^5	43
Potassium	3.0×10^5	3.6×10^5	20
Calcium	1.9×10^6	4.5×10^6	137
Magnesium	3.5×10^5	9.4×10^6	169
Iron	1.1×10^4	2.4×10^4	118
Sulfate	3.3×10^4	6.0×10^4	82

* For, 1927 we included the total forest area (2,563 ha) that was later impounded by beaver. From Naiman et al. (1994).

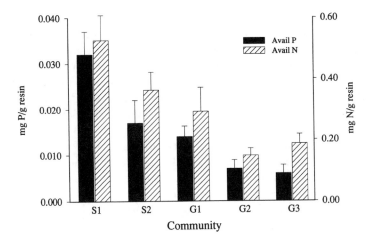

Figure 12-1. Available phosphorus (P) and nitrogen (N) (± 1 SE) in two types of wet meadow communities in abandoned beaver ponds on the Kabetogama Peninsula, Voyageurs National Park, MN. N and P extracted with $1N$ KCL from mixed bed ion-exchange resin (Baker, AGMI-615) placed in the organic horizon over a 7-month period (10/89–5/90). S = sedge-dominated communities; G = grass-dominated communities.

wet meadows, and ponds. Overall, on the peninsula, N available to plants increased from 65 kg km^{-2} to 296 kg km^{-2}.

Microclimate alterations in riparian areas, caused by the presence of beaver, likely include changes in relative humidity, available light, and temperature regimens. For example, in Minnesota, cumulative annual sediment temperatures in beaver-created ponds and wet meadows were 33% greater than in moist meadows and forest soils (Naiman et al., 1991). We expect that forest soils, which are heavily shaded, would have lower temperatures than the more open meadows and ponds, but the relatively low temperatures observed in the dry meadow suggests that sediment temperature may be significantly affected by the high heat capacity of water.

ABIOTIC INFLUENCES: KEY IDEAS AND QUESTIONS

Comprehensive studies from the boreal forest biome document the influences of beaver on hydrological, biogeochemical, and geomorphological processes in streams and riparian forests. The same basic abiotic processes occur in other biomes, but process rates and ecological states may be different. For example, in arid, erosion-prone regions of the west, beaver impoundments may trap sediments at a much faster rate and expansion of wetland areas will have far more important biological implications as compared to boreal regions where wetlands are more common and erosion rates are generally low.

The effect of beaver impoundments on the water table and subsequent alterations to the quantity and timing of water released to streams is also an important issue, particularly in areas where water is scarce and the effect of recolonizing beaver populations on water supplies is of concern. Some evidence suggests that beaver impoundments raise water tables and provide a seasonally steady water supply to streams and riparian vegetation (Apple et al., 1984), but the generality of this observation requires documentation.

The influence of channel morphology on the dynamics of beaver impoundments is not well understood. Channel form affects the long-term abiotic characteristics of abandoned impoundments by significantly altering sediment retention, hydrologic regimens and successional dynamics. A more precise understanding of the variation in beaver influences as a function of channel morphology would allow better predictions of the potential influence of beaver in particular streams or stream reaches.

BIOTIC INFLUENCES

Beaver influence the community dynamics of riparian vegetation (Johnston and Naiman, 1990b), instream benthic organisms (McDowell and Naiman, 1986), fish (Dahm and Sedell, 1986), wildlife (Peterson and Low, 1977; McKelvey et al., 1983), the structural and community diversity of streams and forests (Naiman et al., 1994; Johnston and Naiman, 1990a), and the nutritional value of certain tree species (Basey et al., 1990).

Forest and wetland communities respond to the selective cutting of aspen (*Populus tremuloides*), to altered hydrological regimens associated with dam building, and to biogeochemical changes associated with the formation of hydric soils (Naiman et al., 1988). In boreal forests, the removal of large trembling aspen by beaver leads to a more open canopy structure, an increase in aspen root sucker density, and an increase in the relative importance of nonfood species such as tag alder (*Alnus rugosa*) and white spruce (*Picea glauca*) (Johnston and Naiman, 1990b). Such changes also affect the population dynamics of other herbivores, such as moose, that browse aspen (Pastor and Naiman, 1992).

Diverse wetland plant communities in beaver ponds and meadows can provide variation in structure, habitat, and food supply in riparian systems. On the Kabetogama Peninsula, Minnesota, beaver have converted 13% of forested peninsula during the last 60 years into a mosaic of different types of wetland communities (Johnston and Naiman, 1990c). The riparian system has expanded and now includes 32 different aquatic, emergent, shrub, and forested wetland communities at different successional stages. When a pond drains after beaver abandon a dam, wet meadows form on the newly exposed sediments. Unless the beaver return to reoccupy and reflood a site, these moist meadows appear to be quite stable. Several factors may account for their persistence, including hydric soils that are inhospitable to tree establishment and direct competition from grasses. Whatever the mechanism contributing to the longevity of the meadows, it seems clear that beaver have altered riparian successional dynamics by increasing the areal extent of a stable community type.

In southeast Alaska, ongoing unpublished studies by two of us (M.M.P. and R.J.N.) suggest that beaver create unique communities in the riparian landscape that are biologically and structurally diverse. Of six types of common wetland communities (floodplain forest, wet meadow, beaver pond, lower and upper active channel shelf, and depositional bar), beaver ponds were the most diverse, containing 1.4–2.7 times as many plant species as the other communities (Fig. 12-2). High floral diversity is apparently correlated with high temporal variability of the disturbance (flooding) regimen during the growing season and the spatial heterogeneity of soil substrate, available light, and topography at scales of 0.5–20 m. Together these studies suggest that beaver-influenced riparian systems are floristically diverse because of the high spatial heterogeneity of environmental parameters at multiple scales caused by temporally complex flooding regimens.

Figure 12-2.   Plant species richness and snag (>50 cm dbh) density of three beaver ponds and adjacent riparian forest (each site = 0.25 ha), Tongass National Forest, Chichagof Island, Alaska. Sites: KP = Kadashan pond, KF = Kadashan forest, C1P = Crab Bay pond #1, C1F = Crab Bay forest #1, C2P = Crab Bay pond #2, C2F = Crab Bay forest #2.

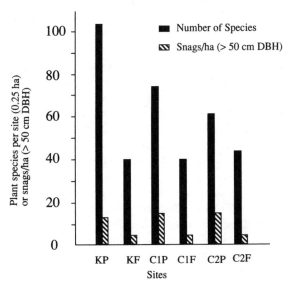

Stream benthic invertebrate communities are also strongly influenced by beaver activities. Decreases in current velocities and the deposition of fine sediments in beaver impoundments allow soft-bottom communities to develop dominated by Tanypodinae and chironomid midges, dragonflies, tubificid worms, and clams (McDowell and Naiman, 1986). Similar communities are found in the slower moving habitats of larger-order streams, but in drainages with streams less than or equal to fourth-order, they represent a unique assemblage. Invertebrate diversity is not increased by beaver activity, but biomass is considerably higher in the ponds (1.3–11.1 g m^{-2}, depending on season) than in unimpounded riffles (0.01–0.6 g m^{-2}) (McDowell and Naiman, 1986).

Similar patterns of biomass and diversity were observed for small mammal populations associated with beaver impoundments (Medin and Clary, 1991). Species diversity of a riparian area and a beaver impoundment were similar, but the small mammal biomass of the beaver pond was much greater, 819 g ha^{-1} vs. 304 g ha^{-1} for the riparian area. There were also differences in community composition. Beaver ponds were dominated by herbivores (*Microtus* spp.) and insectivores (*Sorex* spp.), whereas the riparian area was dominated by omnivores (*Peromyscus maniculatus*) and herbivores (*Microtus montanus*).

Beaver enhance structural diversity by felling trees and creating snags (Fig. 12-2) when rising water tables kill trees (Naiman et al., 1994; M.M. Pollock and R.J. Naiman, unpublished). Browsing of trees by beaver removes about 1.4 Mg ha^{-1} yr^{-1} in the 12–15% of the landscape that is affected by beaver herbivory (Johnston et al., 1993). Most of this biomass is not consumed and instead adds to the structural complexity of forests and streams in the form of downlogs and dams. Riparian lowland areas are habitat for large trees. Flooding by beaver creates numerous large snags critical to the diversity and productivity of avian, mammalian and insect communities (Maser et al., 1988). For example, snags with large, well-insulated cavities are important for overwintering bird species such as the pygmy nuthatch (*Sitta pygmaea*) where communal roosting is a survival tactic employed to minimize heat loss (Guntert et al., 1986). These structures may also be critical to the development of avian community foraging behavior and survival (Rendell and Robertson, 1989). In general, beaver activity creates snags, an increasingly rare but important structural element for the survival of numerous mammalian, avian, and invertebrate species.

Finally, beaver also affect the chemical composition of some tree species. *Populus* species appear to have chemical defenses to reduce grazing by beaver. Basey et al. (1990) demonstrated that beaver avoided shoots that regenerated from the stumps of cut trees. A previously unknown secondary compound was isolated from the shoots and shown to be an effective deterrent against beaver grazing when applied to shoots lacking this compound. This negative feedback loop could help to explain why *Populus* and beaver have historically been able to maintain a dynamic equilibrium in riparian forests.

BIOTIC INFLUENCES: KEY IDEAS AND QUESTIONS

Current questions of interest regarding the influence of beaver on biological processes include the relationship between beaver population dynamics and long-term successional dynamics of beaver-influenced habitats, the use of beaver to restore riparian communities in degraded habitats, and the effects of beaver recolonization on the biota of regions where they have been extirpated.

Fish and wildlife managers have typically managed for steady-state populations. However, from what is known about the effects of beaver, it appears that fluctuating beaver populations improve the health and diversity of aquatic systems. Beaver pond abandonment and recolonization is necessary for producing a rich array of productive and diverse communities. The natural variation in beaver population levels and the effect of population fluctuations on the biota of streams and forests has yet to be established.

Widespread trapping of beaver in eastern Oregon in the 1800s apparently initiated hydrologic and geomorphic changes in stream systems that eliminated deciduous riparian forests (Elmore and Beschta, 1987). Prior to trapping pressures, beaver impoundments raised the water tables in these arid systems and increased the effective area of the riparian corridor. Dam failure subsequent to beaver removal channelized streams, causing erosion and downcutting. This resulted in systems with narrow or no riparian vegetation zones, lowered water tables, and highly variable and intermittent stream flows (Apple et al., 1984).

Beaver are slowly recolonizing some of these areas and the intentional use of beaver to restore riparian systems in this region is also being explored (Apple et al., 1984). However, initial colonization by vegetation to a level that can support beaver populations has proven difficult, especially where intense grazing by livestock has precluded the reestablishment of *Salix* and *Populus* species, both critical food supplies for beaver (Jenkins and Busher, 1979). These natural and human-assisted recolonization efforts provide an opportunity to obtain detailed information on the biological components of a drainage before and after beaver colonization. Such experimental data would provide direct evidence for system-level influences by beaver.

IMPACTS OF OTHER ANIMALS

The system-level impacts described in this chapter are not unique to beaver. A wide variety of mammals, arthropods, and nematodes are able to alter ecosystem states and processes (Andersen, 1987a; Lawton and Jones, Ch. 14).

Foraging by arthropods and nematodes on roots (Andersen, 1987a), foraging by bush elephant (*Loxodonta africana*) on acacia (*Acacia*) trees in Africa (Hatton and Smart, 1984), grazing by mammals on the Serengeti grasslands (McNaughton and Georgiadis, 1986; McNaughton et al., 1988), foraging by snow geese (*Anser caerulescens*) in subarctic salt marshes (Jefferies, 1988), and burrowing and foraging by prairie dog (*Cynomys*) and pocket gopher (*Geomys bursarius*) in North America (Holland and Detling, 1990) affect biogeochemical and geomorphological processes, as well as affecting the productivity and composition of vegetation.

A prime example of behavior in an animal other than beaver that strongly influences system level processes is the burrowing of fossorial rodents (Jones et al., 1994). Early in this century in the western United States, nearly 25% of the mammalian species were fossorial rodents representing nearly one-half of the total number of mammals (Grinnell, 1923). In natural settings free-living gopher and ground squirrel move 10,000–90,000 kg ha^{-1} yr^{-1} of subsurface soil to the surface (Andersen, 1987b). These geomorphological and biogeochemical alterations increase resource heterogeneity and species richness as well as alter the successional dynamics of vegetation (Huntly and Inouye, 1988).

Animal–system interactions, as a general phenomenon, are complicated because many animal population cycles occur over long periods (e.g., decades), alterations to the ecosystem are often subtle over short periods (e.g., vegetative replacement or altered soil characteristics), and shifts in biogeochemical cycles or soil formation are not usually detectable over short periods (e.g., years). Nevertheless, animal-induced alterations to the ecosystem often result in a heterogeneous landscape that would not occur under the dominating influences of disturbance (such as fire), climate, or geology alone.

ACKNOWLEDGMENTS

We wish to thank Jennifer Sampson at the Center for Streamside Studies for her comments, criticisms and insights, which substantially improved the quality of this manuscript.

13

ATMOSPHERIC OXYGEN AND THE BIOSPHERE

Heinrich D. Holland

SUMMARY

Atmospheric oxygen is generated photosynthetically at such a rapid rate that the O_2 content of the atmosphere would be doubled in a few thousand years if all the photosynthetic O_2 were allowed to accumulate. It is not. The rate of O_2 production is nearly balanced by the rate of O_2 consumption by respiration. Together these processes constitute a very rapid geochemical cycle. The cycle is not, however, completely closed. A few tenths of a percent of the photosynthetically produced organic matter are buried with marine sediments, and an equal percentage of the photosynthetically produced O_2 remains as the net O_2 generated by the biological cycle. This O_2 is consumed by the oxidative weathering of rocks and by the burning of volcanic gases. The reason why the partial pressure of oxygen in the atmosphere is 0.2 atm seems to be related to a feedback mechanism involving PO_4^{3-}, a nutrient that limits the biological productivity of the oceans.

The early earth almost certainly was clothed in a mildly reducing atmosphere. The earliest organisms probably used photosystem I and hence did not generate O_2. The origination of photosystem II has not been dated, but it probably preceded the major rise in P_{O_2} about 2,100 million years ago. A role of biology in this rise cannot be ruled out but is not necessary. Additional increases in P_{O_2} just before the Cambrian explosion and following the rise of the higher land plants have been suggested but are not yet confirmed. Despite a good deal of research, the evolution of atmospheric O_2 is still rather imperfectly known, but it is quite clear that the O_2 level in the atmosphere has been determined by the interplay of biological evolution and the physical and chemical evolution of the planet. Individual species do not seem to have mattered.

INTRODUCTION

The topic of this chapter is very ambitious. I am taking the entire biosphere as my biological province and the entire planet as my ecosystem. To complete the hubris, I propose to look at the history of the interaction of the biosphere with the oxygen content of the atmosphere during all of earth history. The enterprise is saved by the word limit imposed by the editors. There must, perforce, be a great deal of lumping, and the conclusions must be quite general. But I will not be too apologetic, because I believe that they are useful of themselves and as an extension of those reached in other studies that are reported in this volume.

OXYGEN PRODUCTION AND USE TODAY

The biosphere produces oxygen at a prodigious rate. By the most recent estimate (Melillo et al., 1993) the terrestrial biosphere produces ca. 53×10^{15} g organic carbon (C)/year and hence ca. 140×10^{15} g O_2/year; the marine biosphere produces roughly an equal amount. At this rate the entire inventory of atmospheric O_2, 1.2×10^{21} g, is produced in about 4,000 years. This is an astonishingly short time, but it is not meaningful as an indicator of the history of atmospheric oxygen, because most of the oxygen produced during green plant photosynthesis is consumed almost immediately in respiration. In a very rough way the well-known equation

$$CO_2 + H_2O \xrightleftharpoons[\text{respiration}]{\text{green plant photosynthesis}} CH_2O + O_2 \qquad (13\text{-}1)$$

illustrates this point. Much of the organic matter produced by green plant photosynthesis on land is consumed within a few years, and little remains unoxidized for more than a few hundred years. In the oceans, where the mean life of microorganisms is measured in days, the cycling of C between the atmosphere and the biosphere is even more rapid.

The rate of photosynthesis on land depends more on the CO_2 content of the atmosphere than on its O_2 content. However, at very high O_2/CO_2 ratios photorespiration can approach or even exceed photosynthesis. The rise in the CO_2 content of the atmosphere due to fossil fuel burning will probably increase terrestrial photosynthesis (Melillo et al., 1993), especially where this is not limited by the availability of water, light, and plant nutrients. In most of the oceans photosynthesis is limited by nutrient availability, and an increase in the CO_2 content of the atmosphere will almost certainly affect marine photosynthesis only slightly.

The make–break cycle of photosynthetic O_2 is nearly closed (see Fig. 13-1). Very little organic material escapes being respired. The net annual increase in the O_2 content of the atmosphere due to biological processes is therefore very

Figure 13-1. The biological parts of the carbon cycle. The C content of the several reservoirs is in Gt (1 Gt = 10^{15} g); the data are largely from compilations by Sundquist (1985, 1993).

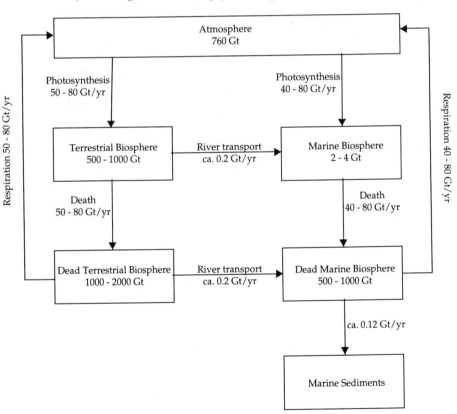

small compared to the rate of O_2 generation by photosynthesis. If we were some-how to stop green plant photosynthesis altogether, O_2 production would cease. During the succeeding centuries the available biomass would disappear, and consumers and decomposers—including *Homo sapiens*—would almost certainly die off. However, the O_2 content of the atmosphere would decrease only slightly, because the oxidation of the entire biosphere would consume only a few percent of the O_2 content of the atmosphere. There are overwhelmingly good reasons to preserve the biosphere, but maintaining atmospheric O_2 is not one of them.

The fact that the make–break cycle of photosynthetic O_2 is not absolutely closed has important long-term consequences for the O_2 content of the atmos-phere. Organic matter is buried with sediments. Most of the burial takes place at sea, where sediments contain 0.5–1.0% organic C on average. The annual burial rate of organic C is ca. 1.2×10^{14} g/year. This is only about 0.3% of the amount of C generated annually in the oceans by green plant photosynthesis. The remaining 99.7% is respired. Most of it is decomposed in the upper few hundred meters of the oceans. The oxidation of organic matter depends some-what on the O_2 content of the atmosphere. In the absence of atmospheric O_2, no O_2 would be dissolved in seawater, fish could not exist, and the decompo-sition of organic matter would be almost entirely bacterial. The rate of bacte-rial decomposition of organic matter seems to depend only slightly on the O_2 content of seawater. In areas where the O_2 content of seawater is low, bacte-ria use NO_3^- and then SO_4^{2-} as oxidants.

Perhaps somewhat surprisingly, the burial of even such a small fraction of the total quantity of organic C is highly significant for the oxygen balance of the atmosphere. Equation (13-1) shows that the synthesis of 1 mol of CH_2O liberates 1 mol of O_2. If this mol of CH_2O is subsequently respired back to CO_2 and H_2O, the mol of O_2 is consumed, and there is no net gain of atmos-pheric O_2. If, however, the CH_2O is buried with marine sediments and is not respired, the mol of O_2 is not consumed and can remain in the atmosphere. The burial of 1.2×10^{14} g organic C corresponds to the burial of 1.0×10^{13} mol of C. An approximately equal number of mols of O_2 are added to the at-mosphere. The O_2 gain of the atmosphere therefore amounts to ca. 3×10^{14} g O_2/year. If all of this O_2 were allowed to accumulate indefinitely in the at-mosphere, it would take only ca. 4 million years to double the O_2 content of the atmosphere. That is obviously a very long time on a human scale, but it is very short on the geological time scale; the earth is, after all, 4,500 mil-lion years old. This raises the question why the O_2 content of the atmosphere is as small as it is. The rate of burial of organic matter is not abnormally high today. The quantity of O_2 generated by the photosynthetic production of the C that was buried with sediments during the last 400 million years alone is therefore a hundred times greater than the present O_2 inventory of the atmosphere.

The excess O_2 must have been used up somehow, and it does not take very long to identify the two major O_2 sinks: oxidative weathering and the burning of volcanic gases. A rather large number of elements are oxidized during the weathering of rocks exposed at the surface of the earth. Of these, only C, sulfur (S), and iron (Fe) are quantitatively important for the O_2 budget of the atmosphere. Organic matter that was buried with ancient sediments can be brought back up to the surface and reoxidized to CO_2. This is probably accomplished by bacteria. Sulfide can be oxidized to sulfate, and Fe^{2+} can be oxidized to Fe^{3+}. The rate of O_2 loss due to oxidative weathering is roughly equivalent to three quarters of the O_2 gain by the burial of organic matter, sulfides, and Fe^{2+} minerals in modern sediments. The remaining quarter is lost by the oxidation of volcanic gases, which always contain H_2, CO, and SO_2. It is difficult to determine these loss rates precisely, but the best estimates of the gains and losses of O_2 indicate that they are approximately equal, and that the O_2 budget of the atmosphere is nearly in balance (see for instance Holland, 1978).

The nature of coal deposits indicates that the O_2 content of the atmosphere has varied rather little during at least the past 350 million years. Coal deposits always seem to contain the remains of some charcoal. This shows that forest fires have occurred as long as there have been forests. Because it is virtually impossible to start fires when the O_2 content of the atmosphere is < ca. 15% (i.e., 0.15 atm), the presence of charcoal indicates that P_{O_2} has been $\geq$ ca. 0.15 atm since the earliest days of coal formation (Watson et al., 1978). Most coal deposits also contain the remains of large trees. Their presence indicates that the frequency of fires has been sufficiently low and the interval between fires sufficiently long, so that trees have been able to attain considerable heights. The frequency and the intensity of fires increases with the O_2 content of the atmosphere. This sets a somewhat uncertain upper limit of 0.30–0.35 atm on the O_2 content of the atmosphere during the past 350 million years. These observations show that O_2 gains by organic matter burial with sediments has been essentially balanced by O_2 loss due to weathering and due to the oxidation of volcanic gases. Short term imbalances between O_2 production and O_2 use cannot, of course, be ruled out.

The current situation is summarized in Fig. 13-2. The upper curve shows that at today's O_2 pressure of 0.2 atm, the rate of O_2 input, dM_{O_2}/dt, is ca. 3×10^{14} g/year. The lower curve shows that at the present value of P_{O_2} the O_2 output is also ca. 3×10^{14} g/year. The net rate of change in the O_2 content of the atmosphere is shown by the middle curve. As the O_2 input rate today is close to the O_2 output rate, the net change of the O_2 content of the atmosphere is zero or close to zero. This could be a happenstance, but the near-constant value of P_{O_2} during a time that is 100 times the residence time of O_2 in the atmosphere makes a happenstance seem rather unlikely. It is much more likely

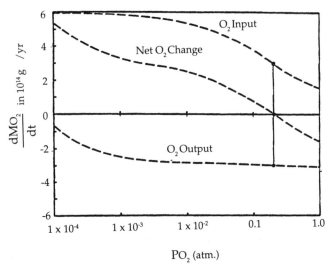

Figure 13-2. Input, output, and net change of atmospheric O_2 in the preindustrial era as a function of the O_2 content of the atmosphere. (Modified from Holland, 1991.)

that PO_2 is and has been controlled by a feedback system that has kept PO_2 at or close to 0.2 atm.

The O_2 output rate is currently only a weak function of PO_2. The oxidation of volcanic gases is independent of PO_2 unless the O_2 content of the atmosphere is essentially zero. The rate of O_2 loss by weathering is also a weak function of PO_2, because at the present O_2 level the oxidation of organic C, S^{2-}, and Fe^{2+} in rocks exposed to weathering is nearly complete. The feedback control mechanism therefore depends almost certainly on the linkage between PO_2 and the O_2 input rate.

There are at least two such connections. One is the burial efficiency of organic matter with marine sediments. It had been suggested that this was higher in anoxic than in oxygenated areas of bottom waters. However, recent results (Calvert et al., 1992; Pedersen et al., 1992) and a detailed analysis of earlier data (Betts and Holland, 1991) showed that the burial efficiency of organic matter is only weakly correlated with the O_2 content of sedimentary environments. The marine chemistry of phosphate is probably a stronger connection between atmospheric O_2 and the rate of O_2 input in the atmosphere. PO_4^{3-} is one of the major limiting nutrients in the oceans. In many parts of the oceans photosynthesis stops when PO_4^{3-} is exhausted or nearly exhausted in the photic zone (see for instance Broecker and Peng, 1982). Most of the PO_4^{3-} that is built into organisms during photosynthesis is released during respiration and becomes available for the next generation of photosynthesizers. Some of the PO_4^{3-} that is sequestered by organisms is removed from the oceans as a constituent of organic matter that is buried with marine sediments. This organic matter has a complex fate. Some of it is bacterially destroyed. The decomposition products, including PO_4^{3-}, tend to diffuse back into the water column. However, in reduced sediments some of the released PO_4^{3-} is precipitated en route as apatite $[Ca_5(PO_4)_3(OH, F)]$ (Ruttenberg and Berner, 1993). In oxidized sediments some of the PO_4^{3-} is removed as a constituent of a $Fe(OH)_3$–$FePO_4$ solid solution (Sherwood et al., 1987; Ruttenberg, 1990).

On a geological time scale the rate of removal of PO_4^{3-} from the oceans cannot exceed the rate of PO_4^{3-} input. The maximum rate of organic C burial with marine sediments is therefore equal to the rate of PO_4^{3-} input to the oceans times the average C/P ratio of permanently buried organic matter. The actual rate of burial of organic matter is less than this, because PO_4^{3-} is removed from the oceans by other mechanisms as well. At present roughly one third of the river input of PO_4^{3-} is removed as a constituent of buried organic matter. The other two thirds are largely removed as a constituent of apatite and $Fe[(OH)_3PO_4]$. These proportions are a function of the oxidation state of sediments, of the oxidation state of the marine water column, and hence of the O_2 content of the atmosphere. In the absence of atmospheric O_2 marine sediments would be anoxic and the removal of PO_4^{3-} from the oceans as a constituent of $Fe[(OH)_3PO_4]$ would almost certainly be negligible. More PO_4^{3-} would then be available for burial with organic matter. The rate of O_2 production via photosynthesis followed by organic matter burial would therefore be greater than it is today. Conversely, at O_2 pressures significantly greater than the present 0.2 atm, sediments that are now anoxic would be oxygenated, more of their iron would be present as Fe^{3+}, and hence more PO_4^{3-} would be removed from the oceans as a constituent of $Fe[(OH)_3PO_4]$. Less PO_4^{3-} would then be available to be buried with organic matter, and the rate of O_2 production would be less than under present-day conditions. The PO_4^{3-}-O_2 connection therefore serves as part of an intriguing negative feedback system that tends to stabilize the O_2 content of the atmosphere. Unfortunately, too little is known about the quantitative aspects of the PO_4^{3-}–O_2 connection to assess the strength of this P_{O_2} control mechanism.

In the proposed control mechanism as explained above, the biology is highly aggregated and the control mechanism has been explained without regard to species. The phytoplankters have not been specified, nor have the decomposers in the food chain been identified. The control mechanism has also bypassed all of physical oceanography. Perhaps all this will turn out to be justified, if PO_4^{3-} availability exerts a strong control on the burial rate of organic matter and if PO_4^{3-} availability is strongly tied to the O_2 content of the atmosphere.

ATMOSPHERIC OXYGEN IN THE DISTANT PAST

In the absence of life the earth would almost certainly have had a mildly reducing atmosphere (Kasting, 1987). The most important reason for this is that volcanic gases are mildly reducing. Part of the hydrogen in these gases would have escaped from the top of the atmosphere, and part of the remainder would have combined with CO and CO_2 to form organic compounds under the impetus of photochemical reactions and in lightning discharges; but there would almost certainly have been a significant H_2 pressure in the atmosphere, perhaps on the order of 10^{-3} atm, and the O_2 pressure would have been vanishingly small. If the first living organisms used photosystem I, they made use of H_2 and H_2S in their environment to reduce CO_2 to organic compounds. In a very simplified manner these reactions can be written in the form

$$CO_2 + 2H_2 \xrightarrow{\ h\nu\ } CH_2O + H_2O \tag{13-2}$$

and

$$CO_2 + 2H_2S \xrightarrow{\ h\nu\ } CH_2O + H_2O + 2S \tag{13-3}$$

Some genetic evidence suggests that the earliest organisms were thermophiles. Perhaps they lived in the vicinity of hot springs, where H_2S is abundant. At any rate, they did not generate molecular oxygen, and the atmosphere almost certainly stayed mildly reducing.

This situation changed with the origination of photosystem II, but it may not have changed dramatically. As long as the net rate of O_2 production was less than the rate of input of volcanic H_2 and CO, the atmosphere probably remained mildly reducing. There may well have been a long time lag between the origination of photosystem II and the appearance of a sizable quantity of atmospheric O_2. This is suggested by the geologic history of the oxidation state of soils and sediments. Today, the chemistry of soils is strongly influenced by the highly oxygenated nature of the atmosphere. In Hawaii and in other tropical areas where basalts are common, many soils are rusty red, because Fe^{2+} in the basalts is oxidized to Fe^{3+} during weathering, and is precipitated almost immediately as $Fe(OH)_3$ and/or $FeO(OH)$. This seems to have been true in all or nearly all soils that were formed during the past 1,850 million years (Holland, 1994). However, in soils that are older than 2,200 million years, Fe^{2+} in parent basalts was not oxidized to Fe^{3+} but was washed downward together with Ca^{2+}, Mg^{2+}, Na^+, K^+, and the other cations that were released during weathering. The Fe^{2+} was precipitated in part as one or more Fe^{2+}–silicate minerals in the lower parts of these soils. The nonoxidation of Fe^{2+} sets an upper limit of about 0.002 atm on the O_2 pressure during soil formation more than 2,200 million years ago. Between 2,200 and 1,850 million years ago P_{O_2} apparently rose to values at least as high as 0.030 atm. Several other lines of evidence agree with such a large change in P_{O_2}. A number of minerals that are oxidized rapidly during weathering today were quite resistant to weathering more than 2,200 million years ago, and the oxidation products of minerals containing Fe^{2+} become abundant in sedimentary rocks between ca. 2,100 and 2,200 million years ago.

The proposed, rapid change in P_{O_2} between 2,200 and 2,100 million years ago might be due to the origination of photosystem II at that time, but there is little paleontological evidence in favor of this hypothesis. Fig. 13-2 suggests an alternative explanation. Early in its history the Earth was undoubtedly more active. Approximately 3,000 million years ago the amount of heat generated by the radioactive decay of U, Th, and ^{40}K was greater by about a factor of 2 than it is today. Volcanism, mountain building, and sea floor spreading were all more intense. Hence the rate of oxygen generation required to maintain the O_2 level in the atmosphere at its present level was significantly greater than today. However, solar insolation may well have been lower, the mean annual temperature no higher, and the chemistry and flux of river water roughly equal to today's. The PO_4^{3-} input to the oceans may have been no greater than it is today. If so, all or nearly all of the river PO_4^{3-} would have been required to balance oxygen consumption by weathering and the burning of volcanic gases. This could be done, but only if the O_2 content of the atmosphere was much lower than 0.2 atm. If the O_2 output rate in Fig. 13-2 were increased by a factor of 2 or 3, the equilibrium value of P_{O_2} would be shifted to much lower values. If the O_2 output rate increased even more, there would be no stable equilibrium point, and P_{O_2} would be essentially zero. If this interpretation is correct, the rise in P_{O_2} about 2,100 million years ago was not due not to biological innovations but was simply a consequence of the evolution of the earth as a planet.

There does seem to have been a biological consequence to the rise in P_{O_2} ca. 2,100 million years ago. *Grypania*, the earliest known eukaryotic fossil, stems from this period (Runnegar, 1994). *Grypania*, like all known eukaryotes, was probably unable to fix atmospheric N_2. In contrast, some of the cyanobacteria knew how to do so, and it is likely that the ability to fix N_2 was present in the common ancestor of all cyanobacteria. This suggests that combined nitrogen was in very short supply on the early Earth. The absence of the ability to fix N_2 in autotrophic eukaryotes suggests that NO_3^- and/or NH_4^+ have been sufficiently available since 2,100 million years ago, so that the ability to fix N_2 was no longer an absolute necessity. This is rather what one might have expected. Inorganic N_2 fixation in the atmosphere increases with increasing atmospheric oxygen. Much of the fixed N_2 rains out into the oceans and is available as a nutrient. The problem of NO_3^- availability therefore became less acute ca. 2,100 million years ago. Another problem also became less pressing at that time. With the rise in P_{O_2} stratospheric ozone became an effective shield against solar UV radiation (Levine, 1991). Organisms that developed since then have therefore not needed as effective a genetic repair mechanism as earlier organisms.

The biological explosion at the end of the Proterozoic and the beginning of the Phanerozoic Eon some 540 million years ago has been ascribed to a further increase in PO_2 at that time. There is little evidence for such an increase in the sedimentary record. However, the tools to detect biologically significant but quantitatively minor increases in PO_2 are not yet available, and the question is still moot. The same is true for another proposed increase in PO_2 following the population of the continents by higher land plants roughly 400 million years ago.

ACKNOWLEDGMENTS

The author wishes to thank E.A. Zbinden, C.R. Feakes, A.W. Macfarlane, and A. Danielson for their many contributions to the oxygen story, A.H. Knoll for stiumulating and enlightening discussions, and NASA for support under NASA Grant NAGW-599 to Harvard University.

APPROACHES

Given the diversity of reciprocal interactions between species and ecosystems seen in Chapters 3–13, how should we tackle integration of population/community ecology and ecosystem ecology? Despite Brown's vision of populations, communities, and ecosystems as Complex Adaptive Systems (Ch. 2), integration is unlikely to be achieved within the framework of one grand unifying theory, at least not in the near future! On the other hand there is a substantial diversity of methodological and conceptual approaches to integration that can be used, and these are brought together in the next 14 chapters. They include modelling and scaling, that is, how to aggregate functions, processes, or entities in space and time; how to use both trophic and nontrophic concepts; how to address constraints of stoichiometry and energetics that apply to both species and ecosystems; how to incorporate evolution; and how to directly couple processes such as population dynamics and material cycles.

Many organisms create or modify habitats. These "ecosystem engineers," as they are called by Lawton and Jones (Ch. 14), directly or indirectly modulate the availability of resources to other species, by causing physical state changes in biotic or abiotic materials. The authors provide a general conceptual model for understanding these effects, identify where their impacts should be greatest, show how engineering is different from concepts such as keystone species, and identify questions that can be asked about organisms as ecosystem engineers.

In contrast, there are clearly important effects on ecosystems of trophic interactions among species, and three chapters address some of these aspects. Estes (Ch. 15) focuses on the diverse array of taxa that constitute top-level carnivores. Discovering their effects on populations and ecosystems by controlled manipulations is often impractical, but valuable insights can be gained from "natural experiments"—comparative analyses of ecosystems in which they are present or absent, including situations where human activities have eliminated top predators. Bengtsson et al. (Ch. 16) present a novel model for examining the effects of such top predators at much smaller scales, focusing on decomposer food webs in soil. They outline a theory of carbon dynamics, and argue, contrary to general consensus, that food web structure may affect decomposition rates, even if trophic interactions are donor-controlled. Martinez (Ch. 17) rounds of this trilogy of food web chapters by discussing "ecosystem webs" that depict trophic interactions between all organisms within a spatially defined area. He argues that who eats whom is the most centrally organizing concept in ecology, and briefly considers extensions to embrace other types of interactions, including mutualisms, engineering (Lawton and Jones, Ch. 14), and the concept of "flow chains" presented by Shachak and Jones in Chapter 27.

The next four chapters are all concerned with the fundamental question of how to model the effects of species on ecosystems and vice versa without becoming lost in overwhelming detail. In essence, the problem is one of aggregation—how to simplify by grouping species with similar ecologies, at appropriate spatial and temporal scales.

Gurney et al. (Ch. 18) make the nature of the problem very clear, before going on to show how relatively simple models that are remarkably free from detailed species biology can do a good job of predicting ecosystem production and nutrient cycling in sea-lochs (fiord-like ecosystems). They point out that the success of these models (as compared to models based on individual species that are not very successful) may be because sea-lochs are relatively open systems, driven by major abiotic forces; ecosystems in which species interactions play a greater role may prove less tractable.

Some of the difficulties are highlighted by Turner and O'Neill (Ch. 19). As several earlier chapters show (e.g., Chs. 1, 8–10), population and ecosystem processes are heterogeneous in space and time, and Turner and O'Neill point out that when we try to simplify and model heterogeneous processes, the scale-dependent effects of aggregation are not well understood. They use a spatially explicit model of large ungulates in Yellowstone to empirically examine the effects of aggregating aspects of species biology and environmental controls.

They conclude that it is not possible to predict, *a priori*, what can and cannot be aggregated, and that decisions about aggregation will be very much system and question dependent. Even successful aggregation of processes will be reliable only if dynamic thresholds are not crossed, if keystone species are not eliminated, and if feedback loops remain intact in the models.

In contrast Schimel et al. (Ch. 20) show that individual-based models of plant physiology can be readily scaled up to an entire canopy of mixed species to predict ecosystem-level productivity and nutrient cycling, probably because for all species these processes are tightly constrained by environmental factors; other processes may require detailed specification of physiological traits at the species or population level. These conclusions are reinforced by Rastetter and Shaver's (Ch. 21) model of five species of plants competing for two elemental resources. A naive, aggregate model with "average" parameter values fails to predict changes in element concentrations and ecosystem productivity. But they then go on to show how the aggregated models can be iteratively calibrated to markedly improve their performance.

In sum, the problem of system aggregation is certainly difficult, but it is not completely intractable. There are no universal rules, but there are a growing number of contingent generalizations and ways forward for linking species and ecosystems in models that are not hopelessly complex.

To a nontrivial extent the ability to aggregate will depend on the similarity of the role of species in ecosystem functioning. Functionally redundant species should be easier to aggregate. Of course, no two species are identical (Brown, Ch. 2), but many do have similar effects on particular ecosystem processes.

Frost et al. (Ch. 22) explore the theme of species redundancy and complementarity in lake zooplankton. They attempted to predict the degree of functional redundancy in this assemblage when the lake was subject to experimental acidification, using initial observations on population fluctuations in the unperturbed system. Although acidification caused loss of species, biomass of several invertebrate groups was maintained by compensatory increases in a few taxa; in other words, ecosystem function was maintained. However, the ability of the system to compensate was not predicted solely on the basis of its behavior prior to acidification; redundancy in one set of circumstances did not allow predictions to be made about redundancy in a different set of circumstances. This is basically Brown's point—every species is different, and although we may gather species into functional groups for one purpose, a different purpose (or question) will require a different grouping. The modellers in Chapters 18–21 are presenting much the same message.

The next four chapters focus on the major constraints of stoichiometry, energetics, and evolution that operate on ecosystems and/or species, and that define the "boundary conditions" within which linkages must operate. Sterner (Ch. 23) shows how differences in the elemental composition of species have profound implications for, and partially control, nutrient cycles in aquatic ecosystems and vice versa. Stoichiometry is a potentially very powerful tool for coupling species and community dynamics with material cycling. Wedin (Ch. 24) uses this approach to understand the roles of herbivores, decomposers, and fire ("consumers" that each have different requirements for, and effects on, carbon and nitrogen) in determining the fate of the bulk of primary production in grassland ecosystems across the globe.

Ecosystem energetics and species energetics have both received substantial attention, and it seems reasonable to suppose that energy might be a universal linking currency (a major theme of Brown's essay, Ch. 2). Yet DeAngelis (Ch. 25) points out that this does not seem to be the case. Instead he shows how energy is a link but not a common currency by developing an explanation for empirical correlations between species richness and available energy, at scales that vary from local to continental. Increased available energy favors the success of specialized strategies, and interacts with disturbance and heterogeneity at local scales to set limits to local diversity. Some simple, albeit speculative, scaling also explains species richness–energy curves over larger geographical areas.

Evolutionary arguments form the central theme of Chapter 26 (Holt). Evolutionary thinking has played little or no role in the development of ecosystem theory, and yet the organismal traits mechanistically responsible for many ecosystem processes result from evolutionary dynamics, and these dynamics are, in turn, constrained by ecosystem processes. Obvious but poorly explored links include adaptive interpretations of resource acquisition and utilization, and contingency and rarity as limits on predictability.

The last chapter in this section, by Shachak and Jones (Ch. 27), shows how some of the complexities of nature that have been addressed in the previous chapters can be represented, interrelated, and integrated. They argue that ecosystems consist of multiple flows of multiple currencies, interacting at multiple scales. Using an example from the Negev Desert, they weave together interactions between terrestrial isopods, hydrology, and soil erosion into "ecological flow chains" comprising an "ecological system," and use this approach to identify critical linkages and the scales over which they operate.

14

Linking Species and Ecosystems: Organisms as Ecosystem Engineers

John H. Lawton and Clive G. Jones

SUMMARY

Ecosystem engineers are organisms that directly or indirectly modulate the availability of resources to other species, by causing physical state changes in biotic or abiotic materials. In so doing they modify, maintain, and create habitats. Autogenic engineers (e.g., corals, or trees) change the environment via their own physical structures (i.e., their living and dead tissues). Allogenic engineers (e.g., woodpeckers, beavers) change the environment by transforming living or nonliving materials from one physical state to another, via mechanical or other means. Here we define and explain engineering, and provide a classification and some general, conceptual models of the processes involved. We show how organismal engineering is related to human engineering and to ecological concepts such as keystone species. We then identify the factors scaling the impact of engineers. The biggest effects are attributable to species with large per capita impacts, living at high densities, over large areas for a long time, giving rise to structures that persist for millennia and that modulate many resource flows. We argue that all habitats on earth support, and are influenced to some degree, by ecosystem engineers, and we raise some general questions about organismal engineering that are worth pursuing.

INTRODUCTION

Ecology textbooks summarize the important interactions between organisms as intra- and interspecific competition, predation, parasitism, and mutualism. Conspicuously lacking from this list is the role that many organisms play in the creation, modification, and maintenance of habitats, although particular examples have been extensively studied (Thayer, 1979; Naiman et al., 1988). In general, however, population and community ecology has neither defined nor systematically identified and studied the role of organisms in the creation and maintenance of habitats. We will call the process *ecosystem engineering* and the organisms responsible *ecosystem engineers*. Here we define, explain, and classify ecosystem engineering, show how it differs from related concepts, and identify some questions for future research. In this short chapter we can only touch on a few aspects of engineering (see Jones et al., 1994 for a more extensive treatment).

DEFINITION AND TYPES OF ENGINEERS

Ecosystem engineers are organisms that directly or indirectly modulate the availability of resources (other than themselves) to other species, by causing physical state changes in biotic or abiotic materials. In so doing they modify, maintain, and/or create habitats. The direct provision of resources by an organism to other species, in the form of living or dead tissues, is *not* engineering. Rather, it is the stuff of most contemporary ecological research. *Autogenic engineers* change the environment via their own physical structures, that is, their living and dead tissues. *Allogenic engineers* change the environment by transforming living or nonliving materials from one physical state to another, via mechanical or other means.

There are numerous examples of engineering [see, for example, Levinton (Ch. 3); Giblin (Ch. 4); Silver (Ch. 5); Huntly (Ch. 8); Anderson (Ch. 10); Pollock et al. (Ch. 12)], but all can be assigned to one or more of five possible types or cases within a general conceptual model (Fig. 14-1). Beaver, *Castor canadensis*, and their dams, illustrate the use of this classification.

Beaver make dams that alter hydrology, sediment, and organic matter retention, nutrient cycling, decomposition and plant and animal community composition and diversity (Naiman et al., 1988; Pollock et al., Ch. 12). They therefore conform to case 4 in Fig. 14-1, that is, they are allogenic engineers, that take materials in the environment (in this case trees, but in the more general case, any living or nonliving material), and turn them (engineering them) from physical state 1 (living trees) into physical state 2 (dead trees in a beaver

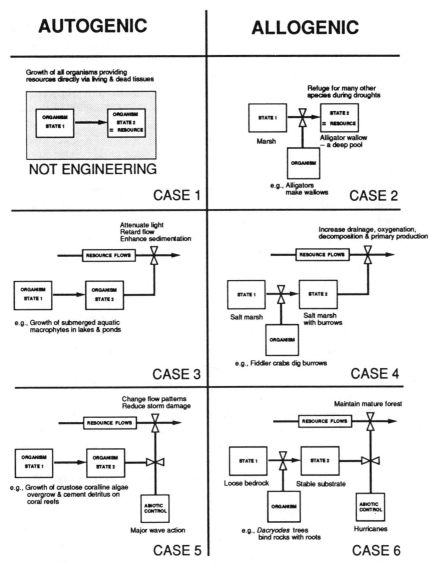

Figure 14-1. Conceptual models of autogenic and allogenic engineering by organisms. (From Jones et al., 1994.) For definitions and examples see text. The symbol $\bar{\mathrm{X}}$ defines points of modulation. For example, allogenic engineers transform living or nonliving materials from state 1 (raw materials) to state 2 (engineered objects and materials), via mechanical or other means. The equivalent (state 2) products of autogenic engineering are the living and dead tissues of the engineer. These products of both auto and allogenic engineering then modulate the flow of one or more resources to other species (cases 2–4) or modulate a major abiotic controller (e.g., fire), which in turn modulates resource flows (cases 5–6). Case 1, the direct provision of resources by one species to another, is not engineering, and involves no modulation of resource flows. Examples of organisms are from Jones et al. (1994) and references therein.

dam). This act of engineering creates a pond, and it is the pond that has profound effects on a whole series of resource flows used by other organisms. The critical step in this process is the transformation of trees from state 1 (living) to state 2 (a dam). This transformation then modulates the supply of other resources, particularly water, but also sediments, nutrients, etc. A critical characteristic of ecosystem engineering is that it must change the availability (quality, quantity, distribution) of resources utilized by other taxa, *excluding the biomass provided directly by the population of allogenic engineers.* Engineering is not the direct provision of resources in the form of meat, fruits, leaves, or corpses. Beaver are not the direct providers of water, in the way that prey are a direct resource for predators, or leaves are food for caterpillars.

Now consider the autogenic equivalents of beaver (Fig. 14-1, case 3). Simple examples are the growth of a forest or a coral reef. The production of branches, leaves, or living coral tissue does not, in itself, constitute engineering. Rather, it conforms to case 1 in Fig. 14-1 (the direct provision of resources). However, the development of the forest or the reef results in physical structures that do change the environment and modulate the distribution and abundance of many other resources. This modulation constitutes autogenic engineering. Trees alter hydrology, nutrient cycles, and soil stability, as well as humidity, temperature, wind speed, and light levels (Holling, 1992); corals modulate current speeds, siltation rates, and so on (Anderson, 1992). Clearly, the numerous inhabitants of these habitats are dependent on the physical conditions modulated by the autogenic engineers, and on resource flows that they influence but do not directly provide; without the engineers, most of these other organisms would disappear.

We can illustrate our arguments further by considering the simplest kind of allogenic engineering (case 2, Fig. 14-1). Various organisms make holes in tree trunks and branches, some quickly (woodpeckers), others more slowly (rot fungi). They transform wood without holes into wood with holes, and indirectly provide resources for other creatures, nesting and roosting cavities for birds and bats for instance. The holes are the resource, not the organisms that make them. Notice that if some of the holes fill with water (Kitching, 1983), the little ponds so created are examples of case 4 engineering, and are conceptually identical to beaver dams, albeit with less ecological impact! When natural hollows and cavities at branch junctions and root bases form as a tree grows (and then fill with water—"phytotelmata," Fish, 1983) they are examples of case 3 (autogenic) engineering. So the same type of engineered structure (water-filled holes) can result from both autogenic engineering and allogenic engineering. However, it is not universally the case that plants are allogenic, and animals are autogenic engineers. Rates of physical and chemical weathering of rocks into soil (allogenic engineering) are enhanced by algae, bacteria, and higher plants as well as by animals such as chitons and snails

(Bloom, 1978; Krumbein and Dyer, 1985; Shachak et al., 1987; Jones and Shachak, 1990).

Autogenic (case 5) and allogenic (case 6) engineering are situations where the autogenic or allogenic engineers modulate very powerful abiotic forces, for instance fires, storms, and hurricanes, that are fundamental modulators of the distribution and abundance of resources, rather than actual resources. Fire provides a particularly interesting case. It is logical, albeit unconventional, to regard the production of combustible living and dead biomass as autogenic engineering (case 5). Different species of plants produce different qualities and quantities of living and dead fuel, modulating the magnitude, intensity, and duration of fire and, in turn, profoundly altering the supply of resources for many other species (Christensen, 1985; Dublin et al., 1990). Plants also act as autogenic (case 5) and allogenic engineers (case 6) to modulate powerful abiotic forces. In Puerto Rico *Dacryodes excelsa* trees are able to withstand hurricanes because their extensive roots and root grafts bind and stabilize bedrock and superficial rocks; this species therefore dominates tropical mountain forests where hurricanes are common (Basnet et al., 1992). Similarly coral reefs modulate the wave action of storms (Anderson, 1992).

HUMAN ANALOGUES

The parallels between ecological and human engineers are very close. Humans are tool-using organisms that specialize in engineering. Many human activities, from dam-building and skyscraper construction to forest clearance and the dredging and canalization of water courses, conform exactly to cases in Fig. 14-1. Construction of nesting boxes for birds and hives for bees are examples of case 2. Ploughing by farmers and the construction of dams and reservoirs by water engineers provide examples of case 4. Building harbors and sea walls to reduce storm damage from waves are examples of case 6. Humans also mimic autogenic effects, using tools to construct glass houses and build air-conditioning plants (mimicking case 3), and by bulldozing fire breaks to counteract fire (mimicking case 5).

SOME RELATED CONCEPTS

The idea that organisms can alter the physical structure of their environment, with impacts on their own and other populations, is not new. For example, Huntly and Inouye (1988) explicitly describe pocket gophers *Geomys bursarius* as "soil engineers" because of their role as earth-movers. Gophers are, indeed, excellent examples of allogenic (case 4) engineers. What is new is the general conceptual framework defined and developed here. Not surprisingly, several authors have developed related concepts, some of which we briefly review next.

In marine benthic environments the activities of large burrowing animals play a dominant role in determining the physical structure of sediments, altering habitat suitability for other species (Rhoads and Young, 1970; Thayer, 1979; Lopez and Levinton, 1987; Meadows and Meadows, 1991; Levinton, Ch. 3; Giblin, Ch. 4). Rhoads and Young (1970) called the process "trophic amensalism" when large deposit feeders create unstable sediments, restricting the presence of suspension feeders and attachment by sessile epifauna. "Trophic amensalism" is actually nontrophic and another good example of case 4 allogenic engineering brought about by bioturbation of sediments.

Meadows and Meadows (1991) and Meadows (1991) review the environmental impacts of animal burrows and burrowing animals, and Hansell (1993) the ecological consequences of animal burrows and nests. Many of these artifacts (e.g., meiofaunal burrows, magapode nests, termite mounds, and mole rat colonies) have landscape level effects, and serve to concentrate and redistribute resources for other species—that is, they are classic examples of case 4 engineering. Meadows (1991) points out that there are "underlying similarities between the impact of [burrowing] animals from different terrestrial and aquatic habitats on environmental change and modification," and Hansell (1993) recognizes that the "services and substances of the builders create a new range of habitat niches which can be exploited by a wide variety of specialists" and suggests that "the presence of nest builders and burrowers can . . . significantly contribute to species diversity in habitats." The examples provided by both authors all conform to either case 2 (other species use the nests and burrows) or case 4 (species respond to changes in distribution and abundance of resources).

The importance of animal artifacts is recognized by Dawkins (1982) as an example of species' extended phenotypes. He also points out that not every example of what we are now calling engineering can be regarded as an extended phenotype, because impacts on the environment are of no consequence to the engineer's fitness and hence are not subject to natural selection. A good example would be water-filled footprints made by an ungulate. The distinction between engineering that is subject to natural selection (because it is an extended phenotype) and engineering that is not ("accidental" engineering) appears to be unimportant in terms of its shorter-term ecological consequences; all types of engineering modify and modulate resource flows for other organisms. But there may be interesting longer-term differences, particularly in the nature of the feedback loops that operate on "extended phenotype" vs. "accidental" forms of engineering (Holt, Ch. 26).

By definition, keystone species (Paine, 1969; Krebs, 1985; Daily et al., 1993) have major effects on species composition and other ecosystem attributes. The critical links are usually regarded as trophic but also frequently involve engineering, for example via disturbance. In a frequently cited example, removal of sea otters (*Enhydra lutris*) leads to an increase in sea urchins (*Strongylocentrotus* sp.) and hence to the disappearance of kelp beds, which in turn changes wave action and siltation rates, with profound consequences for other inshore flora and fauna (Estes and Palmisano, 1974; Estes, Ch. 15). Kelp are autogenic engineers (case 3); removal of kelp by urchins is, among other things, allogenic engineering (case 4). In other words, in this familiar example, the species traditionally regarded as the keystone (sea otter) has major effects because it changes the impact of one engineer (urchin) on another (kelp), with knock-on effects on other species in the web of interactions.

Krebs (1985) concludes his textbook review of keystone species as follows: "Keystone species may be relatively rare in natural communities, *or they may be common but not recognised* (our italics). At present, few terrestrial communities are believed to be organised by keystone species, but in aquatic communities keystone species may be common." We believe that such views probably reflect a consensus among ecologists, because we have failed to recognize the role of ecosystem engineers as keystone species. It is trite, but true, that a forest is a forest because it has trees, which not only provide food and living space but that also autogenically engineer the forest climate, and modulate the flows of many other resources. Our views are therefore very close to Holling's (1992) *Extended Keystone Hypothesis*, in which he argues that "all terrestrial ecosystems are controlled and organised by a small set of key plant, animal, and abiotic processes that structure the landscape at different scales." We would add two points. First, a critical, but not exclusive controlling mechanism is some form of engineering; and second, we believe that keystone engineers occur in virtually all habitats on earth [see Jones et al. (1994), for examples], not just terrestrial ones.

SCALES OF ENGINEERING IMPACT

The impact of an ecological engineer depends on the spatial and temporal scale of its actions. Six factors scale the impact. They are: (1) Life time *per capita* activity of individual organisms; (2) population density; (3) the spatial distribution, both locally and regionally, of the population; (4) the length of time the population has been present at a site; (5) the durability of constructs, artifacts, and impacts in the absence of the original engineer; and (6) the number and types of resource flows that are modulated by the constructs and artifacts, and the number of other species dependent on these flows.

Thus, the most obvious engineering impacts are attributable to species with large *per capita* effects, living at high densities, over large areas for a long time, giving rise to structures that persist for millennia and that affect many resource flows (for example, mima mounds created by fossorial rodents; Cox and Gakahu, 1985). Autogenic engineers may also have massive effects; as Holling (1992) succinctly states: "To a degree, . . . the boreal forest 'makes its own weather' and the animals living therein are exposed to more moderate and slower variation in temperature and moisture than they would otherwise be." Boreal forest trees have large *per capita* effects on hydrology and climatic regimens, occur at high densities over large areas, and live for decades. But their impacts as autogenic engineers may have a relatively short memory if the forest is logged.

Organisms with small individual impacts can also have huge ecological effects, providing that they occur at sufficiently high densities over large areas, for sufficient periods of time. Burrowing meiofauna (Reichelt, 1991; Levinton, Ch. 3; Giblin, Ch. 4) and bog-forming *Sphagnum* mosses are good examples. Accumulated *Sphagnum* peat may persist for thousands of years after the death of the living moss (Tansley, 1949).

Ecological engineers may also enhance and speed up large-scale physical processes, including geological erosion and weathering (Yair and Rutin, 1981; Krumbein and Dyer, 1985; Hoskin et al., 1986; Shachak et al., 1987; Jones and Shachak, 1990). Worldwide, but especially in the tropics, heavily undercut coastal cliffs of sedimentary rock are apparently being eroded by tides and storms. In fact the process is greatly accelerated by two groups of engineers, both with low *per capita* effects, but very abundant. Cyanobacteria (*Hyella* spp.) bore the rock and are food for chitons which rasp away the rock to reach them, apparently speeding up coastal erosion by an order of magnitude or more (Krumbein and Dyer, 1985).

QUESTIONS

We finish with a list of open questions.

Are there any ecosystems on earth that have not been physically engineered by one or more organisms to a significant degree? A cautious, preliminary answer is no; we currently cannot identify any habitat on earth that is not engineered in some way by one or more species (Jones et al., 1994).

How many species (or what proportion of species) in various communities have a clearly defined and measurable impact as engineers? Is it 10% or 0.1%? What are the relative frequencies of the five classes of engineering identified in Fig. 14-1, say in terms of the numbers of species acting as engineers? Cases 5 and 6 are presumably rather rare; but how much rarer are they than the other types of engineering?

Are the most physically structured ecosystems (or subsystems, e.g. soil or sediments) the ones in which engineers are most important?

We know of very few field manipulation experiments designed to quantify the impact of ecosystem engineers by removing or adding species. Studies by Bertness (1984a,b, 1985) are rare, but excellent examples, prompting us to ask how many other species are impacted by engineers in any ecosystem. What happens to species richness if we remove or add engineers?

How does the persistence of the products or effects of engineering influence population, community, and ecosystem processes? If engineers make long-lived artifacts, then their effects will usually, also, be long lived. But ephemeral products can also have long-term impacts. For example, fecal pellets produced by marine zooplankton (Dunbar and Berger, 1981; Fowler and Knauer, 1986) decompose relatively quickly, but not before they have sunk into the deep ocean, removing nutrients from surface waters for millennia. In many cases the impacts of engineers are ephemeral, operating on time scales shorter than, or similar to, the lifetime of the organism itself (e.g., the nests of small passerine birds). But in other cases, the engineers leave monuments with impacts that extend many lifetimes beyond their own—mima mounds, termite nests, buffalo wallows, beaver dams, peat, and so on. These persistent effects must greatly slow down rates of ecological change, and impose considerable buffering and inertia on ecosystems, contributing to population, community, and ecosystem stability, resistance, and resilience (Pimm, 1984).

How much of the effects of keystone species are due to engineering vs. trophic effects? Earlier, we speculated that few keystone effects are purely trophic; is this hypothesis correct? How do engineering and trophic relations interact?

How should we model engineering? There is, in principle, no reason why we cannot write down the equation: $d\text{mayfly}/d\text{beaver} = F(x, y, z)$, where mayfly populations respond to changes in beaver numbers on long time scales, and where the response is affected by various key variables, including feedbacks to beaver from other components in the engineered habitat. Interesting theoretical questions center on the generation time of the engineer, the half-life of whatever it is that is engineered, the rate of restoration of nonengineered habitat, the generation times of impacted species, and their various interactions. There are intriguing problems of nested time scales, delayed responses, donor control, and long chains of indirect interactions, that might usefully be explored using relatively simple models.

Extending these arguments, there is no reason in principle why processes driven by engineering should not be coupled to the rich diversity of trophic linkages to create not simply descriptions and models of food webs (Pimm et al. 1991), but of *interaction webs*, that more accurately reflect interactions in communities and ecosystems (Martinez, Ch. 17).

Last, but by no means least, this chapter attempts to define and classify the phenomenon of physical engineering. We recognize that organisms can also be chemical and transport engineers. But definition and classification of this and other forms of engineering are merely a small beginning, not an end. What predictions might we make that would not have been made without the conceptual framework provided here, or something akin to it? We do not, currently, know the answers to this, or any other of the questions raised in this final section. But if the notion of organisms as ecosystem engineers results simply in an accumulation of "just-so" stories, it will not have been particularly useful.

ACKNOWLEDGMENTS

This work was supported by the Mary Flagler Cary Charitable Trust and the NERC Centre for Population Biology, U.K. Numerous colleagues have allowed us to sharpen and refine our ideas about engineering by discussing problems with them, and have generously provided us with examples and references; to all of them we express our sincere thanks. This is a contribution to the program of the Institute of Ecosystem Studies.

15

TOP-LEVEL CARNIVORES AND ECOSYSTEM EFFECTS: QUESTIONS AND APPROACHES

James A. Estes

SUMMARY

Top-level carnivores comprise a diverse array of taxa and occur in most natural ecosystems. This chapter considers three main questions: What are the ecological and evolutionary effects of top-level carnivores, how general are they, and under what conditions do they occur? To answer these questions one first must determine whether a food web is under top-down control, and ultimately, the species, systems, and circumstances for which top-down control exists. For systems affected by top-down forces, the cascading effect of carnivory on autotrophs is expected to vary with the order of trophic complexity. Given this conceptual framework, three more specific questions are explored. First, how general are the direct and indirect effects of top-level carnivores? Second, what is the breadth of trophic influence from the direct impact of carnivory? Broadly ranging effects are expected for systems in which (1) top-level carnivores occupy odd-numbered trophic levels and (2) the food web is strongly linked. (3) what are the evolutionary consequences of trophic cascades on species lower in the food web? These questions can best be answered via manipulative or natural experiments in which the presence or absence of carnivores varies in space or time. Natural experiments are advocated as the only practical means of addressing these questions for many species and systems, in particular those in which the key players are large and highly mobile, or those in which the manipulation of predators has legal, political, or social ramifications.

INTRODUCTION

Top-level carnivores are represented by a wide variety of taxa and occur in most natural ecosystems. The broad questions I address in this chapter are: What are their ecological effects and how do these effects vary across taxa or

Figure 15-1.   Hypothetical distributions of the relative importance of carnivores among taxa or ecosystems. *Hatched bars*—uniform distribution; *closed bars*—distribution under keystone predator hypothesis; *diagonal bars*—normal distribution. As some species of carnivores are known to have important effects in some systems, all distributions must begin with species of maximum relative importance. (Figure redrawn from Mills et al., 1993.)

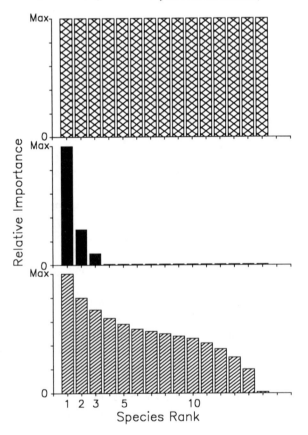

ecosystems (e.g., Fig. 15-1)? My focus on top-level carnivores is based not so much on the *a priori* assumption that they have stronger community- and ecosystem-level influences than do species lower in the food web, as it is on my belief that understanding the influences of these carnivores poses particular challenges to ecologists.

The effects of top-level carnivores may be limited to what are traditionally viewed as community-level properties (e.g., diversity, abundance, and structure of the component populations) or they may extend to such ecosystem-level properties as energy and nutrient flow, or production at lower levels. Top-level carnivores are known or suspected to have important community- and ecosystem-level effects in some cases, although whether the known cases represent the rule or exceptions for other species and systems is uncertain. This uncertainty will be difficult to resolve because (1) meaningful field studies are hard to do under the best of circumstances, (2) so many top-level carnivores and the systems in which they live have been disturbed by human activities, and (3) ecologists have not been highly motivated to take on the problem, except for the most manageable species and ecosystems. My treatment of the subject focuses on two simple yet fundamental questions: What would we like to know about the ecological importance of top-level carnivores, and how should we go about learning it?

THE QUESTIONS

1. *How often and under what circumstances do top-level carnivores have important ecosystem-level effects?* At least two conditions seem necessary for important effects to occur. One is that the food web must be strongly linked (MacArthur, 1972; Paine, 1980), especially between the carnivores and their prey; the other is that the community must be under top-down control (Hunter and Price, 1992; Bengtsson et al., Ch. 16).

2. *How do we recognize these conditions and are they predictable from the characteristics of species, food webs, or ecosystems?* At present there is no theoretical basis for resolving these questions. Strong (1992) argued that top-down control is most prevalent in aquatic systems, which indeed provide the best examples (e.g., Kerfoot and Sih, 1987; Carpenter, 1988; Power, 1990b; Kitchell, 1992). However, a possible reason for the prevalence of aquatic examples is that generation times for species with low trophic status (especially autotrophs) are typically short in aquatic systems, thus permitting predator-induced changes to be seen over short periods of time. In contrast, generation times of the dominant plants in most terrestrial systems range from decades to millennia. The typically short generation times of autotrophs in aquatic systems also raise the possibility that the cascading influences of top-level carni-

vores, although prevalent and demonstrable in these systems, are transient effects as opposed to stable states of community organization (Ginzburg and Akçakaya, 1992).

Another reason that Strong's (1992) dichotomy is not compelling is that top-down forces *are* important in some terrestrial ecosystems. For instance, the introduction or natural colonization of mammalian herbivores on predator-free islands has altered their vegetation (e.g., Klein, 1968; Coblentz, 1980); megaherbivores are known have landscape-level effects in Old World tropical forests (Owen-Smith, 1988); and there are numerous examples of seed predation having important population- and community-level effects (Janzen, 1971; Brown et al., 1979). Examples in soil ecosystems are discussed by Bengtsson et al. (Ch. 16). Top-down control does not establish that top-level carnivores *commonly* control the structure and organization of terrestrial ecosystems, although it is a necessary condition for such interactions.

3. Is the number of trophic levels odd or even? For strongly linked food webs that are under top-down control, the influence of top-level carnivores may vary depending on whether the number of trophic levels is odd or even (Fretwell, 1987). That is, in odd-numbered systems the effect of top-level carnivores should be to enhance autotrophs through the reduction of herbivory whereas in even-numbered systems it should be to enhance the intensity of herbivory. This prediction has been confirmed in river (Power, 1990b, Ch. 6) and kelp forest (P.D. Steinberg, J.A. Estes, and F.C. Winter, unpublished) ecosystems.

4. What is the depth and breadth of food web effects by top-level carnivores? The array of possible influences by top-level carnivores on strongly linked food webs is truly vast. For instance, Zaret and Paine (1973) suggested that the introduction of peacock bass to Lake Gatun, Panama, altered the food web so as to increase the threat of human malaria, and Duggins et al. (1989) demonstrated that growth rates of suspension feeding invertebrates in the Aleutian archipelago are influenced by food web linkages stemming from sea otter predation on herbivorous sea urchins.

5. How general are the effects of top-level carnivores? Many ecological interactions have been conclusively demonstrated, but the extent to which any of these can be generalized is uncertain. The effects of top-level carnivores are no exception. Variation in the effects of top-level carnivores, even in tightly linked food webs, is undeniable. The important questions are, how extensive are these effects and what conditions modulate them?

Those interactions closest to the top of the food web should be the most generalizable in space and time because they present fewer possibilities for the intercession of a weak link, the influence of alternate food web influences, or other idiosyncratic effects inherent in linkages involving longer chains of events. This is true for the trophic cascade among sea otters, sea urchins, and kelp. That is, sea otter predation has reduced the number and biomass of herbivorous invertebrates nearly everywhere people have looked (Estes and Duggins, 1994), whereas the secondary interaction between herbivores and macroalgae is less predictable (Foster and Schiel, 1988). Other interactions in this food web also should be increasingly less general with trophic distance.

6. *What are the evolutionary consequences of top-level carnivores on species of lower trophic status?* If a top-level carnivore exerts demographically significant forces on other members of the food web over ecological time, then these effects could have selective influences over evolutionary time. Even so, the interpretation of ecological patterns on evolutionary time scales can be problematic (Gould and Lewontin, 1979) because of the singular nature and limited resolution of the historical record. A necessary condition for any such interpretation is that the key players have a history of association with one another. Although the selective importance of carnivory has not been well studied, evolutionary responses to top-down selective forces have been demonstrated or proposed (Vermeij, 1977; Steneck, 1983; Hay, 1984; Estes and Steinberg, 1988; P.D. Steinberg, J.A. Estes, and F.C. Winter, unpublished).

APPROACHES

In answering any of the questions posed above, it is always helpful to know who is eaten by whom as this can help identify which trophic interactions, from the set of possible combinations, are worth studying (a daunting number in species-rich ecosystems). Information on population abundance, biomass, nutrient and energy content, and feeding rate also can be useful in judging which interactions are most important. By themselves, however, food web descriptions of this sort can be misleading. For instance, we might mistakenly infer that a predator–prey interaction is unimportant if the prey is rare in the predator's diet. It might be that the prey are vulnerable to predation and thus rare in the predator's presence, but become abundant when the predator is absent. Similarly, a large and productive prey population may comprise the dominant element of a predator's diet whereas losses to predation are of little or no consequence to the dynamics of the prey population. The former circumstance could be misinterpreted as evidence for a weak linkage whereas the latter circumstance could be misinterpreted as evidence for a strong one. In general, food web descriptions are useful for generating questions and hypotheses, but not conclusions.

Key to understanding the effects of top-level carnivores is the ability to cast the system into a dynamic perspective, that is, to observe how it responds to changes in carnivory. This requires that top-level carnivores be added to or removed from the system. Such additions or removals can be achieved by way of manipulative or natural experiments. The experimental manipulation of carnivores, perhaps best known from Paine's (1974) work on sea stars in mussel beds, is appealing because of its replicability (thus permitting statistical hypothesis testing) and because it allows one to control for potentially confounding sources of variation. This approach has since been used in other studies of the top-down effects of consumers. However, the experimental manipulation of top-level carnivores has several obvious limitations. This approach is most applicable for systems with key players that are sessile or weakly motile, which probably explains why so little is known about the effects of large, mobile carnivores. In addition, generation times (or related life history characters) of the key players must be short enough for demographic responses by lower trophic forms to be seen within the "window of patience" of a human observer. Few investigators are willing to initiate ecological experiments that are unlikely to produce results during normal funding cycles, a restriction that excludes many natural ecosystems (if not most of them) from manipulative study. Finally, the limited spatial and temporal dimensions on which predator manipulations can be implemented also limit the extent to which their results can be generalized.

Natural experiments (i.e., spatial or temporal comparisons made around the presence or absence of carnivory), although often lacking the control and replicability of purposeful manipulations, are probably the most practical and generally useful means of assessing the importance of many top-level carnivores. The manipulation of predators is often impractical for a host of reasons, not the least of which are social, economic, ethical, and legal. In contrast, comparisons of community structure and ecosystem function based on natural or human-caused variation in the distribution and abundance of carnivores are often easy to make and can be extremely informative in understanding the importance of top-level carnivores on ecological and evolutionary time scales. Unfortunately, this approach has been underutilized, perhaps because of the dominating recent influence by manipulative experimentalists on the methodological philosophy of community ecology.

Effects of Carnivores on Ecological Time Scales

The ecological effects of top-level carnivores (or any species for that matter) often can be inferred from patterns of community structure seen in association with meso-scale extinctions or recolonizations. This approach is exemplified by studies of sea otter predation in the North Pacific Ocean. The top-level carnivore (sea otters in this instance), once broadly distributed across the North Pacific, was hunted to near extinction (Kenyon, 1969). Following protection, the surviving remnant colonies grew and spread into a series of fragmented populations. By knowing the history and distribution of this species, its diet, and several features of the food web, it was possible to hypothesize what the top-down effects of sea otter predation might be and then test these hypotheses by contrasting areas with and without sea otters in space or time. This approach has been repeatedly taken (see VanBlaricom and Estes, 1988; Riedman and Estes, 1990; Kvitek et al., 1992; Watson, 1993 for reviews) and the resulting data permit a reasonable assessment of the generalities, exceptions, and geographical variation in the influence of sea otter predation.

Can the same approach be applied to other species and systems? Perhaps, but in order to know where to look and what to look for, one first must know something about the species' history and the food web's structure and function. Reintroductions, or localized population reductions or extinctions, are two particularly useful means of judging the ecological importance of top-level carnivores. For instance, the proposed reintroduction of gray wolves to Yellowstone National Park, or the restocking or protection of predatory fishes in lakes and rivers, would provide marvelous opportunities to make such assessments. Similarly, the removal of coyotes from areas of the western United States, wolves from parts of Alaska, or arctic foxes from islands of the North Pacific Ocean and Bering Sea, all could be used to understand better the importance of these species in their respective ecosystems, either as natural players or exotics.

Effects of Carnivores on Evolutionary Time Scales

Comparisons of systems within which the presence, absence, or abundance of carnivores varies in space or time also can be used to assess the selective effect of carnivory on evolutionary time scales. Here the comparisons must be made over larger scales of space and time, and the foci of comparison should be characteristics of taxa rather than those of populations, communities, or ecosystems. For instance, the ecological effects of top-level carnivores most often should be manifested in the abundance and population structure of species lower in the food web. The evolutionary consequences of these ecological forces should appear in such species-level characters as life history, morphology, behavior, and physiology. The selective effects of carnivory over evolutionary time are largely unstudied [but see Vermeij (1978) Palmer (1979) for examples].

CONCLUSIONS

Static food webs, no matter how long or hard we look at them, are apt to tell us little about the importance of top-level carnivores. Experiments are needed in which the structure and behavior of these systems can be observed when top-level carnivores are added or taken away. Manipulative experiments are useful when the populations in question are effectively closed, either because of their own limited mobility or boundaries to dispersal. Unfortunately, many top-level carnivores are highly mobile, which together with political, economic, and ethical issues, often precludes these species from being experimentally manipulated. The use of natural experiments to compare habitats with and without carnivores is advocated as an approach that often can be achieved by designing the comparison around such things as fragmented populations, re-introductions, and animal damage control programs. Nearly all of the (few) experimental studies of top-level carnivores have failed to look much beyond the direct impact on their prey. In the future these comparisons should be made with a more open view to the possibility of indirect food web effects on ecological time scales and selective effects on evolutionary time scales.

Few generalizations can be made about the effects of top-level carnivores, either within or among species or systems. The ability to generalize is not likely to come easily. It probably will require an inductive approach, in which recurrent properties of species, populations, and ecosystems under the influence of top-level carnivores are gradually identified from the accumulated results of many individual studies.

ACKNOWLEDGMENTS

I thank Jim Brown, Suzy Kohin, Scott Mills, Rick Ostfeld, Mary Power, and Peter Steinberg for sharing ideas with me and providing comments and assistance on an early draft of the manuscript. I owe a special debt to the sea otter for teaching me that carnivores can do more than just eat sheep.

16

FOOD WEBS IN SOIL: AN INTERFACE BETWEEN POPULATION AND ECOSYSTEM ECOLOGY

Jan Bengtsson, David Wei Zheng,
Göran I. Ågren, and Tryggve Persson

SUMMARY

We propose that the population dynamics of soil organisms can be linked to ecosystem processes by studies of food webs. Key questions concern the types of trophic interactions occurring between functional groups, and if consumers influence their resources in detritus-based food webs. Carbon and nitrogen dynamics are tightly linked in terrestrial ecosystems, suggesting that classic food web theory without "productivity feedbacks" probably is insufficient to analyze the relation between food web structure and ecosystem processes. A theory of carbon dynamics and soil food webs is outlined. We show that food web structure may influence decomposition rate, even if the trophic interactions are donor-controlled.

INTRODUCTION

In most terrestrial ecosystems, most of the carbon (C) fixed by plants usually reaches the soil, rather than being grazed and respired by herbivores. Therefore it is essential to understand the processes regulating the decomposition of organic matter to understand nutrient dynamics in ecosystems such as forests, grasslands, and agricultural land. In the past, decomposition and nutrient cycling have been treated mainly as problems at the ecosystem level, with comparatively little effort put into studies of the organisms doing the work. Recently, several authors have emphasized the need for explicit consideration of processes at the population and community levels for a more mechanistic understanding of soil processes (e.g., Anderson, 1988b; Verhoef and Brussaard, 1990; Parmelee, Ch. 11; Wolters, 1991). The situation is comparable in studies of lake ecosystems (e.g., Carpenter and Kitchell, 1988).

The purpose of this chapter is to outline a theoretical framework to describe how soil organisms may affect decomposition and nutrient cycling in forest ecosystems and to present some empirical studies relevant to this question. Our proposition is that rates of decomposition and nutrient cycling and population dynamics of soil organisms can be linked by studying soil food webs and interactions between functional groups. These may be single species or groups of species with similar ecology and a broadly similar function with regard to ecosystem processes.

SOIL FOOD WEBS AND ECOSYSTEM PROCESSES

The major part of the biomass in soils is usually found at the lowest trophic level as fungi and bacteria. Soil animals at higher trophic levels can influence decomposition and nutrient cycling in many ways. Among these are (1) grazing on microorganisms or roots, (2) indirect effects on grazing by predation and competition, (3) comminution or restructuring of the substrate for microorganisms, (4) translocation of nutrients or substrates (Anderson, 1988b, Ch. 10; Parmalee, Ch. 11; Wolters, 1991). We will focus on the first two mechanisms, that is, the kind of interactions treated by classic population dynamics theory and current food web theory. However, the latter two may be just as important in many ecosystems, an obvious example being earthworm activity.

Soil food webs are species-rich and complex with a large proportion of "between-trophic-level omnivory." Along food chains (energy channels sensu Moore and Hunt, 1988) deriving from different sources, such as bacteria, fungi, or roots, it may nevertheless be possible to distinguish clearly defined trophic levels (sensu Oksanen et al., 1981). However, for several reasons, traditional food web theory with its emphasis on consumer–resource interactions (e.g., Oksanen et al., 1981; Pimm, 1982; Abrams, 1993) is probably insufficient for analyzing detritus-based soil food webs. In this theory, resources enter at the bottom of the food web from outside the system, and their dynamics are not related to the structure of the food web. The dynamics of detritus and decomposers need to be coupled to ecosystem productivity through the cycles of C and nutrients, primarily nitrogen (N), in terrestrial ecosystems (DeAngelis, 1992).

On short time scales, bacterial and fungal activity can be energy- or nutrient-limited. Microorganisms act both as mineralizers and immobilizers of nutrients and compete with plants for nutrients in short supply. Their activity is usually located to waxing and waning "hot-spots," over which animals may integrate by feeding and movement. Animals may immobilize nutrients when feeding on resources with much higher C/nutrient ratios than themselves, but generally they contribute to nutrient mineralization. On longer time scales, the dynamics of N determines plant productivity and detritus input to the soil. If there are any long-term effects of soil animals on N cycling, these may be related to whether the composition of the decomposer community influences long-term vegetation dynamics.

Several examples exist of animals and food web structure affecting nutrient cycling and decomposition rate, but mainly from the laboratory. Experiments with one or a small number of species have shown that grazing by collembola or millipedes on fungi, or bacterial grazing by protozoans or nematodes, can increase nutrient mineralization, although intense grazing pressure may reduce microbial activity (e.g., Clarholm, 1985; Allen-Morley and Coleman, 1989; Verhoef and Brussaard, 1990). In other experiments large parts of the food web have been deleted (Persson, 1989; Setälä and Huhta, 1991) or a food web has been built up progressively (Coûteaux et al., 1991). These studies suggest that gross food web structure influences nutrient mineralization and often also C mineralization rate. Field studies are less common. Although animals usually contribute to nutrient mineralization, effects vary from suppression to enhancement depending on food web composition, habitat and other factors (Ingham et al., 1989; Faber and Verhoef, 1991; Beare et al., 1992).

The detrital food web is a classic textbook example of donor-controlled consumer–resource interactions, where consumer density has no effect on the rate of resource renewal (Pimm, 1982; Begon et al., 1990). However, this proposition has seldom been examined, and two things should be pointed out: First, focussing on direct interactions between consumers and resources may leave out important feedbacks in the soil system (cf. below). Second, the concept of donor-control is poorly understood, and it is likely that it is an interaction that requires rather special circumstances (e.g., DeAngelis et al., 1975).

If all interactions in soil food webs are donor-controlled, an increased input of detritus is expected to lead to increased biomasses at all higher trophic levels [DeAngelis, 1992; see Eqn. (16-3)]. This would also be the case if consumer–resource interactions are completely ratio-dependent (Ginzburg and Akçakaya, 1992). On the other hand, if consumers affect resource densities (e.g., interactions of the Lotka–Volterra type), an increased detritus input is usually expected to have little effect on biomass at intermediate trophic levels, but the highest trophic level will increase (see Abrams, 1993 for examples and exceptions). Experimental manipulations of a higher trophic level are expected to result in corresponding changes at the next lower level.

Few studies have explicitly addressed these questions in soil food webs. In a long-term laboratory experiment we are examining if field densities of macroarthropod predators [miniature "top-level" carnivores in the sense of Estes (Ch. 15)] affect numbers and community composition of fungivores, and whether this leads to subsequent effects on decomposition and N mineralization. The results from the first sampling after 50 days' duration show that a mixed predator community (gamasid mites, spiders, and cantharid beetle larvae) has significant effects on both total abundances of fungivorous collembola and mites and fungivore species composition, when compared to the no-predator treatment (T. Persson, L. Lenoir, J. Bengtsson, unpublished). This indicates that predator regulation of the fungivore trophic level occurs. Whether it also results in effects on fungal biomass (i.e., a trophic cascade), decomposition rate, and N mineralization is yet to be determined.

Field studies of soil food webs often suggest that consumers have little influence on their resources (Ingham et al., 1989; Beare et al., 1992; Wardle and Yeates, 1993). However, Wardle and Yeates (1993) suggested that the abundance of predators on microbial-feeding nematodes was related to microbial biomass but not prey abundance. This is expected if interactions are reciprocal. Beare et al. (1992) found clear indications that fungivorous microarthropods regulated fungal densities. It has also been suggested that protozoa may regulate bacterial biomass (Clarholm, 1985; Beare et al., 1992), and that microarthropods can regulate densities of organisms at lower trophic positions (Moore et al., 1988).

Studies of soil food webs have usually neglected the larger predators such as ants and carabid beetles. These may play an important role for ecosystem processes on larger temporal and spatial scales, as they may average out the faster dynamics at lower trophic levels. An example of such a top predator is wood ants (*Formica* spp.) in boreal forests. By translocating and concentrating detritus and nutrients to their mounds, they may in themselves have substantial effects on ecosystem processes (cf. Lawton and Jones, Ch. 14; Anderson, Ch. 10). Also, if forest soil food webs are not completely donor-controlled, top-down effects on lower trophic levels by ant predation are possible. These are largely unexplored questions that need to be examined.

To summarize, soil animals and food web structure can both contribute to decomposition rates and nutrient cycling. To understand these processes and mechanisms in more detail and to predict what will happen when conditions change, future studies need to be focussed on the *interactions* between functional groups. The types of interactions between components in soil food webs are important, as theoretical predictions regarding ecosystem processes probably differ depending on the forms of these interactions (see following section).

THEORY OF FOOD WEBS AND ECOSYSTEM PROCESSES IN SOILS

One of the reasons why few ecologists have tried to link population dynamics and food webs with ecosystem processes is the lack of a theory relating these levels to each other. Our aim is to formulate a theory incorporating the population dynamics of simple food webs and the ecosystem-level dynamics of C and N. This is similar to DeAngelis (1992), but we will explicitly relate soil food web dynamics to a general theory of C and N dynamics in terrestrial ecosystems formulated by G.I. Ågren and E. Bosatta (e.g. Ågren and Bosatta, 1987; Bosatta and Ågren, 1991). Initially we will focus on C dynamics in food webs where the interactions are donor-controlled. The choice of donor-control interactions was motivated by their simplicity, which provides a starting point for theoretical analyses, and because of the widespread conception that detrital food webs are donor-controlled (e.g., Pimm, 1982).

Consider a simple system consisting of a detritus compartment X_1, a compartment of primary decomposers X_2 (for example, bacteria or fungi), microbial grazers X_3 feeding on the primary decomposers, and a predator compartment X_4. Assume that the fluxes F_{ij} from resource i to consumer j are donor-controlled, and that all consumers are present in amounts fairly close to their equilibria. Then the functional responses in the consumer–resource interactions can be described by equations of the form

$$F_{ij} = f_{ij} X_i \qquad (16\text{-}1)$$

where f_{ij} is the removal rate of the resource per unit consumer biomass. Under these conditions, the system can be described by the following equations:

$$
\begin{aligned}
dX_1/dt &= I_C - (f_{12}/e_{12})X_1 + \mu_2 X_2 + \mu_3 X_3 + \mu_4 X_4 \\
dX_2/dt &= f_{12} X_1 - (f_{23}/e_{23} + \mu_2)X_2 \\
dX_3/dt &= f_{23} X_2 - (f_{34}/e_{34} + \mu_3)X_3 \\
dX_4/dt &= f_{34} X_3 - \mu_4 X_4
\end{aligned}
\qquad (16\text{-}2)
$$

where X_i is the amount of C in each compartment, I_C is the input of C to detritus from outside the decomposing system, f_{ij} is the net incorporation of C by X_j feeding on X_i, e_{ij} is the efficiency (production to assimilation ratio) of X_j

when feeding on X_i, and μ_i is the nonpredation mortality of compartment X_i. At steady state,

$$X_1^* = I_C/\Psi$$
$$X_2^* = X_1^* f_{12}/(f_{23}/e_{23} + \mu_2)$$
$$X_3^* = X_2^* f_{23}/(f_{34}/e_{34} + \mu_3) \qquad (16\text{-}3)$$
$$X_4^* = X_3^* f_{34}/\mu_4)$$

where

$$\Psi = (1/e_{12} - p_2)\, f_{12}$$
$$p_2 = (p_3 f_{23} + \mu_2)/(f_{23}/e_{23} + \mu_2)$$
$$p_3 = (f_{34} + \mu_3)/(f_{34}/e_{34} + \mu_3).$$

At the ecosystem level, using one detritus and one decomposer compartment, $X_1^* = I_C/k$, where k is the specific decomposition rate (Ågren and Bosatta, 1987). Hence Ψ can be identified as the specific decomposition rate (mass loss) but with details about the soil food web included. In donor-controlled systems, a general format for Ψ can be given through the use of a set of p-factors,

$$p_j = (p_{j+1} f_{j+1} + \mu_j)/(f_{j+1}/e_{j+1} + \mu_j) \qquad (16\text{-}4)$$

where the p-factor for the highest trophic level in the chain is 1 (Zheng, 1993).

Equation (16-4) shows that the p-factor for each component in each chain is a function of the interactions with the trophic level above it, the p-factors of the upper trophic levels, and the nonpredation mortality. The decomposition rate of detritus Ψ is directly related to the values of the p-factors at the microbial level [cf. Eqn. (16-3)]. These p-factors can be viewed as a measure of the efficiency of the decomposers—the higher the microbial p-factor, the less efficient the decomposer community is and the lower the decomposition rate.

These results apply to food webs both with and without omnivory by summation over all chains c in the food web. The specific decomposition rate can then be expressed as

$$\Psi = \Sigma\,(1/e_{cj} - pc_j)f_c \qquad (16\text{-}5)$$

where $e_{c,j}$, $p_{c,j}$, and f_c refer to the microbial or detritivore components immediately above the detritus in each chain (Zheng, 1993).

In donor-controlled systems, Ψ decreases linearly with $p_{c,j}$ [Eqn. (16-5)], and $p_{c,j}$ usually decreases with the addition of a new component in the food chain. Hence, the number of trophic levels may influence the decomposition rate (and the steady-state amount of detrital C), even though all consumer–

resource interactions are donor-controlled. This result is unexpected in view of traditional food web theory, and is mainly a result of food web structure influencing intrasystem cycling. Also, the addition of new food chains or energy channels to a decomposition system will increase decomposition rate, given that consumers do not influence each other's consumption rates or mortalities. These theoretical results coincide with several laboratory studies (e.g., Allen-Morley and Coleman, 1989; Setälä and Huhta, 1991).

Hence, a connection between decomposition rate, a process at the ecosystem level, and "population" dynamics in simple food webs has been achieved. It remains to be seen how the theory can be extended to more general types of trophic interactions, such as Holling (Monod) and Lotka–Volterra functional responses. In these nonlinear systems, the relationships between decomposition rate and trophic structure are unlikely to be as simple as those presented above.

We aim to develop this theory further by including a variable substrate quality (Bosatta and Ågren, 1991), modification of the substrate and microbial activity by animals (e.g., communition), and resource competition. Furthermore, because animal effects on N dynamics usually are more prominent than effects on decomposition rate, it is essential to incorporate food web structure into the ecosystem-level theory of C and N dynamics formulated by Ågren and Bosatta (1987) and Bosatta and Ågren (1991).

ACKNOWLEDGMENTS

J. Faber, D.L. DeAngelis and an anonymous reviewer commented on the manuscript. This study was financed by grants from NFR (J.B.), SJFR (T.P.), and NUTEK (D.Z., J.B.; grant to H. Lundkvist).

17

Unifying Ecological Subdisciplines with Ecosystem Food Webs

Neo D. Martinez

SUMMARY

Ecosystem food webs equitably depict the trophic interactions between all organisms within a spatially explicit unit of the earth. Membership and trophic interactions within ecosystem food webs are determined for explicit time periods using explicit linkage criteria. These webs are shown to be useful tools for merging ecological subdisciplines and linking species to ecosystems. This utility stems from the fact that who eats whom appears to be the most central organizing concept in ecology. Such centrality allows ecosystem food webs to integrate information from nearly all ecological subdisciplines into comprehensive portraits of ecological organization. Quantitative analysis of these portraits allows prediction of the trophic structure of populations, communities, and ecosystems. Beyond trophic interactions, ecosystem webs may be used to address other types of interactions. Such webs also emphasize explicitness, completeness, and consistency but may be based on criteria such as which organisms create habitats for each other (i.e., ecosystem engineering webs) or which organisms have mutually beneficial symbiotic relationships (i.e., ecosystem mutualism webs).

INTRODUCTION

What is the fate of ecological subdisciplines such as population, community, and ecosystem ecology? A look to the development of physical sciences suggests at least two disparate possibilities. Our earliest physical science diverged into physics and chemistry because largely separate frameworks proved most useful for understanding and predicting physical and chemical phenomena. In contrast, early and essentially separate research on mechanics and heat was unified by the discovery that both heat and mechanical forces are different manifestations of energy. While this volume espouses ecological subdisciplinary unification, it is possible that our subdisciplines will successfully diverge. The value of unification vs. divergence is best evaluated based on the analytical and predictive power achieved by developing along either path.

If divergence prevails, there is still an important role for interdisciplinary work. In the physical sciences, physical chemistry occupies this role by linking physics and chemistry through the explanation and prediction of physical-chemical phenomena that cannot be explained by either discipline alone. If unification prevails, there must be a unifying framework that allows explanation and prediction of the phenomena studied within subdisciplines. Given either possible trajectory, ecosystem food webs demonstrate how species and ecosystems along with their associated subdisciplines (e.g., population, community, and ecosystem ecology) are linked. Whether research on ecosystem food webs will provide a unifying framework or merely demonstrate links between distinct subdisciplines is an open question. Here, progress in these directions is discussed.

Most authors in this book embrace unification through either a theoretical or pragmatic approach. The theoretical approach specifies an ecological system from an ecosystem perspective in terms of mass flows and from a community perspective in terms of organisms and their populations. Once these specifications are made, the theoreticians typically map relations between the two approaches. The pragmatic approach specifies a research problem primarily from one ecological perspective but then strategically employs concepts from another perspective that helps to address the original problem. Both approaches reduce the possibility of divergence by linking advances within one subdiscipline to advances in other subdisciplines.

The ecosystem web approach is different from the theoretical and pragmatic approaches because the original system specification is a rarefied hybrid of the ecosystem and community perspectives. Neither quantitative estimates of mass flows nor of species' populations are a necessary part of specifying an ecosystem web. The utility of avoiding these traditional perspectives is demonstrated by the current use of food webs to construct and disprove ecological theories as well as successfully predict patterns in data that synthesize observations from many subdisciplines.

THE CONCEPT OF ECOSYSTEM FOOD WEBS

Tansley (1935) introduced the ecosystem concept and considered ecosystems to be "the basic units of nature." Likens (1992) reified this concept by defining an ecosystem "as a spatially explicit unit of the Earth that includes all of the organisms, along with all components of the abiotic environment within its boundaries." The ecosystem concept is widely thought to be the most important concept in ecology (Cherrett, 1989). Focusing the concept toward food webs, Martinez (1991a) defines an ecosystem food web as a network of trophic interactions "with relatively even emphasis on all trophic groups within an explicitly bounded ecosystem, and employing explicitly defined criteria for designating links." Even emphasis prevents ecologists from arbitrarily restricting food webs to only include organisms within a narrow taxonomic range. Following a tradition of ecosystem research, boundaries are chosen based on empirical and operational tractability rather than on problematic notions of community or population boundaries (Yodzis, 1993). Explicit linkage criteria facilitate objective and reproducible observations (Peters, 1988; Lawton, 1989). Omission of taxa and linkage criteria have been one of the most serious impediments to progress in food web research (Strong, 1988). Use of ecosystem food webs reduces these impediments (Martinez, 1991a).

The ecosystem concept and the definition of ecosystem food webs specifies that these food webs are comprehensive networks of producing, consuming, and decomposing organisms linked by trophic relationships within explicit spatial and temporal boundaries. Practically all ecological subdisciplines conduct some degree of fundamental research on how organisms are consumed or consume and subsequently respire other organisms' biomass. The observation of ecosystem food webs requires synthesis of these type of data from a wide variety of subdisciplines. For example, a lake ecosystem food web may synthesize observations from ecosystem, community, population, microbial, benthic, phytoplankton, zooplankton, and vertebrate ecologists. The pervasive nature of trophic interactions and the highly synthetic nature of ecosystem food webs suggests that "who eats whom" may be the most central organizing concept in ecology.

OBSERVATIONS AND PREDICTIONS
CONCERNING ECOSYSTEM FOOD WEBS

The utility of comprehensive descriptive food webs is not without its critics (see Paine, 1988). Just because the idea of collecting and collating certain data is attractive, it may not be scientifically useful to analyze such observations. However, the discovery of robust and successfully predictive constraints among nearly all populations, communities, and ecosystems, combined with the delineation of mechanisms responsible for those patterns, could greatly advance and synthesize ecology (Martinez, 1994). These types of discoveries are important because the lack of synthesis and prediction seriously impedes the maturation of ecology as a scientific discipline (Peters, 1991).

One of the most confounding aspects of describing ecosystem food webs is deciding what constitutes a "species." This problem hinges on how much of what type of similarity among organisms is enough to warrant aggregating the organisms into a "species." When deciding among such criteria, preference should be given to a definition that is agreed on and is relatively easy to apply consistently. Biological species, or groups of potentially interbreeding organisms that are reproductively isolated from other populations, are natural candidates because they are the most widely accepted and applied groups of organisms in ecology. However, application of sex-based grouping criteria to bacteria, fungi, plants, and other organisms included in ecosystem food webs that can reproduce asexually or hybridize readily is problematic and often impossible. The concept of trophic species, or groups of organisms that share the same predators and prey (Briand and Cohen, 1984), is an alternative to biological species that has gained wide acceptance among trophic ecologists (Cohen et al., 1990; Pimm et al., 1991).

Martinez (1994) discusses how criteria used to define trophic species and their interactions that comprise ecosystem food webs can be consistently applied to virtually all organisms. Unlike biological species, the definition of trophic species is both agreed upon (Pimm et al., 1991 and many others) and can be consistently applied to organisms regardless of their sexual status. Also, because organisms typically engage in trophic interactions more frequently than sexual interactions, data that specify trophic interactions are easier to collect than data on reproductive interactions and therefore trophic species is a more operational concept than biological species. The most pragmatic approach may be to resolve organisms to the finest taxonomic level possible without going beyond their Latin binomials and then aggregate to trophic species while retaining knowledge of the taxonomic diversity within each trophic species (Martinez, 1991a).

Another important aspect of describing ecosystem food webs is deciding what constitutes a "link" (Martinez, 1988, 1991a, 1993a; Paine, 1988). The variable and undefined linkage criteria of most food webs are asserted to be a fatal flaw in food web theory because the resultant variable quality of trophic information may confound identification of patterns in ecosystem food webs (Paine, 1983, 1988; Polis, 1991, 1994). However, since successfully predictive constraints in food web structure that are robust to variable resolution of species and linkages appear to have been discovered (Martinez, 1993b), the consequences of variably defining links appears to have been overstated. Successfully predictive constraints suggest that trophic ecologists generally observe similar links despite variable criteria for defining when an organism consumes another organism. Still, agreement on consistent explicit linkage criteria would aid analyses of food web properties (Cohen et al., 1993). Martinez (1991a) assigned links "between taxon 'A' and taxon 'B' whenever an investigator believes that 'A' is likely to eat 'B' during a typical year." Eschewing "belief" in favor on more operational criteria would be an important improvement of this definition.

The criticism of variable linkage criteria appears to be especially inapplicable to directed connectance, a simple measure of the number of links in ecosystems and communities. In systematic analyses that varied the numbers of species of several food webs and the criteria for establishing links between them, Martinez (1991a,b, 1993b) demonstrated that connectance is one of the most robust of the commonly observed food web properties. The largest food web to date, the 93-trophic species food web of Little Rock Lake, WI, has a directed connectance [directed connectance equals links/species^2 (Martinez, 1992)] equal to 0.118 (Martinez, 1991a), which means that about 10% of possible links including cannibalism are realized. In an attempt to demonstrate the wide variability and scientific irrelevance of connectance data, Paine (1988, Table 3) described 10 intertidal food webs with a mean directed connectance of 0.11 (SD = 0.05). Contradicting his own argument, Paine's estimates are comparable to previously cited means of 0.11 (SD = 0.03, n = 5, Martinez, 1992) and 0.103 (SD = 0.025, n = 50; Martinez, 1993a). The connectance of a recent high quality food web of the island of St. Martin containing 42 trophic species is 0.116 (Goldwasser and Roughgarden, 1993).

The constrained variability of connectance supports the hypothesis that food webs have been useful in discovering a synthetic, robust, and useful pattern among aquatic and terrestrial populations, communities, and ecosystems. Ecosystem food webs should play an even stronger scientific role inasmuch as they are more rigorously defined than food webs have been in the past. Constrained connectance has been used to successfully reject an early hypothesis titled the "link-species scaling law" (Martinez, 1992, 1993a) which asserts that links increase roughly linearly with species (Cohen et al., 1990; Pimm et al, 1991). The fact that the constant connectance hypothesis was advanced (Martinez, 1988, 1991a,b, 1992) before recent corroborative food web observations (Goldwasser and Roughgarden, 1993; Martinez, 1993a) demonstrates that the hypothesis may be one of the very few general quantitative hypotheses that successfully predict the structure of populations, communities, and ecosystems irrespective of their terrestrial or aquatic status.

Roughly constant connectance of 0.1 means that the trophic function of species' populations within communities and ecosystems is constrained to eat, and to be eaten by, a mean of roughly 10% of the species within their system. This pattern suggests that ecosystem scientists who design studies to estimate flows of trophic energy between species should be prepared to examine a number of flows equal to $0.1(\text{species})^2$. Populations of species who eat >10% may be considered trophic generalists whereas those that eat <10% may be considered trophic specialists (Martinez, 1993b). Without such a distinction research on causes and consequences of trophic specialization in ecological systems is inhibited (Strong et al., 1984). The connectance constraint provides a tractable and widely applicable distinction between trophic specialists and generalists. For example, even among ecosystems with different species richness, the tendency for links to equal $0.1(\text{species})^2$ allows marine specialists to be compared with terrestrial specialists to test whether trophic specialists are generally more or less likely to be more abundant (in terms of biomass, species, or individuals) than trophic generalists. Another intersubdisciplinary hypothesis that can be tested is the question of whether species that are relatively vulnerable (>10% of species in its system consumes the species) have more variable populations and larger fractions of biomass.

Again, the demonstration that mechanisms studied within a wide variety of ecological subdisciplines are responsible for the general structure of ecosystem webs would do much to merge ecological subdisciplines. For example, investigation of ecosystem webs may suggest that natural selection has caused most species to be specialists although some species, such as nonspecific insectivores, are extremely general. Although few in species number, the generalists may be found to have both significantly more population variability and significantly higher fractions of an ecosystem's biomass. Discovering such a trade-off between diversity, rarity, and trophic specialization on the one hand, and population variability, biomass, and trophic generalization on the other, could merge ecological subdisciplines by demonstrating general relationships between variables that different subdisciplines emphasize.

OTHER ECOSYSTEM WEBS

Following Lawton and Jones' (Ch. 14) discussion of ecosystem engineers (Jones et al., 1994), another ecosystem web that could be defined is an ecosystem engineering web. Among several possibilities, these webs could depict which species create habitats for which species within an explicit volume of nature. Hypotheses that could be tested with ecosystem engineering webs include whether or not the average fraction of species for which species create habitats is fixed among a wide variety of ecosystems. Further tests may reveal that quantitatively predictable fractions of species in an ecosystem are "generalist" habitat creators that create habitat for a fixed fraction of species in the ecosystem. Although this is only speculation, it highlights the scientific possibilities of employing consistent, explicit, and synthetic criteria to conduct research at the intersection of our various subdisciplines. However, if interactions in engineering webs are defined as existing between "organisms that directly or indirectly modulate the availability of resources to other species, by causing physical state changes in biotic or abiotic materials" (Jones et al., 1994), all possible interactions may be realized because all organisms appear to modulate the availability of CO_2 and O_2 to all other organisms. As with ecosystem food webs, the scientific utility of such observations will have to be critically evaluated.

Another possible ecosystem web could depict which organisms have mutually beneficial symbiotic relationships with one another. These mutualism webs along with engineering webs will have to face the problem of defining species and interactions in a robust and operational manner.

WEBS COMPLEMENT CHAINS

Ecological flow chains (Shachak and Jones, Ch. 27) synthesize ecological subdisciplines in a manner analogous to ecosystem food webs with a focus on mechanistic chains rather than phenomenological webs. Much of ecology proceeds by employing the concept of causal chains to analyze the dynamics of a small subset of interactors within ecosystems. It is widely appreciated that specification of a small subset is necessary to make ecology tractable to theory [e.g., analytical and simulation modeling, (Hall and Day, 1977)] and to experimentation. This small subset is first specified and then the many possible interactions between the components of the subset are reduced to those most important for explaining the behavior or structure of interest. Paine's (1980) and Pimm's (1982) studies of inter- and intraspecific interactions are classic examples. Components are almost always fewer than a dozen functional types and the interactions between them are almost always reduced to a very few strong or causally important interactions. This allows powerful interpretations of models, experiments, and simulations. However, if one desires more comprehensive ecological research that explicitly includes a nontrivial fraction of biodiversity in an ecosystem, one must embrace the much greater number of interactors and interactions emphasized by Tansley and Likens.

Shachak and Jones' (Ch. 27) discussion of ecological flow chains represents an important development of the chain concept across ecosystem and community/population subdisciplinary boundaries. Shachak and Jones advocate analyses of chains that explicitly interconnect material flows with species abundances. These analyses serve to merge insights that typically remain within separate disciplines while maintaining a detailed understanding of interactions among species and ecosystems.

Although it may be tempting to simply expand the notion of chains that contain detailed information to webs that maintain this information while explicitly delineating the wide variety of organisms in ecosystems, such expansion may be analytically intractable (Hall and Day, 1977) and may be less productive than a strategically compromised approach that trades the details of causal interactions for inclusiveness of hundreds of functionally different components and thousands of simple interactions. Giving up detailed information of how a few components are linked in chains may allow for more general descriptive information that identifies components and their interactions within webs.

The web approach is complementary to the chain approach. The chain approach generates important understanding of dynamically important links between only a very small, and sometimes trivial, fraction of the components that make up ecosystems. Webs embrace the diversity of components but make few if any distinctions between the importance of different links. The power in the complementarity may lie in the possibility of combining both perspectives to provide more independent tests of each perspective's insights and hypotheses. It is not clear that the mechanisms delineated in chains will be relevant to the more inclusive ecosystem food web structure. Certainly, hypotheses that such mechanisms (e.g., trophic cascades) affect ecosystem food web structure will be important hypotheses to test. However, as suggested earlier, the hypothesis that other mechanisms such as natural selection and community assembly constraints are responsible may be more plausible.

A type of complementarity is beginning to emerge within trophic ecology. Research on the strength of causal interactions has suggested that interaction strength of the sort that causes trophic cascades tends to decrease as chains become more reticulate and functionally speciose (Strong, 1992). Research on food web structure has found connectance to be relatively constant among webs with widely varying numbers of species (Martinez, 1992, 1993a). A complementarity between these findings can be seen by employing May's (1974) classic mathematical assertion. This assertion states that, for stable randomized food web models, interaction strength must decrease if the number of species increases and connectance remains constant. The main point of this complementarity is not that these findings and assertions are corroborative, and therefore are in some sense true. The main point is that diverse perspectives, such as those represented by chains and webs, that can address similar questions add power to scientific methodology. Such power allows us to arrive more quickly at an acceptable approximation of truth.

ECOLOGICAL PERSPECTIVES, SCALES, AND SUBDISCIPLINES

Central to linking species and ecosystems or merging subdisciplines is an understanding of the entities being linked. Although the title of this book suggests a linking of natural entities, deeper inspection reveals the goal to meld the historically different perspectives ecologists use to study nature. Population biologists emphasize the population fluctuations of certain species in ecological systems. Community ecologists emphasize the relative importance of the distribution and abundances of species. Ecosystem ecologists emphasize the stocks and flows of energy and materials. The search for a vehicle to merge these perspectives is a central concern of this volume.

Adherents of the cell–organism–population–species–community–ecosystem–landscape hierarchy argue that the difference between ecological subdisciplines merely represents a difference in the scale of interest. With regards to scale, this hierarchy is false and misleading (Allen and Starr, 1982; O'Neill et al., 1986). Each entity along this hierarchy may occupy either larger or smaller time and space scales than the entity on either side of it. For example, an organism (e.g., pitcher plants) may contain entire communities. A cell has DNA sequences that vary on vastly longer time scales than the dynamics of ecosystems. The difficulty of consistently arranging the entities along a spatial or temporal hierarchy of scale suggests that the differences between these entities, and their associated subdisciplines, are more fundamental than mere differences of scale. As with physics and chemistry, the difference between ecology's subdisciplines concerns which phenomena are of interest; not the scale at which entities are presumed to operate.

The discussions of Tansley (1935) and Likens (1992) suggest that "ecosystem" has taken on an unfortunately parochial connotation. While Likens and Tansley have emphasized inclusiveness, subsequent practice has tended to narrow the application and meaning of ecosystem science to "stock and flow" ecology. Material stocks and flows are a narrowly defined subset of ecological interactions and miss much of what Tansley and Likens have defined as ecosystems that includes community and population structure. Population and community ecologists study many interactions outside the typical purview of ecosystem ecologists and analyses of these interactions have much to contribute to the understanding of the structure and dynamics of organisms within explicit boundaries.

From this perspective, the question of how to link species and ecosystems is answered by recognizing that species and ecosystems are inseparable and that the illusion of separateness comes largely from the different terms for the same entities used by different subdisciplines. For example, plants are considered a plant community by community ecologists while "stock and flow" ecologists consider the plants to be producers. The simple answer of how to link these two entities is to realize that they are the same thing with different names and used for different scientific purposes. However, realization of the deeper urge to merge ecological subdisciplines requires more than semantic reorganization. It requires systematic formalization of relationships between fundamental tenets embraced by ecological subdisciplines.

CONCLUSIONS

The definition of ecosystem food webs incorporates several of the most fundamental characteristics that population, community, and ecosystem ecologists use to define their research. Predator–prey relationships have long been central to population biology. Species diversity is central to community ecology. Explicit boundaries, inclusiveness, and trophic completeness defined by material (e.g., live biomass) flow is central to ecosystem ecology. Trophic relations are also important to evolutionary and physiological ecologists as evidenced by research from how beak structure affects a bird's diet to the effects of secondary plant chemicals on herbivory. Ecosystem food webs incorporate each of these subdisciplinary fundamentals into a reproducibly observable and quantifiable ecosystem portrait. Such portraits are an important first step in analyzing relationships between ecological organization and function (e.g., diversity, stability, and biomass). The mechanisms responsible for the structure of ecosystem food webs appear likely to include evolutionary, behavioral, and physiological aspects of organisms as well as community assembly rules and the environmental constraints of ecological systems (Martinez, 1992). The discovery of how such mechanisms generate universal and quantifiably predictable patterns in ecological organizations could do much to illustrate relationships among fundamental aspects of population, community, ecosystem, and evolutionary ecology.

ACKNOWLEDGMENTS

Discussions with Jennifer Dunne, Marc Fischer, John Lawton, Clive Jones, John Harte, and Kristina Stinson contributed many important ideas to this manuscript. Jennifer Dunne greatly improved the manuscript by editing early versions. Support was provided by NSF Grants BIR-9207426 and DEB-9208122.

18

COUPLING THE DYNAMICS
OF SPECIES AND MATERIALS

William S.C. Gurney, Alex H. Ross,
and Niall Broekhuizen

SUMMARY

After analyzing the failure of early attempts to model energy and material flow
in ecosystems, we consider the influence of individual-based modelling on understanding of population and ecosystem dyamics. We conclude that such models are necessarily very data-hungry, and suggest that taken together with the difficulty of measuring many body effects, this implies that models based on detailed
representations of individual properties are unlikely to prove widely useful. We
then show that a sparsely formulated nutrient/energy flow model is capable of
reproducing the observed annual cycle in a well-defined, open, ecosystem (a sea
loch). That this success is systematic rather than fortuitous is shown by the ability of the same model, with appropriately changed environmental parameters,
to predict the differences in annual cycle within a group of related systems.

A HISTORICAL PERSPECTIVE

The idea that considering energy fluxes through an ecosystem ought to yield
insight into how it works has its genesis in the considerable advances made
possible in the physical sciences by the laws of conservation of energy and

mass. During the 1960s and 1970s, the view that this success could readily be extended to the life sciences led to the expenditure of many thousands of man-years on grandiose simulation models describing ecosystems ranging from trees and ponds to oceans and continents, in terms of energy flow, material flow, or an unholy combination of both.

Conservation laws are at their most useful in the closed systems tradition-ally studied in the laboratories of physical scientists. In the open systems com-monly dealt with by engineers their use requires us to define and measure the fluxes across the system boundaries, a difficulty that normally necessitates very careful choice of those boundaries. Early attempts to formulate energy flow ecosystem models rapidly disclosed that although the inflow of energy to an ecosystem could often be estimated quite well, the outflows were hard even to define, let alone measure.

The observation that the significant energy flows inside ecosystems concern chemically bound energy, and are thus necessarily accompanied by flows of nutrients, led to alternative formulations in terms of elemental nutrients such as carbon (C) or nitrogen (N). Many ecosystems are more nearly closed to nu-trients than they are to energy, and where they are not, nutrient inflows and outflows are often easier to measure than their energetic counterparts. How-ever, either because of the difficulty of drawing suitable system boundaries, or because of the prevailing distaste for "reductionism," most workers still tried to construct highly detailed models of entire ecosystems. Such models typically had many tens of state variables, representing the standing stock of C bound in each species or trophic group, and an unsurprisingly large ac-companying list of parameters. In consequence, they are immensely hard to parameterize, both because of the sheer number of quantities for which val-ues are needed, and also because of the difficulties of interpreting the param-eters themselves in a sufficiently precise manner.

The severity of the parameterization problem becomes especially acute when (as ill luck would frequently have it) the predicted dynamics are highly sensitive to at least some of the parameter values. In the physical sciences such parameters would generally be regarded as fitting parameters, whose precise values should be determined by optimizing agreement with a well-defined tar-get data set. However, in the case of ecosystem models suitable data for such an exercise are almost never available. This brings us to the central reason why we believe the early energy and material flow ecosystem models were such a resounding failure. The objects represented by such models are geographi-cally extended, operate in a highly variable environment, and change state on a time scale of months and years rather than milliseconds. Collecting data on many tens of state variables over such time and space scales is a practical im-possibility. Thus the models were seldom if ever subjected to the virtuous cycle of testing, falsification, and refinement, which is a necessary condition for progress. Instead the culture came to see the very construction of a model as the culmination rather than the start of the process of scientific investigation.

One reaction to the systematic quality of these failures has been either to model simplified systems, such as the CEPEX mesocosms (Andersen and Nival, 1989), to focus on specific processes, such as Spring Bloom initiation (Taylor et al., 1992), or to restrict attention to a subset of ecosystem processes (Steele and Frost, 1977; Cloern and Cheng, 1981; Tett et al., 1986). Such models often have been highly successful. However, their necessarily tight focus, and consequent neglect of locally irrelevant but globally critical nutrient and energy flows, implies that they may not give a good overview of ecosystem function.

It is an evident truism that flows in organisms, populations, and ecosystems are driven by the input of energy and that they carry chemically bound energy and elemental nutrients in characteristic stoichiometric proportions (Sterner, Ch. 23; Wedin, Ch. 24). It is further evident that natural systems do not contravene the basic laws of physics, so the books must always balance. It is hard to see how these considerations cannot be made to yield insights into the dynamics of biological systems. Indeed, our thesis in this chapter is that they are extremely powerful ideas, whose initial failure was largely attributable to inappropriate methodologies.

INDIVIDUAL ENERGETICS

An area in which considerations of material and energy flux have made significant contributions is the study of the dynamics of individual organisms (e.g., Kooijman, 1986; Ross and Nisbet, 1990). Growth and reproduction have been particularly intensively studied in this context, both because of their intrinsic interest, and because they are conveniently studied in the laboratory (e.g., McCauley et al., 1990; Bradley et al., 1991).

Armed with a well validated model relating individual characteristics to energetic considerations, it should be possible to predict population properties on the same basis. Numerous formal methods exist for constructing population models from descriptions of the demographic properties of individuals. These range from the elegant distribution function techniques pioneered by the Dutch structured population group (e.g., Metz and Diekmann, 1980) to explicit representation of all the individuals who compose the population (e.g., Huston, 1992; MacKay, 1992).

To illustrate the high quality of fit that can be obtained when validating models describing the properties of individuals we shall use two models specifically developed to underpin individual based population models (IBPMs). One model describes growth and fecundity in *Daphnia*, and the other describes compensatory growth in fish. Both are two state variable models, concentrating on the flow of C. Both take account of the fact that during periods of starvation the organism burns reserves to meet maintenance costs. The need for two state variables in both cases comes from the recognition that although reserve depletion involves weight loss, a vital group of tissues such as the gut, mouthparts, and skeleton remain unchanged during this process. Thus, the configuration, and feeding capacity of an individual of a given weight will depend critically on whether it got there by growing unchecked from a smaller size, or by growing to a larger size and then being starved.

The *Daphnia* model (Gurney et al., 1990) uses total weight and carapace length as its state variables. It postulates that the first activity to be sacrificed as uptake becomes inadequate for total needs is growth. When the uptake rate becomes too small to meet the costs of reproduction and maintenance, the allocation to reproduction is reduced just sufficiently to allow body weight to be held constant. If uptake falls still further, so that it cannot even meet maintenance requirements, then body reserves (body weight) are burned to meet the deficit, but carapace length remains constant. Although the model is conceptually simple it has a substantial parameter count, because it must describe the allometry of both uptake and maintenance cost as well as the rules governing energy allocation. Fortunately there is a large body of experimental data on our chosen subject (*Daphnia pulex*), and so 16 out of the 18 parameters could be determined from independent studies. We confronted the model with a data set comprising six growth and fecundity curves taken by two experimenters under very different laboratory protocols, and broad characterisations of the growth and fecundity observed in different laboratories by a further two workers. Fig. 18-1 illustrates the degree of agreement achieved.

The fish model (N. Broekhuizen, W. Gurney, A. Jones, and A. Bryant, unpublished) is particularly concerned with the compensatory growth often observed when fish are reintroduced to food after a period of starvation. It uses reserve and skeletal weight as its two state variables, and regards the second of these as a surrogate for length. It postulates that in times of plenty the fish arranges energy allocation so as to maintain its reserve-to-skeletal weight ratio (essentially its "fatness") at an optimum value. It further assumes that both maximum uptake rate and maintenance costs are to an extent under behavioral and physiological control, and can alter in response to changes in reserve ratio. A small reduction in reserves triggers a "hungry" state in which maximum uptake rate is increased, whereas a larger depression takes the animal into a "torpid" state in which both maximum uptake rate and specific maintenance costs are depressed. The model has 14 parameters of which seven were reliably determined from independent sources. We confronted it with length and weight data for groups of aquarium-reared salmonids subjected to repetitive periods of food deprivation and abundance. Five of the seven fitting parameters were held constant over the whole test data set, whereas two were

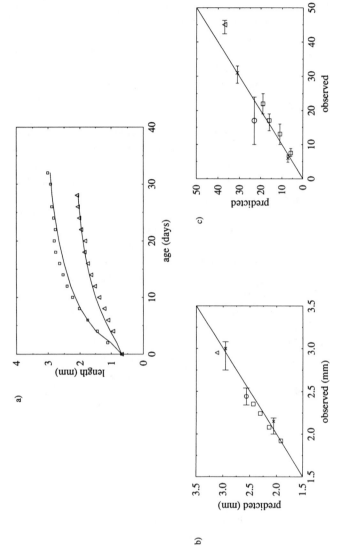

Figure 18-1. Testing the *Daphnia* individual model (Gurney et al., 1990) against *D. pulex* data. (a) Typical growth curves at high (1 mg carbon (C)l⁻¹ —*squares*) and low (0.1 mg Cl⁻¹ —*triangles*) food rations. *Continuous lines* are model predictions, not statistical fits to data. Data are from Taylor (1985) and Taylor and Gabriel (1985). (b) Predicted and observed maximum daphnid lengths. Data are from Paloheimo et al. (1982) —*triangles*, Lynch et al. (1986) —*triangles*, Richman (1958) —*squares*, Taylor (1985) and Taylor and Gabriel (1985)—*crosses*. (c) Predicted and observed maximum brood sizes. Data are from the same sources as *b*.

regarded as experiment specific. Again, the quality of the fit achieved was gratifyingly good, as is illustrated by Fig. 18-2.

Notwithstanding the excellent quality of fit to individual performance yielded by these two models, we found the *population* models based on them to be highly unsatisfactory. Indeed they failed to predict major qualitative features of the observed population behavior in both field and laboratory conditions (Murdoch et al., 1992). Although most methods of model construction place some limitations on the nature of the underlying individual description, these seldom determine the ultimate outcome. It is thus tempting to seek an explanation for our failure in changes in individual behavior brought about by the presence of large numbers of conspecifics, and such "many-body" effects certainly do occur. However, the true roots of systematic failure seem to lie in much more mundane considerations. For example, the dynamic range of conditions explored by laboratory experiments seldom matches the entire range experienced in the field. Certain aspects of individual demography, for example senescence, are difficult and time consuming to observe, and are less well studied than more amenable characteristics such as fecundity. Foraging behavior and mortality due to predation and accident occur on scales that effectively preclude laboratory observation altogether. Thus, although some components of an underlying individual model may be honed to perfection against excellent data, others are often no more than educated guesses informed by anecdotal field observations.

If enough of these difficulties could be overcome it might be possible to determine the remaining "free" parameters from population data without removing all possibility of falsification. However, the likelihood of this seems slight, even in the single-population context, when one remembers that a typical individual model has tens of parameters, a number that may well double when interactions and individual variability are accounted for. Using individual-based methods to construct a useful community model covering tens of interacting populations seems unthinkable.

BOX-MODELS REVISITED

One way to ameliorate the data-hunger associated with individual-based community models is to return to the idea of aggregated functional groups used in the early ecosystem models. Although the antireductionist philosophy underlying many such models led to just the sort of parameter proliferation we are currently trying to escape, this is not inherent in the approach. The broad phenomenological perspective taken by these models makes it essential to characterize each functional group by a few parameters, and they can thus be quite simple provided the level of aggregation is high enough (see also Turner

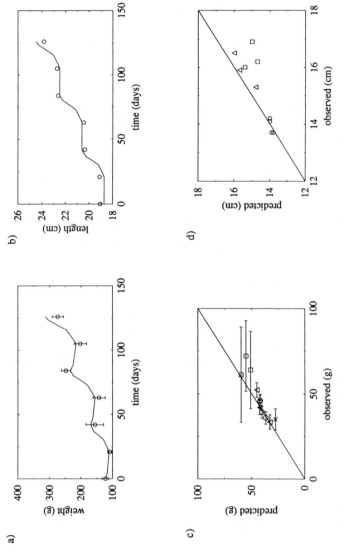

Figure 18-2. Testing the fish individual model (N. Broekhuizen, W. Gurney, A. Jones, and A. Bryant, unpublished) against data on salmonids subjected to repetitive periods of starvation and refeeding. Data are from Quinton and Blake (1990)—*circles*; Weatherly and Gill (1991)—*squares*; Miglavs and Jobling (1989a,b)—*crosses*; Kindschi (1988)—*triangles*. (a, b) Typical growth curves (weight and length) for relatively large fish subject to 3-week starvation/refeeding cycles. *Continuous lines* are model predictions, not statistical fits to data. (c) Predicted and observed final weights for small fish subject to a variety of feeding regimens. (d) The predicted and observed final lengths corresponding to the weights of the fish in c.

and O'Neill, Ch. 19). This view is by no means original, as many strategic models of material cycling and energy flow have been constructed on just this basis (e.g., May, 1973; Nisbet and Gurney, 1976; DeAngelis et al., 1989; see Schimel et al., Ch. 20; Rastetter and Shaver, Ch. 21). However, we now propose to argue that models of this general type can also achieve meaningful comparisons with field data.

The starting point for the work that we shall use as an illustration was a requirement by the Scottish Office Agriculture and Fisheries Department (SOAFD) for guidance in planning a large-scale experimental program in Loch Linnhe, a large sea-loch on the west coast of Scotland. This program was to focus on the potential effect of nutrient enrichment on ecosystem dynamics, and thus planned simultaneous measurements of key nutrients and key elements of the biota over a whole annual cycle. To optimize the insight gained from an exercise involving the deployment of very extensive resources, they required a semistrategic model realistic enough to give clear guidance as to the critical nutrient pathways, but yet intuitive enough for the robustness of its conclusions to be readily assessed.

In view of both the intended purpose of the model, and the time scale on which it had to be constructed, we adopted a conservative "box-model" approach. After discussion with SOAFD Marine Laboratory we concluded that the simplest viable description of the biological pathways for the circulation of nutrients is that shown in Fig. 18-3. Phytoplankton use incoming solar energy to fix dissolved N and C. The primary production is grazed by zooplankton, which are in turn preyed upon by carnivores (mainly jellyfish). Excretion and respiration products are returned directly into the water column, while corpses and fecal pellets fall to the bottom, where they are slowly remineralized and hence recycled. Dissolved C is assumed to be present in excess, so primary productivity is limited by the availability of light and N. Grazing and carnivory produce linked flows of C and N, which must be accounted for if the biomass of these functional groups is to be correctly estimated. However, as primary production in marine ecosystems is almost never limited by the availablity of dissolved CO_2, we need only take account of the recycling of N, both directly and through the sediments.

The set of pathways depicted in Fig. 18-3 greatly resembles a number of models of both freshwater and marine systems that either are, or were for simplicity assumed to be, effectively closed to everything except light (e.g., Nisbet et al., 1991). However, as Fig. 18-4 shows, a sea-loch is very far from being a closed system. The steady inflow of fresh water from the land and saline water from the sea implies that the system is always to some degree stratified. Deep systems, such as Loch Linnhe, have a stable layer of very cold, saline water at the bottom, an intermediate layer into which the incoming tidal water flows, and a low-salinity surface layer that carries terrestrial runoff toward

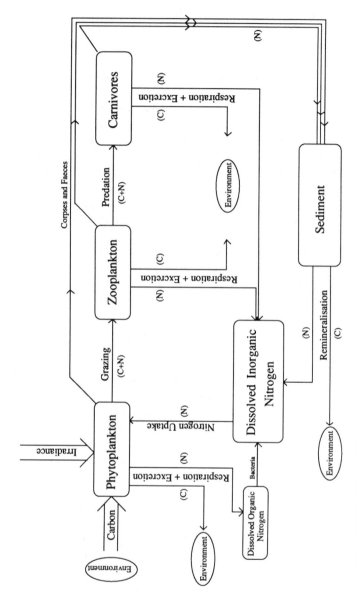

Figure 18-3. The sea-loch model food web. Arrows show flows of carbon (C) and nitrogen (N). The model is described in detail in Ross et al. (1993).

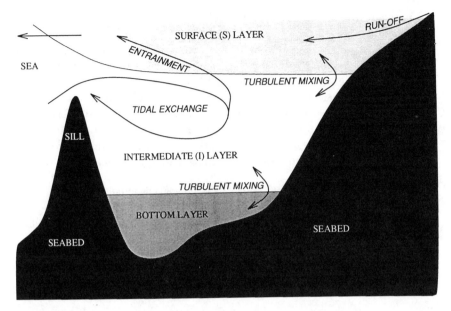

Figure 18-4.   The sea-loch model: physical environment. The shaded areas show the layer structure, and the arrows show water exchanges. The model is described in full in Ross et al. (1993).

the sea. These layers are coupled by turbulent mixing driven by the kinetic energy of both wind and tide, and by direct upwelling of incoming tidal water. Terrestrial runoff carries a heavy load of dissolved organic and inorganic N. Interchange with the sea carries with it dissolved nutrients and primary, secondary, and tertiary producers.

Despite its relative simplicity, this model has 36 parameters describing the biota, 6 parameters describing the internal physical structure and dynamics, 2 parameters describing the exchange rates with the outside world, and 14 driving functions specifying the annual cycle in the appropriate parts of the environment. However, there are some important redeeming features. The physical characteristics of many Scottish sea-lochs have been the subject of intensive study, and the relevant parameters are well known (Edwards and Sharples, 1986). The species structure of the aggregated trophic groups is relatively simple, and the organisms concerned are common and frequently studied. We were thus able to obtain good estimates of all but 10 of the biological parameters (see Ross et al., 1993 and references therein). The annual cycles of irradiance and water temperature have been measured in the close vicinity, and are thus well known. The biological conditions in the external sea are less well known, but we were able to determine some driving functions from measurements made in the area during cruises by SOAFD personnel. The rest we inferred from data from a not dissimilar area on the west coast of Ireland (Roden et al., 1987).

Perhaps unsurprisingly in view of the small number of nonlinear pathways it contains, the dynamic behavior of the model depends on the parameter values in a very smooth and continuous fashion. We were thus able to conduct an extensive numerical investigation of its behavior, with particular emphasis on understanding the mechanisms underlying any predicted effects. Our conclusions turn out to be highly robust to quite large changes in the values of almost all the parameters, and, in particular, to be very insensitive to all the parameters whose values we do not know. Our main conclusions are:

- Nutrient recycling through the sediments is dynamically unimportant.
- The system's nutrient dynamics resemble a chemostat, with nutrient import/export rates proportional to the difference between internal and external standing crops.
- Specific primary productivity is light limited for virtually the whole season.
- During the productive part of the season, primary production is grazing limited.
- Nutrient standing crops are determined from the top down.

Armed with these insights, it was clear that the SOAFD study needed to concentrate on the biota, and particularly on secondary production more than is often the custom in such programs. Moreover, as external conditions play a crucial role in both the nutrient and the biological dynamics it was essential to pay considerable attention to conditions in the sea close to the outer sill of the loch, that is in the Firth of Lorne.

A CONFRONTATION WITH OBSERVATION

As part of the exercise of convincing ourselves that the sea-loch model was behaving sensibly, we combed the literature for data on the annual cycle of similar fjordic systems. We did not, at this stage, expect to be able to obtain quantitative agreement with such data, even if it was complete enough for such a test to be possible. We hoped for our predicted average abundances to be in the right ball park, and for the predicted annual cycle to have qualitatively correct properties. Considerable persistence finally enabled us to unearth a suitable data set (Ryan et al., 1986; Roden et al., 1987) covering standing crops of nutrients, primary and secondary producers in Killary Harbour, a (loosely) fjordic inlet in the west of Ireland. More digging also yielded measurements of the nutrient and biological conditions in near inshore waters close to (although not at) the entrance to the fjord (Roden, 1984). Fig. 18-5 shows a side-by-side comparison of the annual cycle observed in Killary Harbour, and that predicted by our model.

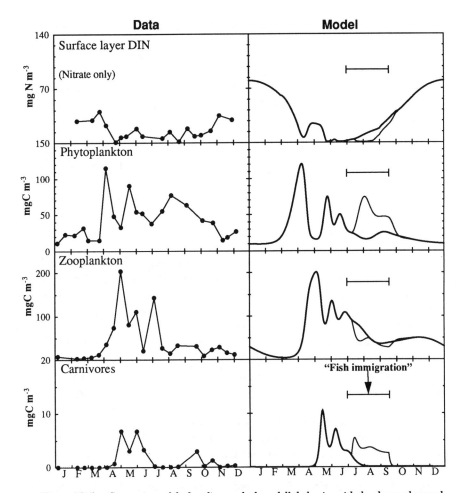

Figure 18-5. Comparison of the baseline sea-loch model's behavior with the observed annual cycle in Killary Harbour. The model, its parameters, and the sources for the test dataset are described in Ross et al. (1993). *Heavy lines* show unmodified model predictions; *thin lines* show model predictions including immigration of fish larvae (see text). DIN = dissolved inorganic nitrogen.

In the early part of the season the predictions for the primary and secondary producers are both qualitatively and quantitatively good. However, the data show an autumn phytoplankton bloom that the model does not reproduce. This failure is due to a late summer collapse in (model) carnivore population, which allows the (model) zooplankton to increase and hold down the standing stock of primary producers. Although this result is quite sensitive to the assumed maintenance cost of the carnivores, one of our unknown parameters, we could find no value that gave an acceptable fit to both the early- and the late-year data.

We "solved" the problem by postulating a plausible (but unobserved) immigration of fish larvae at the appropriate time of year.

The rather unexpected success of this model test exercise led us to question whether we couldn't achieve rather more with this model than we have previously thought likely. The data from the Loch Linnhe experimental program were not yet available, but by digging in the most unlikely places we were able to obtain partial data sets on three more fjordic systems with quite distinct hydrographical characteristics.

- Loch Airdbhair—a shallow quickly flushed embayment in the extreme north of Scotland
- Loch Etive—a deep, two-basin system with a very slowly flushed inner basin
- Loch Creran—similar in conformation to Killary, but deeper and more slowly flushed.

We postulated that the species composing the functional groups would be the same for all four components of our extended data set, and hence that it should be possible to achieve a qualitative fit to the whole body of data with a single set of biological parameters, and using (known) physical parameters to define the differences between systems.

In its original form, the model failed this test quite badly. However, before making a serious start on upgrading its performance we made some simplifications. We treated the lowest layer of the water column and the sediment as a single nutrient storage compartment, and also removed the rather arbitrary "fish immigration" introduced to induce an autumn bloom. However, despite the dramatic gains in simplicity it offered, we did not go to the ultimate extent of removing nutrient dynamics, since dissolved organic nitrogen (DON) and dissolved inorganic nitrogen (DIN) measurements formed an important part of our extended data set, and the validity of the test that they offered was in no way compromised by the top-down mechanism by which nutrient levels are determined.

The simplified model was dynamically identical to its predecessor; thus our detailed knowledge of the way in which the model parameters influenced different features of the yearly cycle now enabled us to effect a series of improvements in its performance. First, the DON measurements revealed that our parameterization of the DON remineralization process was incorrect. Second, our predictions of the bloom concentrations of phytoplankton in the three additional lochs were all considerably too high. This proved to be due to the omission of self-shading from our model—a neglect that was unimportant in the turbid conditions of Killary Harbour, but that significantly affected the predictions in the clearer Scottish conditions. Third, further literature searches suggested that our failure to predict an autumn bloom in Killary Harbour could be attributed to neglect of additional zooplankton losses attributable to production of resting eggs during early autumn.

With the aid of these three improvements the model predictions came a great deal closer to the full test data set, but significant discrepancies remained. The time at which the spring bloom was initiated showed a significant variation between systems that was not reproduced by the model, and the predicted dynamics during the grazing controlled period after the spring bloom showed unrealistic instabilities. Investigations suggested that the second of these problems could be counteracted by many stabilizing mechanisms, the most plausible of which we felt to be the aggregation of zooplankton grazers onto high concentrations of their prey. The first problem arose from the assumption that the phytoplankton vertical distribution is uniform, whereas in reality it is both nonuniform and changes significantly with season. Such effects are not readily incorporated in box models, but we were able to represent them in a crude way, by assuming a step-function distribution whose characteristic depth changed according to grazing pressure. With these modifications the final model predicted trajectories that are in excellent agreement with the test data set. This is illustrated in Fig. 18-6.

Finally, to demonstrate the validity of our earlier conclusion that nutrient standing crops are set from the top down, and play no part in determining the dynamics of these systems, we constructed a super-simplified model. The structure of this three-state-variable model is shown in Fig. 18-7. We discard all nutrient dynamics and assume that specific primary productivity is always light limited. In consequence we can also simplify our description of the physical structure to a single layer flushed by tides and runoff. Fig. 18-8 shows that this model produces a fit to the non-nutrient parts of the test data set which is indistinguishable from that of its more complex progenitor.

METHODOLOGICAL CONSIDERATIONS

Despite being very similar in conception to earlier box modelling approaches, and having a rather intimidating parameter count, the sea-loch model appears to have succeeded in passing tests that many earlier such constructs would almost certainly have failed. Why then is this model different?

Our belief is that there is no single answer to this question. The system being modelled is certainly simpler than many of those attempted earlier, both in terms of the number of functional groups and linkages that it contains and in the species complexity of those functional groups. This undoubtedly helps to make the behavior of the model less sensitive to key parameter values, and facilitates the systematic investigation of its properties. The dominance of most functional groups by a very small number of species also makes the definition of the parameters much clearer, and helps identification of appropriate experimental values in the literature.

The nature of the sea-loch dynamics is probably also an important contributory factor in making this model successful where many similar prede-

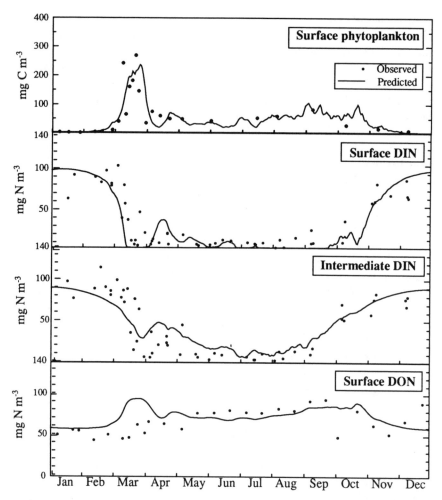

Figure 18-6.   Comparison of the fully developed sea-loch model's behavior with the observed annual cycle in Loch Creran. The model, its parameters and the sources for the test dataset are described in Ross et al. (1994). DIN = dissolved inorganic nitrogen; DON = dissolved organic nitrogen.

cessors have failed. Most of these failures have been in closed systems, whereas the sea-loch receives fresh water and nutrients from the land and exchanges water, nutrients, and biota with the external sea. However, the system boundaries are well defined and of limited geographical extent. The fluxes across them are thus relatively easy to measure. In return for a moderate increase in complexity, the presence of external forcing brings many beneficial effects. In contrast to closed systems, where the highly nonlinear processes of nutrient uptake and recycling play a key dynamic role, nutrient concentrations in sealochs exercise no control over the biota, rather they are set by the require-

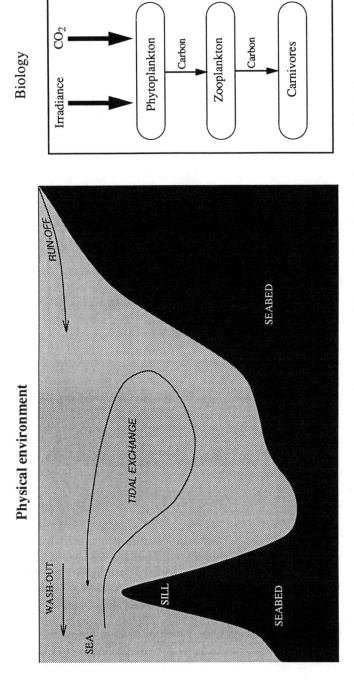

Figure 18-7. The single-layer, three-state-variable sea-loch model, illustrating both the physical system and food web structure.

191

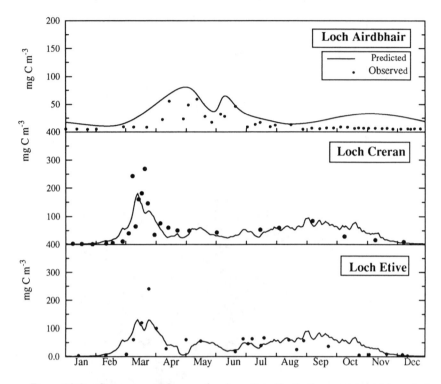

Figure 18-8. Comparison of the annual cycle predicted by the three-state-variable sea-loch model (Fig. 18-7) with observed annual cycles in Lochs Ardbhair, Creran, and Etive. The model, its parameters, and the sources for the test dataset are described in Ross et al. (1994).

ment to balance external supply and internal demand. The link with the outside sea also reduces the role played by overwintering effects, which our model barely represents at all, and which would be vital in determining the early part of the annual cycle in a closed system.

A third important consideration is the availability of adequate individual and process level data to parameterize the model. Even a model such as ours, which was constructed with some attention to minimizing the parameter count, has far too rich a repertoire of behaviors for fitting procedures to be successful when very many of the parameters are unconstrained. Awareness of this problem caused us to take considerable care in seeking a match between the scale and complexity of our chosen representation and the available parameterization data. The quality of our initial fit to the Killary Harbour data set (Fig. 18-5) attests to the success of this strategy.

Our ability to further improve the model, and ultimately achieve a quantitatively validated system representation, was dependent on the coincidence of two further vital factors: first the availability of a large body of data against which the model could be tested and refined, and second a robust and intuitive baseline model.

An essential characteristic, which the available data fortunately satisfied, is that the measured quantities must be close analogs of those predicted by the model. A further feature that we found particularly helpful was that our test data set contained yearly cycles from a number of generically similar, but parametrically distinct, systems. We were thus able to use consistency of pattern between systems to distinguish systematic deviations between predictions and observations from those caused by noise or errors in the observations.

A further requirement for the unambiguous identification of meaningful discrepancies between observation and prediction is a model whose behavior changes smoothly and continuously in response to changes in its parameters. Unless this requirement is satisfied it is difficult to distinguish deviations caused by inaccuracies or uncertainties in parameter values from those that indicate a structural fault in the model. Failure to discriminate between these two conditions will either lead to a (probably divergent) process of inappropriate stuctural modifications, or to futile parameter fitting exercises.

Once a structural failure has been identified, the utility of a model whose behavior is intuitively understandable becomes apparent. Only if the mechanism leading to a particular abberation can be identified is it possible to construct improved hypotheses about how the particular aspect of the system being modelled really works. The testing and falsification (or acceptance) of such modifications is the key driving force behind the steady improvement of a model. Clearly, this process cannot proceed beyond the point at which all remaining systematic discrepancies are comparable with the noise and inconsistencies in the test data set—a condition that Fig. 18-8 suggests our sea-loch model has almost reached.

Thus, our overall conclusion is that there are certainly systems for which energy and materials-based ecosystem modelling is a viable methodology. Our belief is that necessary conditions for success are the existence of an adequate body of applicable data, and a baseline model whose behavior is robust and intuitive. These two conditions imply the possibility of an iterative process of falsification and modification that we believe to be a (perhaps the) vital element in successful modelling.

ACKNOWLEDGMENTS

We gratefully acknowledge many helpful conversations with SOAFD staff, especially Mike Heath, Steve Hay, and Eric Henderson, and with members of the University of Aberdeen Field Station staff, especially John Ollason and Andy Bryant. We received vital financial support from Scottish Office Fisheries and Agriculture Department, Ministry of Agriculture Fisheries and Food, and the European Commission (MAST).

19

EXPLORING AGGREGATION IN SPACE AND TIME

Monica G. Turner and Robert V. O'Neill

SUMMARY

Population and ecosystem processes are heterogeneous in both time and space, and every ecological study requires some level of aggregation or abstraction. Aggregating organism or environmental dynamics is challenging because the processes occur at a variety of spatial and temporal scales, and the scale-dependent effects of aggregation are not well understood. We used a spatially explicit individual-based simulation model of winter foraging and survival of free-ranging ungulates in northern Yellowstone National Park to explore effects of aggregation in space, in time, and of individual animals on model predictions. Aggregation in space was examined by (1) varying the heterogeneity represented in forage abundance across the landscape and (2) eliminating spatial heterogeneity in the accumulation of snow. Results suggest that any aggregation that averages the broad-scale patterns of forage biomass availability underestimates ungulate survival. Aggregation in time was examined by varying the temporal grain used to simulate snow accumulation through the winter. Ungulate survival was not sensitive to this temporal grain, probably because the response remained linear within the range explored. Aggregation of individuals was done by varying the number of individuals contained within ungulate groups assumed to contain identical individuals. Aggregating across individuals was reasonable for small group sizes but led to substantial under-estimates of survival for large group sizes. The effect of aggregation on an ecosystem or population parameter is a function of the question asked and a specified spatial and temporal scale. Even successful aggregation of processes will be reliable only if dynamic thresholds are not crossed, if keystone species are not eliminated, and if feedback loops remain intact.

INTRODUCTION

Whenever ecologists study populations or ecosystems, phenomena of interest must be abstracted from the vast complexity of the world. For example, the study of a particular bird population requires decisions about what attributes of the vegetation (e.g., species, phenology, vertical or horizontal structure, age), other species with which it interacts (e.g., competitors, predators, or prey), and ecosystem-level processes (e.g., net primary production) are suffi-ciently important to warrant inclusion. Similarly, the study of ecosystem-level processes (e.g., production, decomposition) requires decisions about what species or species groupings must be included. Every ecological study, be it empirical or theoretical, requires some level of aggregation. To understand a population, what ecosystem phenomena must be disaggregated? To understand an ecosystem, what population phenomena must be aggregated?

Hierarchy theory (Allen and Starr, 1982; O'Neill et al., 1986) offers some guid-ance in determining what can be disaggregated: a study or model should include all the components and interactions that affect the process of interest and oper-ate on the same spatial and temporal scales (but see Shachak and Jones, Ch. 27 for operation on multiple scales). O'Neill et al. (1986) suggest that processes that are three orders of magnitude larger or smaller than the process of interest can be safely aggregated. That is, those operating over much larger spatial or tem-poral scales act as constraints on the dynamics of a process, whereas those op-erating over much finer scales occur so rapidly that their dynamics are perceived as static. Although this rule-of-thumb provides some guidance, a lot of complex-ity still remains within that range of six orders of magnitude, and it is herein that ecologists come face to face with the problem of aggregation.

Examining aggregation effects and scale dependencies is complicated by the various ways in which processes can be aggregated. First, processes can be ag-gregated in scale, that is, with grain and/or extent in space and time. Grain refers to the finest level of spatial or temporal resolution used within a study, whereas extent refers to the size of the study area or the duration of the study (Turner et al., 1989a,c). Second, the spatial and temporal scales can remain constant, but groups of organisms, species, or similar processes might be aggregated for particular functions (e.g., Gurney et al., Ch. 18; Schimel et al., Ch. 20). This type of aggregation can occur within or across biological levels of organization. Prior studies suggest that we can expect different results from these different types of aggregation (see next section); hence, it is difficult to develop a taxonomy of simple rules that can be applied to process aggregation.

Given the nature of ecological complexity, how then do we abstract what is important? What are the implications of particular choices of abstraction and aggregation? Are there any general rules that can guide the aggregation process? In this chapter, we first examine some results from previous studies of aggregation. Then, we use a spatially explicit simulation model of winter foraging by large ungulates to explore the implications of aggregation in space, in time, and of individuals on projected ungulate survival and habitat use. Aggregation in space is examined in two ways: (1) by varying the heterogeneity represented in forage abundance across the landscape and (2) by eliminating spatial heterogeneity in the accumulation of snow. Aggregation in time is examined by varying the temporal grain used to simulate snow accumulation through the winter. Aggregation of individuals within a feeding group is explored by varying the number of individuals from single animals to groups of 80. From these and other examples, we draw some general implications for process aggregation.

APPROACHES TO AGGREGATION

Problems with aggregating dynamic components of a complex system have long been discussed by ecologists (O'Neill, 1973, 1979; Cale and Odell, 1979, 1980) and economists (Chipman, 1975, 1976). Most of the work has been done on mathematical models (Zeigler, 1976; Iwasa et al., 1987, 1989) but the principles can be applied to the abstractions necessary for any ecological study (Rastetter et al., 1992a; Rastetter and Shaver, Ch. 21).

If one is dealing with "near normal" conditions (i.e., equilibrium) and simple responses to environmental change (i.e., linear dynamics), then reasonable criteria exist for aggregation. One can, for example, aggregate "functionally redundant" species [i.e., parallel configurations and similar turnover rates (O'Neill and Rust, 1979; Gardner et al., 1982)]. The aggregated model is not exact, but the errors are bounded and acceptably small relative to other sources of error.

It may also be possible to isolate one part of the system [i.e., decomposable matrix (Simon and Ando, 1961)] and consider it as a simple component. Almost all ecologists have performed this type of implicit abstraction when they lumped several thousand species (not to mention phyla and kingdoms) in a little box labelled "decomposition" and assumed it would "do its thing" stably during the course of an experiment.

These types of aggregation are reliable as long as radical changes do not occur in the behavior of the aggregated components. Clearly, if one crosses dynamic thresholds, eliminates keystone species, or breaks feedback loops, the assumptions are violated and the aggregations can lead to serious error (Cale et al., 1983).

The situation becomes significantly more difficult if a study requires aggregating food webs [i.e., series configurations (Gardner et al., 1982)] or lumping complex nonlinear responses. Nevertheless, some progress has been made in approximating global responses to CO_2 and temperature increases, while retaining fine-scale understanding of the processes (Sinclair et al., 1976). Approaches include explicit integration of fine-scale models across measured spatial and temporal heterogeneity (King et al., 1989). In simple cases, an analytical solution can be obtained directly. In other cases, the fine-scale model can be considered to be stochastic in space and time. Solutions of the model for larger scales can be approximated by expanding the moments of the stochastic model (Rastetter et al., 1992a). For very complex models, Monte Carlo simulation provides a solution (King et al., 1991).

For limited, well-defined research objectives it is sometimes feasible to use a calibration approach (Rastetter et al., 1992a). Monte Carlo runs of the fine-scale model are made over the required range of conditions. The output is regressed against the environmental conditions of interest (e.g., temperature and rainfall). The regression model is then used to calculate large-scale responses, with known uncertainty. This approach is strictly limited to the range of conditions considered in the regression analysis.

Research on model aggregation has clearly demonstrated the dependency of any abstraction on scale and research objectives. The aggregation may be adequate at one scale (e.g., annual dynamics on a landscape) and false at another (e.g., daily dyamics on a 1-ha plot). The annual/daily difference may be obvious but other changes of scale are not (Cale and O'Neill, 1988). The abstraction is necessarily specific to the immediate objectives and even subtle changes in objective can invalidate an abstraction (Bartell et al., 1988). A water quality model may aggregate fish species as functionally redundant. This aggregate may well be valid, but the model is now invalid for studying the impact of pollutants on a specific game fish.

OVERVIEW OF THE NORTHERN YELLOWSTONE PARK MODEL

The Northern Yellowstone Park (NOYELP) model simulates the search, movement, and foraging activities of individuals or small groups of elk (*Cervus elaphus*) and bison (*Bison bison*) (Turner et al., 1994). The model was developed to explore the effects of fire scale and pattern on the winter foraging dynamics and survival of these free-ranging ungulates in northern Yellowstone National Park, Wyoming, U.S.A. The 77,020-ha landscape is represented as a gridded irregular polygon with a spatial resolution of 1 ha. Grasslands cover most of the northern range and are dominated by big sagebrush (*Artemisia tridentata*), bluebunch wheatgrass (*Agropyron spicatum*), and Idaho fescue (*Festuca idahoensis*).

The model simulates daily forage intake as a function of an animal's initial body weight, the absolute amount of forage available on a site, and the depth and density of snow. Energy balances are computed daily, with energy gain a function of forage intake and energy cost a function of baseline metabolic costs and travel costs. When the energy expenditures of an animal exceed the energy gained during a day, the animal's endogenous reserves are reduced to offset the deficits. Simulations are conducted with a 1-day time step for a duration of 180 days, approximately November 1 through April 30. The model has been used to examine the effects on ungulate survival and habitat use of fire size, fire pattern, winter weather conditions, and initial ungulate numbers (Turner et al., 1994).

EXPLORING AGGREGATION IN SPACE AND TIME: EXAMPLES USING THE NOYELP MODEL

The NOYELP model was used to explore the effects of several different types of aggregation on simulated elk survival and habitat use. Because we focus only on aggregation effects, several parameters were held constant across all simulations. We simulated the winter of 1988–1989, the first winter following the extensive fires that affected Yellowstone National Park and a winter of average severity. These fires affected 22% of the northern range study area, and no forage was present in burned areas during this first post-fire winter. All simulations use the actual snow accumulations observed in, 1988–1989 and begin with 19,270 elk and 600 bison. Observed winter mortality of elk was between 38 and 43% (Singer et al., 1989), and the baseline prediction of the model results in 40% mortality (Turner et al., 1994). The total abundance of forage across the landscape was also held constant in the simulations, although its spatial heterogeneity was varied.

Aggregation Across Space I: Forage and Habitat Heterogeneity

Biomass abundance varies spatially across a landscape. When lines are drawn on a map to delineate different habitat types, assumptions are often made about the characteristics of the vegetation within those habitats. The NOYELP model includes six different vegetation types (dry, mesic, moist, and wet grasslands; canopy forest; aspen stands) whose spatial distribution was obtained from the Park's geographical information system. The abundance of prewinter forage in each of these six habitats was quantified during the fall of 1990 (Wallace et al., 1994) by field sampling. The model uses the mean and 95% confidence interval obtained from these data to distribute forage to each grid cell such that the mean forage abundance within each habitat type matches the mean obtained from the data, but any given grid cell within a habitat type has a forage value assigned from the distribution.

To explore the effects of aggregating across this spatial heterogeneity, a series of three successive aggregations was conducted. First, the burn pattern and the spatial distribution of the six unburned habitats was maintained, but the within-habitat heterogeneity in forage abundance was removed by assigning to each grid cell the mean forage abundance for its habitat type (Tab. 19-1, "no within-habitat heterogeneity"). Second, the burn pattern was preserved, but the between-habitat heterogeneity in the unburned area was removed (Tab. 19-1, "no between-habitat heterogeneity"). The forage biomass assigned to each grid cell was obtained by dividing the total biomass on the landscape by the number of unburned grid cells and assigning this mean value to each unburned grid cell. Finally, all heterogeneity in forage abundance was removed by completely eliminating the burn pattern (Tab. 19-1, "no heterogeneity"). Each grid cell in the landscape was assigned the same forage value, obtained by dividing the total landscape biomass by the number of grid cells.

Table 19-1. Levels of forage biomass assigned by habitat in the baseline model and each of three successive aggregations.

Aggregation	Habitat Type	Forage Biomass (kg/ha)	
		Mean	SEM
Baseline	Burned	0	—
	Dry	520	59
	Mesic	631	139
	Moist	1122	166
	Wet	2259	266
	Forest	333	85
	Aspen	659	101
No within-habitat	Burned	0	—
heterogeneity	Dry	520	0
	Mesic	631	0
	Moist	1122	0
	Wet	2259	0
	Forest	333	0
	Aspen	659	0
No between-habitat	Burned	0	0
heterogeneity	Unburned	729	0
No heterogeneity	All	568	0

The removal of within-habitat heterogeneity had little effect on elk survival (Fig. 19-1a). However, aggregating habitats such that between-habitat heterogeneity was eliminated underestimated survival of elk cows and bulls by 30% (Fig. 19-1a). Removal of all spatial heterogeneity in forage abundance further underestimates elk survival by approximately another 20%. The decreases in survival with the loss of between-habitat heterogeneity are due, in part, to energy costs associated with increased daily movement. The mean distance travelled in a day for elk cows increases sooner (in February rather than in March) when there is no between-habitat heterogeneity (Fig. 19-1b). Animals are forced to move more when they are not obtaining sufficient forage to meet their daily requirements, and hence expend more energy. If we examine the proportion of animals present at the end of each month (Fig. 19-1c), we note that mortality begins during February for all scenarios but is much more rapid when the between-habitat heterogeneity in forage abundance is reduced.

These results illustrate for wintering ungulates the importance of maintaining coarse-scale variability in forage biomass when conducting a study/simulation at this spatial and temporal scale. Assuming animals respond to the mean landscape biomass would be misleading; the distribution across the landscape of areas of high and low biomass enhances ungulate survival.

Aggregation Across Space II: Snow Conditions and Topography

The depth and density of snow also varies spatially across the landscape. In the baseline simulations, the depth and density of snow on each grid cell are assigned by modifying a projection that assumes a flat site by the slope and aspect of the cell. A set of coefficients was obtained by sampling snow depth and density across a wide range of topographic positions in northern Yellowstone (Turner et al., 1994). To eliminate heterogeneity in snow, simulations were conducted in which each baseline projection was applied to each grid cell; that is, snow conditions were applied homogeneously across the landscape.

The homogeneous snow conditions resulted in substantial underestimates of ungulate survival (Fig. 19-2a). The proportion of elk cows surviving, for example, declined from ~70% to ~20%. The mean distances travelled per day by elk cows increased dramatically in January in the absence of snow heterogeneity and remained elevated as compared to the baseline simulation for the rest of the winter (Fig. 19-2b). Some changes in the frequency of use of different habitats occurred (Fig. 19-3a), with dry habitats receiving more use and moist habitats less use. Because fewer animals survived and foraging was more difficult in general, the end-of-winter biomass present within each habitat generally increased with homogeneous snow (Fig. 19-3b).

These results illustrate that the spatial heterogeneity of snow conditions, especially the maintenance of snow-free or low-snow areas, is important for

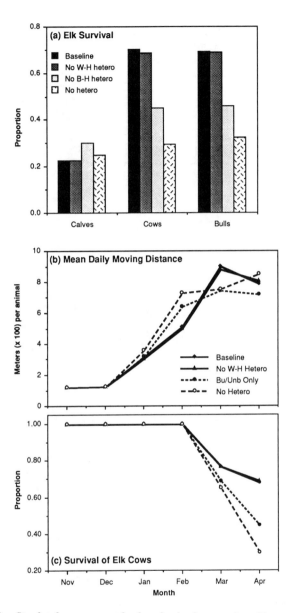

Figure 19-1. Simulated responses to the three levels of aggregation of forage heterogeneity across the landscape. (a) Overall survival of elk cows, calves, and bulls. (b) Mean distance moved per day for elk cows at the end of each month. (c) Survival of elk cows through time (legend as in *b*). "Baseline" indicates no aggregation; "No W-H hetero" indicates no heterogeneity within each habitat; "No B-H hetero" indicates no heterogeneity between habitats, but the burned vs. unburned pattern remains; and "No hetero" indicates no heterogeneity among grid cells (see Tab. 19-1).

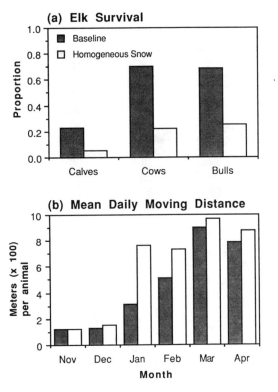

Figure 19-2.   Simulated responses with and without spatial heterogeneity of snow depth and density. (a) Elk survival. (b) Mean distance moved per day for elk cows at the end of each month (legend as in *a*).

estimating winter ungulate survival. Although the mean snow accumulation across the landscape will certainly be an index of the overall winter severity, homogeneity across space cannot be assumed in a study at this spatial and temporal scale. Of course, this does not negate the use of mean snow accumulation in some other analyses, for example, statistical models of ungulate mortality across many years as a function of mean snow conditions.

Aggregation in Time: Interval Used for Snow Conditions

Snow varies not only across space but also through time. Snow accumulation data are available on a monthly basis from a snow course sampled on approximately the first of each month through the winter. However, snow conditions are actually changing at a finer temporal scale. In the baseline runs of the model, a snow interval of 3 days is used. The snow conditions for two sequential months are compared, change between them is assumed to be linear, and the changes are then parcelled out in 10 equal 3-day increments. To examine the effect of aggregating across the changes in snow conditions within each month, we varied the interval from 1 day, giving a linear increase in snow, to 30 days, which is a step function. Note we are varying the temporal grain of the snow simulation, but the temporal extent and underlying temporal pattern of snow through the winter are maintained.

Elk survival was not very sensitive to variation in the interval used to modify snow conditions. Survival declined somewhat when snow was simulated as a monthly step function (Fig. 19-4a), possibly in response to increased daily moving distances in April (Fig. 19-4b). These results illustrate that the temporal grain of snow within the range of 1 to 30 days was not particularly important. This may occur because there is a linear response to snow conditions within that time scale and no qualitatively different response occurs. However, some deviations from the baseline simulation did occur at a 30-day grain size, suggesting that larger aggregations would not be appropriate.

Figure 19-3.   Frequency of habitat use by elk (a) and remaining end-of-winter biomass by habitat (b) with and without spatial heterogeneity of snow depth and density (legend as in *a*).

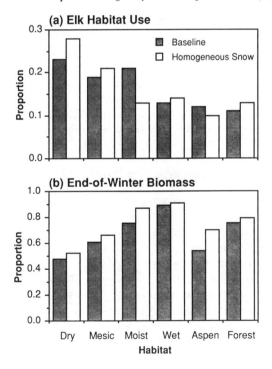

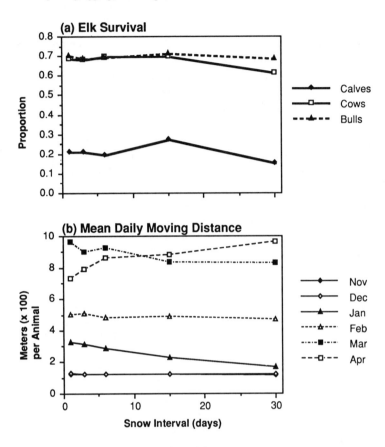

Figure 19-4.   Simulated ungulate responses to changing the temporal grain of the snow simulation from 1 to 30 days. (a) Overall elk survival. (b) Mean daily moving distance of elk cows at the end of each month.

Aggregation of Individuals

The final aggregation experiment we conducted was to vary the size of groups that are simulated together from single individuals to groups of 80. When a group is simulated, the group size is used as a multiplier for all foraging and energetic relationships. For example, if the maximum daily foraging rate for an individual elk cow is 5 kg per day, then the same rate for a group of 10 elk cows is 50 kg. Of course, if the forage intake required is greater than the forage present at a site, the group obtains less than its daily requirement and must move to another site to meet its needs.

Elk survival was sensitive to group size. Survival of elk calves began to decline at a group size of four animals and reached zero for a group size of eight (Fig. 19-5a). Survival of adults, however, declined rapidly between group sizes of 20 and 40. This decline in survival corresponds to a substantial increase in the mean daily moving distances (Fig. 19-5b). Above group sizes of 10, moving distances increase even in early winter, resulting in increased energetic costs. Both simulated habitat use (Fig. 19-5c) and end-of-winter biomass (Fig. 19-5d) remaining in each habitat show fluctuation with group size.

The results from varying ungulate group sizes illustrate effects of aggregating across a level of biological organization. Although sex and age differences were maintained in this aggregation, individuals were lumped under the assumption that all were identical and that simply multiplying the dynamic processes by the number of individuals would maintain the biological relationships intact. This assumption was acceptable with small groups of individuals (≤ 4 for calves, ≤ 10 for cows and bulls) but not with large groups. At the spatial and temporal scale of the model, aggregating across individuals led to serious error.

DISCUSSION

The effect of aggregation on an ecosystem or population parameter clearly is a function of the spatial and temporal scales at which the parameter is being studied and of the question being asked. Describing the effects of aggregation cannot be done in the absence of a question of interest. For example, aggregating the spatial heterogeneity of forage biomass between habitats had a substantial effect on simulated survival of wintering ungulates studied over a single season and using a daily time step. Within-habitat heterogeneity was not important, and we aggregated the multitude of species into a single "biomass" component. However, aggregating in this way would not be suitable for a study of wintering ungulates over a short time span (e.g., days) and within a single habitat because critical interactions would be missing.

The aggregations or abstractions made in ecological studies or models are often represented as unstated assumptions. For example, in the model presented here, we assume that species-by-species differences in forage quantity and quality can be aggregated, but sex and age differences of the two ungulate species cannot. We aggregate when we expect, either by hypothesis or through empirical data, that the aggregated entities do not differ qualitatively for the process of interest. Here, we assumed that the individual grasses and forbs could simply be summed, without changing the fundamental effect on ungulates. However, summing all elk and bison could not be done because the energetic demands differ substantially among males and females and between juvenile and mature animals. The important point is this: for ALL ecological

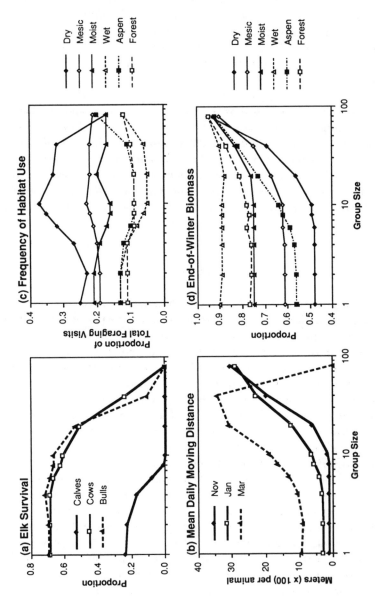

Figure 19-5. Simulated responses to varying the group size of elk used in the simulations from 1 to 80 animals. (a) Overall elk survival. (b) Mean daily moving distance of elk cows in November, January, and March. (c) Frequency of habitat use. (d) End-of-winter biomass by habitat.

206

studies or models, a tremendous amount of aggregation has occurred, even though it may not be stated as such.

The most important conclusion to be drawn from the studies of aggregation can be summarized in the principle of the "Title Translator." By this principle, one takes the title of a paper, such as, "An Analysis of X," and translates the title into: "One specific analysis of X, over very limited time/space scales, using a set of limiting assumptions, mostly unstated, for a specific research objective, and not precluding other, completely different ways of looking at X."

We cannot offer a simple conclusion on process aggregation, but we suggest an approach toward determining what can and cannot be aggregated in an ecological study:

1. Ask the question. Whenever the question is changed, then both the spatial/temporal scale and the level of abstraction also change.

2. Specify the spatial and temporal scales of interest, both in terms of grain and extent. Again, the level of aggregation or abstraction will vary as the scale of the question is changed.

3. Aggregate the processes that occur much more slowly than the processes of interest and consider these to be constraints on the process of interest.

4. Aggregate the processes that occur much more quickly than the processes of interest and consider these to be "noise" to the process of interest.

5. Identify the interacting factors that influence the process of interest. To determine whether any more aggregation can be done at this scale, address the following:

   a. Is there functional redundancy among some components or processes? In our example, there was functional redundancy among forage species, but not among different habitat types or different age/sex classes of ungulates.

   b. Are there feedbacks at the particular spatial and temporal scale of interest? If not, then the processes may be aggregated or ignored. In our model, for example, the temporal extent is 180 days, a single winter. Therefore, processes such as reproduction, nutrient feedbacks, and decomposition were not addressed, but these would likely be unacceptable omissions from a mulityear consideration.

   c. Are there nonlinearities, or threshold dynamics? If so, it may be possible to aggregate below and above a threshold, but not across it. Spatially, for example, there may be thresholds in connectivity for a particular organism or process (e.g., Gardner et al., 1987; Turner et al., 1989b; Gardner and O'Neill, 1991). Above the threshold, the abundance of the organism might be proportional to the abundance of habitat because the organism can access all the habitat. Below the threshold, however, the abundance of the organism might be related to the probability of success of long-distance dispersal, a qualtitatively different relationship.

Even successful aggregation of processes will be reliable only if dynamic thresholds are not crossed, if keystone species are not eliminated, and if feedback loops remain intact. If any of these occur, either through aggregation or through fundamental changes in the system, then the aggregation will lead to serious error (Cale et al., 1983).

This approach does not presume to be a cookbook recipe for aggregation nor does it presume to be exhaustive. Rather, the steps listed above illustrate the critical dependence of aggregation effects on both the question and scale of interest in a study. We suggest that thoughtful consideration of each of these steps will help answer questions about what processes can and cannot be aggregated in studies linking populations and ecosystems.

ACKNOWLEDGMENTS

We appreciate technical assistance on the model modifications from Yegang Wu. Comments on the manuscript from Scott Pearson, Richard Flamm, and an anonymous reviewer were helpful. This work was supported by the U.S. National Park Service and U.S. Forest Service through a research contract from the University of Wyoming-National Park Service Research Center; and by the Ecological Research Division, Office of Health and Environmental Research, U.S. Department of Energy, under Contract No. DE-AC05-84OR21400 with Martin Marietta Energy Systems, Inc. This is Publication No. 4243, from the Environmental Sciences Division, Oak Ridge National laboratory.

20

Aggregation of Species Properties for Biogeochemical Modeling: Empirical Results

David S. Schimel, V.B. Brown,
K.A. Hibbard, C.P. Lund, and S. Archer

SUMMARY

In many biogeochemical models, plant species are aggregated such that only one generic plant type is represented. If multiple species are present, each "species" or functional type is represented as a collection of physiological traits. We recently have been exploring the physiological responses of co-occurring plant species of a variety of growth forms to determine how many separate physiological types are needed in order to capture the dynamics of net primary production, decomposition, carbon (C) storage and nitrogen (N) availability. We measured species-level photosynthetic responses, canopy light extinction, leaf N and lignin, and soil nutrient processes. In both the prairie and the shrub-savanna we found that photosynthetic responses scale with light within the plant canopy such that for calculation of C gain, the canopy can be modeled as a single unit, regardless of the vertical distribution of species. Nutrient cycling and C storage are, however, very different in herbaceous areas compared to wooded areas, such that shrub-dominated areas have different soil C levels and N mineralization rates from grassland areas. In Texas, shrub-dominated areas have higher soil C and N mineralization. The dominant woody plant in the Texas site is an N fixer. Significant areal expansion of this plant, documented since the mid-1800s, has undoubtedly affected regional patterns of N cycling and net primary production (NPP) (similar to results of Vitousek and Walker, 1989, from Hawaii) . The dominant woody plant in Kansas is not a fixer and invasion (which is locally common due to fire suppression) results in losses of N availability and stored C. We suggest that some ecosystem processes may be modeled successfully with highly aggregated models, particularly models of processes that are highly constrained by environmental factors. Other processes will require detailed specification of physiological traits at the species or population level. The degree of species aggregation in ecosystem models should be regarded as a research problem rather than a quasi-ideological problem.

INTRODUCTION

It is now well known that models of biogeochemical cycling in terrestrial systems are sensitive to the representation of physiological traits of the primary producers (Holland et al., 1992). Key attributes include those that affect either net primary production, or the allocation of NPP among plant tissue types with different biochemical and nutrient composition (Melillo et al., 1984; Schimel et al., 1991; Holland et al., 1992). In an important study of this type, Holland and co-workers (1992) showed sensitivity of C and N dynamics to population-level variations in allocation responses to herbivory. Because many biogeochemical models do not explicitly represent species or their dynamics, two issues are important: (1) plant attributes must be represented in aggregated parametric fashion when not explicitly simulated at the species level; and (2) the implications of aggregating species or population-level attributes for simulation of biogeochemical processes must be evaluated as part of model sensitivity analyses. In addressing the first point, it is important to understand and identify processes that may be robustly represented by simplified paramaterizations, neglecting taxa, as opposed to processes that exhibit a high degree of species specificity. In addressing the second point, it is important to identify the range in species compositions for which model predictions are reasonably robust. In this chapter, we discuss results from a recent study attempting to link species, community and ecosystem processes governing photosynthesis and C storage. In particular, we discuss preliminary data showing how the vertical allocation of N for photosynthesis corresponds to the light gradient in a multi-species community. We also examine the role of species level processes on ecosystem C storage at longer time scales. Finally, we discuss several "modes" whereby species changes can affect ecosystem processes and how these processes may be captured in simple models.

APPROACH AND RESULTS

Canopy Photosynthesis

A number of workers have shown that N is allocated vertically within the canopy of individual plants so as to optimize the utilization of light. This results in a pattern of decreasing N concentrations with decreasing height and incident photosynthetically active radiation (PAR) within a plant canopy (Field, 1983; Sellers et al., 1993). We showed (Schimel et al., 1991) that this same response was expressed in multispecies swards of grasses in Tallgrass Prairie. Specifically, N concentrations and light-saturated photosynthetic rates were considerably higher at the top of the grass canopy than near the bottom, and this gradient was steeper in dense compared to sparse canopies. This latter attribute supports the dependence of the vertical allocation response on light attenuation, as the light gradient is not extreme in sparse canopies, whereas light extinction approaches 100% at the bottom of dense stands. In the Tallgrass, there is considerable horizontal heterogeneity in species composition, but the canopy is usually monospecific vertically: no overstory–understory structure normally occurs. Thus, the allocation response is a within-individual response, which is however consistent between adjacent individuals of the different dominant species.

We have recently begun to explore canopy processes in a system with considerable vertical stratification of species, a subtropical thorn woodland in south Texas (La Copita: Archer et al., 1988). The site is an experiment station of Texas A & M University. In this system, an overstory of Mesquite (*Prosopis glandulosa*) occurs with an understory composed of a number of shrub species (*Zanthoxylum fagara*, *Celtis palida*, *Ziziphus obtusifolia* and *Condalia obovata* are common). We sampled the overstory and understory in regular vertical increments within species for light attenuation, leaf area index and mass, foliar N, light-saturated photosynthesis, and dark respiration. We were thus able to construct vertical sections of leaf area, foliar biomass and N, light extinction, and gas exchange.

In summary, the results showed the following:

Light extinction. On the average, light extinction could be described by an exponential decay $I(z) = I(o)e^{-kLAI}$ where $I(z)$ and $I(o)$ are radiation at height z and at the top of the canopy, respectively, LAI is the leaf area index, and k is a coefficient with value of approximately 0.4. This value does vary with leaf geometry, but we found no evidence for significant differences between the species within this ecosystem.

Leaf N. The leaf N gradient (in mg/g) was linearly related to the mean light extinction profile (expressed as $I(z)/I(o)$).

Light-saturated photosynthesis (Amax). As has been shown numerous times (Field and Mooney, 1986), Amax was linearly related to leaf N.

Dark respiration. Dark respiration increased with increasing leaf N.

Thus, as mean light levels decrease within the canopy, so too do values of N, Amax, and dark respiration. This is a very similar pattern to that expected within a single plant. This resemblance raises several questions. First, although the daily average light profile is exponential, there are periods when light levels in the understory reach near-full sunlight levels. Why should the subordinate shrubs not have the photosynthetic capacity to respond with high rates of photosynthesis? The answer lies in the dark respiration–N relationship. Since respiratory costs increase as N content increases, the shrubs cannot sustain a positive C balance with only infrequent episodes of high light, if they maintain high respiratory C losses. Thus, they balance low Amax with low dark respiration rates.

As a corollary, the understory shrubs generally have longer growing seasons than the overstory *Prosopis*. That is, whereas the overstory species is drought deciduous and often carries leaves for only relatively short periods of time, the understory contains many broadleaf evergreen species. Thus low instantaneous rates of NPP (photosynthesis less respiration) are balanced by long leaf area duration. While the apparent pattern of N allocation in the multispecies canopy is very similar to that expected within a single plant, this occurs by a change in another process, namely leaf turnover rates. The patterns observed in this study can be understood in terms of the C balance of the component species without invoking selection at the community level. Although we found similar relationships between gas exchange, N allocation, and light extinction in multispecies woodland canopies and vertically homogeneous grasslands, the patterns arise from different mechanisms. The similar scaling of photosynthesis to light in multi- and single-species canopies allows the use of simple models for canopy gas exchange, even in complex canopies.

Ecosystem Carbon Storage

The La Copita study site has recently undergone extensive conversion from grassland to shrubland and woodland (Archer et al., 1988). A significant consequence of this has been a major increase in plant biomass storage in woody tissue (compared to negligible storage in herbaceous tissue) and a near-doubling of soil organic matter. This increase in C storage has occurred as a result of the invasion of *Prosopis*, which is a vigorous N fixer. The increase in N capital makes possible a substantial increase in C fixation and storage over the low-N grassland system. This major change in ecosystem function and element storage occurred as a result of the invasion or spread of the N-fixing species. Our analyses of the scaling of photosynthesis suggest that to first order the species differences along the vertical light extinction gradient can be accounted for by a simple C balance model, without specific recognition of physiological idiosyncracies at the species level. Instead, the gradient of physiological traits is predictable from first principles. However, when considering the ecosystem C balance, the change over approximately 50 years was driven by the spread of a species with a particular physiological trait for N fixation. Given the uneven distribution of dominant species exhibiting N fixation, it would be difficult to predict the ecosystem trajectory from first principles, though if the presence of the fixer were known, its dynamics might be modeled. In other settings, for example, the Tallgrass Prairie, the situation is quite different. When Tallgrass Prairie is invaded by woody plants, few or none of which are N fixers, this often results in decreases in soil organic C storage. Given our current state of knowledge, we cannot model the difference in invasion patterns without recourse to *a priori* knowledge of the biogeography of the woody species.

CONCLUSIONS

We began this chapter by considering the appropriateness of the representation of physiological attributes by "lumped" representations in biogeochemical models [see also Gurney et al. (Ch. 18) and Turner and O'Neill (Ch. 19)]. We sought to find experimental situations and field studies where such questions could be addressed using data as well as model sensitivity analyses. Our results provide two perspectives. First, we suggest that scaling of photosynthesis within canopies can be described by C balance and resource (nutrient) use arguments without significant reference to physiological traits which can be included only if the identities of the participating species are known. That is, the scaling of photosynthesis can be described from first principles with minimal specification of species-specific or growth-form-specific traits. Second, long-term ecosystem C balance is sensitive to the identity of the tree species. The occurrence of the N-fixing trait would have to be known and prescribed at our current level of understanding of the biology and biogeography of N fixation. In a similar situation to woodland expansion in Texas, woodland expansion in Tallgrass Prairie generated opposite results. Holland et al. (1992) documented a similar sensitivity in examining the effects of grazing-tolerant and -intolerant ecotypes of grasses in a grazed setting. The results could be simulated if the physiological traits could be specified, but the existence of the ecotypes could not be predicted from biogeochemical or physiological theory.

We suggest a taxonomy of species-level traits for plants. First are species-level traits that are in a sense predictable from environmental constraints. In the case of photosynthesis, certain traits are predictable, or nearly so, given minimal information concerning environmental conditions within the canopy. Second are traits that are consistent with resource constraints but where multiple strategies exist. Examples of this latter category are allocation strategies between above- and belowground tissue, changes in lignin and lignin/N ratios in leaves (Melillo et al., 1984; Parton et al., 1994; Wedin, ch. 24), and allocation responses to herbivory (Holland et al., 1992). Theory exists to predict the ranges of these variables, but not as unambiguously as in the case described for photosynthesis (Chapin et al., 1990). For properties such as allocation and tissue chemistry, modeling approaches range from explicit dependance on species composition (Pastor and Post, 1986) to completely parametric approaches (Parton et al., 1987), and simulation of allocation responses to resources (Rastetter et al., 1992b; Rastetter and Shaver, Ch. 21). Third are species attributes that essentially must be prescribed external to resource and environmental responses. The presence of N fixation is the clearest example of such a trait (Vitousek and Walker, 1989). Flammability and serotiny may also fall into this category. Although conditions conducive to N-fixing or flammable species may be predicted from environmental conditions (Wedin, Ch. 24), species with the appropriate traits must be present, or, on longer time scales, must evolve.

While a polarized debate has revolved around on the one hand the omission of species as explicit entities in ecosystem models, and on the other hand the importance of species-specific properties for ecosystem processes, recent research makes it clear that both extreme views are flawed. Clearly, reduced-form models collapsing species properties into resource space are appropriate for certain problems (Gurney et al., Ch. 18; Rastetter and Shaver, Ch. 21). Equally clearly, certain problems require explicit or nearly explicit inclusion of properties at the species level (e.g., Vitousek and Walker, 1989; Holland et al., 1992). The intellectual conditions for the species-ecosystem debate now clearly exist to consider the role of species in ecosystem processes, and biogeochemical processes in community and population dynamics as research topics, rather than as ideological issues.

ACKNOWLEDGMENTS

Research described in this chapter was supported by the EOS Program of NASA. C.P.L. was supported by a DOE Fellowship. The senior author thanks the second through third authors for permission to discuss unpublished results from their graduate work. The authors thank Beth Holland, Chris Field, and Paul Barnes for valuable discussions and Becky Riggle for assistance with analyses. The National Center for Atmospheric Research is sponsored by the National Science Foundation.

21

Functional Redundancy and Process Aggregation: Linking Ecosystems to Species

Edward B. Rastetter and Gaius R. Shaver

SUMMARY

A simple model of multiple-element limitation for acclimating vegetation was used to examine the problem of process aggregation at the ecosystem level. The model was applied to five species competing for two elemental resources. Because of differences in the abilities of the species to compete for the two resources, their relative abundances change in response to changes in the availability of the resources.

A naive, aggregate model of ecosystem-level productivity and biomass was developed using equations identical to those in the species model, but with an "average" parameterization. This aggregate model failed to predict changes in element concentrations and ecosystem productivity and biomass in response to changes in resource replenishment rates.

The aggregate model was then sequentially calibrated to ecosystem-level productivity and biomass data, generated with the species model, for several levels of resource availability. An analysis of parameters derived from these calibrations revealed strong quantitative relationships between the calibrated parameters and the model variables. These relationships revealed how within-

and between-species processes are manifested at the ecosystem level and suggested mechanistically interpretable equations for the aggregate behavior of the ecosystem. These equations were used to develop a corrected aggregate model that faithfully simulates changes in element concentrations and ecosystem productivity and biomass

INTRODUCTION

Ecosystems are composed of species that are often lumped into functional groups that carry out somewhat redundant processes (e.g., primary producers). However, within functional groups, species differ quantitatively in how they perform these processes (e.g., shade-tolerant vs. shade-intolerant primary producers). Because of these differences among species, it is difficult to derive an aggregate representation of their function at the ecosystem level.

To address this problem, we have adapted a simple model of multiple-element limitation for acclimating vegetation (Rastetter and Shaver, 1992) to examine the behavior of a group of plant species competing for two elemental resources. The individual species in this model differ in their uptake kinetics for the two elements. We also use a naive model of the aggregate behavior of this group of plants. We exercise both models to determine what types of perturbations result in the failure of the aggregate model to represent the behavior of the multiple-species model. The aggregate model fails when the perturbation results in significant changes in the relative abundances of the competing species. We then use a "sequential-calibration" procedure to identify changes in the structure of the aggregate model that improve its performance.

SPECIES MODEL

Rastetter and Shaver (1992) developed a model for simulating responses of monospecific stands of vegetation to changes in the availability of two essential elements. The vegetation biomass in this model is assumed to be composed of only two elements and acclimates to changes in the availability of these elements by adjusting the relative capacities for their uptake. We made four modifications to this model. First, we modified it to simulate several species competing for the two elements. Second, we used a more complex equation relating uptake to biomass to allow species extinction at low resource concentrations [Eqn. (21-6) below]. Third, we added a random immigration term for each species [I_j in Eqn. (21-2)]. Finally, we added a proportional loss term for each element resource so that the nonlimiting resource did not build up infinitely in the environment [$p_i E_i$ in Eqn. (21-1)].

The model equations are as follows:

$$dE_i/dt = R_i - p_i E_i - \sum_{j=1}^{n} U_{ij} \tag{21-1}$$

$$dB_{ij}/dt = U_{ij} - m_{ij} B_{ij} + I_j \tag{21-2}$$

$$dV_{1j}/dt = -a_j \ln\{B_{1j}/(B_{2j} Q_j)\} V^*_j \tag{21-3}$$

$$dV_{2j}/dt = -dV_{1j}/dt \tag{21-4}$$

$$U_{ij} = S_j V_{ij} E_i/(k_{ij} + E_i) \tag{21-5}$$

$$S_j = B_{*j}^2/(1 + B_{*j}^{4/3}) \tag{21-6}$$

$$B_{*j} = B_{1j} + B_{2j} \tag{21-7}$$

$$V_{*j} = V_{1j}; \quad \text{if } B_{1j}/B_{2j} > Q_j$$
$$= V_{2j}; \quad \text{if } B_{1j}/B_{2j} < Q_j \tag{21-8}$$

where t is time, n is the number of species, E_i is the amount of element i in the environment, B_{ij} is the amount of element i tied up in the biomass of species j, U_{ij} is the uptake rate of element i by species j, S_j is the surface area active in element uptake for species j, V_{ij} is the maximum rate of uptake of element i per unit S_j, R_i is the replenishment rate of element i to the environment, I_j is the immigration rate for species j expressed in terms of biomass accrual, Q_j is the optimum ratio of B_{1j} to B_{2j}, k_{ij} is the half-saturation constant for the uptake of element i by species j, p_i is the proportional loss rate of element i from the environment, m_{ij} is the proportional biomass loss rate of element i from species j, and a_j is the acclimation rate for species j.

Species in this model might vary in their values of m_{ij}, a_j, Q_j, or k_{ij}. In the present study, we assumed that the five species differed only in their respective values of k_{ij} (Tab. 21-1). For all species, I_j was random with a uniform distribution between 0 and 0.1. Initially, R_1 and R_2 were set at 4.5 and 4.3, respectively. The model was allowed 200 time steps to adjust to these initial conditions and then, at time 0, R_1 was doubled; at time 250, R_2 was doubled; and at time 500, R_1 was halved back to its original value. Because $Q_j = 1$ and $m_{1j} = m_{2j}$ for all species, there is an equal demand for both elements. Therefore, long-term community productivity should be limited by the slower of the two replenishment rates, that is, by R_2 up through time 500 and by R_1 thereafter (Rastetter and Shaver, 1992).

The species acclimated during the initial adjustment period based on their relative ability to take up the limiting element at low concentrations (Fig. 21-1a). Because species 1 had the highest half-saturation constant for the limiting element (k_{21}), it had the lowest abundance at the end of the adjustment period. Species 5, with the lowest half-saturation constant for the limiting el-

Table 21-1. Parameter values and initial conditions

Species Model:
Parameters: for all i and j,

$p_i = 0.1$	$Q_j = 1$	$m_{ij} = 0.1$	$a_j = 0.01$	
$k_{11} = 0.2$	$k_{12} = 0.6$	$k_{13} = 1.0$	$k_{14} = 1.4$	$k_{15} = 1.8$
$k_{21} = 1.8$	$k_{22} = 1.4$	$k_{23} = 1.0$	$k_{24} = 0.6$	$k_{25} = 0.2$

Initial Conditions; for all i and j,

$E_i = 1$	$B_{ij} = 8.6$			
$V_{11} = 0.3$	$V_{12} = 0.4$	$V_{13} = 0.5$	$V_{14} = 0.6$	$V_{15} = 0.7$
$V_{21} = 0.7$	$V_{22} = 0.6$	$V_{23} = 0.5$	$V_{24} = 0.4$	$V_{25} = 0.3$

Aggregate Ecosystem Model:
Parameters: for all i,

$p_i = 0.1$ $Q_a = 1.0$ $m_{ia} = 0.1$
(other parameters as specified in Eqns. (21-9)–(21-16)

Initial Conditions; for all i,
$E_i = 1.0$ $B_{ia} = 43.0$ $V_{ia} = 0.5$ $k_{ia} = 1.0$

---

ement (k_{25}), had the highest abundance at the end of the period. The other species were arranged according to their half-saturation constants between these two extremes. The total community biomass adjusted much more quickly than that of the individual species and remained nearly constant after about 50 time steps despite continued change in the relative abundances of the species (Fig. 21-1b).

When R_1 was doubled at time 0, there was a substantial readjustment of the relative abundances of the five species (Fig. 21-1a). Initially, all species increased in abundance. However, this increase was not sustained and all but species 5 had lower abundances by time 250. The total community biomass also had an initial, transient increase (Fig. 21-1b). However, because R_1 was not the limiting replenishment rate when it was doubled, the community biomass eventually decreased back toward its value at time 0. This occurred because long-term productivity was limited by the slower replenishment rate (R_2). Under these conditions, the equilibrium community biomass should also depend almost entirely on R_2 and should be nearly the same before and after R_1 was doubled (equilibrium was not yet reached by time 250). Thus, any gain in biomass by species 5 as a result of the doubling of R_1 was at the expense of a nearly equal decrease in the biomass of the other four species.

When R_2 was doubled at time 250, all five species had sustained increases in biomass, but their relative abundances remained about the same (Fig. 21-1a). Because R_2 was the limiting replenishment rate at time 250, the total community biomass also had a sustained increase and rapidly approached a new

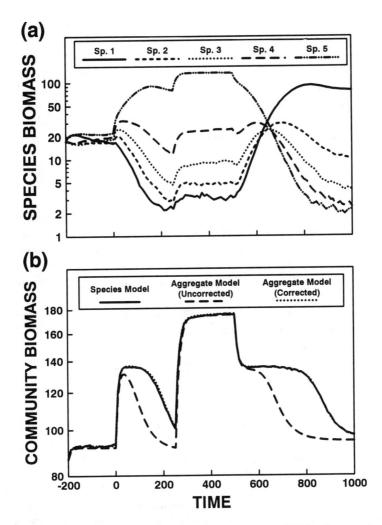

Figure 21-1. Predicted species biomass using a five-species model (a) and community biomass using the five-species model and two aggregate models (b). Changes in biomass result from changes in the replenishment rates of two element resources to the environment. The solid line in *b* is simply the sum of the biomass of the five species in *a*. The uncorrected aggregate model (dashed line in *b*) fails to predict changes in community biomass during periods when species composition is changing dramatically. The corrected aggregate model (dotted line, largely hidden by solid line in *b*) has been altered to account for these changes in species composition.

equilibrium value that was about twice as large as the initial equilibrium value (Fig. 21-1b).

When R_1 was halved at time 500, the community suddenly went from R_2 limitation to R_1 limitation. This resulted in a reversal of the relative abundances for the entire community (Fig. 21-1a). The most abundant species (species 5) became the least abundant and the least abundant species (species 1) became the most abundant. The total community biomass decreased precipitously immediately after R_1 was halved, due to a sudden drop in the biomass of the individual species (Fig. 21-1b). Following this precipitous drop, the community biomass decreased only slightly while the relative abundances of the five species readjusted. Once this readjustment was complete, community biomass decreased more rapidly toward a final equilibrium.

AGGREGATE ECOSYSTEM MODEL

How can species-level processes be linked to ecosystem-level properties? The answer is not straightforward. We will illustrate one approach by developing an aggregate ecosystem model from the species model described above. The objective is to develop a model that does not explicitly represent any of the five species, but still accurately represents total community biomass and productivity.

The aggregate behavior of functionally redundant components is often qualitatively similar to that of the individual components (Rastetter et al., 1992a). Thus, as a first attempt at an aggregate model, we simply used the species model with average values for the two half-saturation constants, an immigration rate equal to the sum of the individual-species immigration rates, and initial biomass equal to the sum of the initial biomasses of the individual species (Fig. 21-1b).

Because both models were similarly constrained by the element replenishment rates and all species have the same Q_j and m_{ij}, equilibrium biomass and productivity for the community are predicted well by the aggregate model. The aggregate model also does well at predicting the transition between time 250 and 500, during which there was not a major shift in community structure. However, it fails to predict the transitions between time 0 and 250 and between time 500 and 1,000, during which there were major changes in community structure.

The aggregate model structure is clearly inadequate for dealing with major changes in community structure. However, it can be sequentially recalibrated to adequately represent any particular community structure. An analysis of how the aggregate parameters change as community structure changes can reveal a great deal about how species-level processes are manifested at the ecosystem level. This information can then be used to restructure the aggregate model.

To implement this analysis, we first used the species model to generate equilibrium data for 100 values of E_1 and E_2 selected to cover their expected ranges. Next, we calibrated the aggregate model to each of these 100 equilibrium data sets by "back calculating" what the values of S_a, k_{1a}, and k_{2a} should be if the aggregate model were consistent with the species model (subscript a denotes "aggregate model"). These three variables were selected because aggregation errors are associated with nonlinearities such as that in the relationship between S_a and B_{*_a} (Rastetter et al., 1992a) and because changes in species composition were expected to be manifested in changes in k_{1a} and k_{2a}. We then searched for relationships between these calibrated variables and other model variables. This search uncovered the following approximate relationships:

$$k_{1a} \approx 0.2 + 1.6E_1 (1 + E_2) / \{E_1(1 + E_2) + E_2(1 + E_1)\} \tag{21-9}$$

$$k_{2a} \approx 0.2 + 1.6E_2 (1 + E_1) / \{E_1(1 + E_2) + E_2 (1 + E_1)\} \tag{21-10}$$

$$S_a \approx n (B_{*_a}/n)^2/\{1 + (B_{*_a}/n^{4/3}\} \tag{21-11}$$

$$n \approx 5 - 4 \{(k_{1a} - k_{2a})/1.6\}^2 \tag{21-12}$$

Equations (21-9) and (21-10) suggest that the ecosystem behaves more like species 1 when E_1 is limiting, and like species 5 when E_2 is limiting. Equation (21-11) suggest that uptake surface area for the whole community is grossly underestimated by Eqn. (21-6) and is better represented by the surface area of the "average" species multiplied by an "effective" number of species present. The "effective" number of species is estimated by Eqn. (21-12) based on the difference between k_{1a} and k_{2a}. When the community is dominated by only one species (species 1 or 5), the difference between k_{1a} and k_{2a} approaches its extreme (1.6), and the effective number of species approaches 1. When the five species are evenly distributed, k_{1a} and k_{2a} are equal, and the effective number of species present is 5. These changes in species composition are not instantaneous. Therefore, we implemented the changes in k_{1a} and k_{2a} by adding a second "acclimation" equation to the aggregate model:

$$dk_{1a}/dt = a_{ka}(k_{1eq} - k_{1a}) \tag{21-13}$$

$$dk_{2a}/dt = -dk_{1a}/dt = a_{ka} (k_{2eq} - k_{2a}) \tag{21-14}$$

where a_{ka} is the acclimation rate of k_{1a} and k_{2a} and k_{1eq} and k_{2eq} are the equilibrium values of k_{1a} and k_{2a} from Eqns. (21-9) and (21-10).

The final correction we made to the aggregate model was to estimate a_{ka} and a_{V_a} (the acclimation rate for V_{1a} and V_{2a}). Unfortunately, nothing can be gleaned about these two parameters from an analysis of the community at equilibrium because at equilibrium Eqns. (21-3), (21-4), (21-13), and (21-14) all equal zero regardless of the acclimation rates. Therefore, we calibrated the model to nonequilibrium data from the original simulations by backcalculating a_{ka} and a_{V_a} from changes in the aggregate values of k_{ia} and V_{ia}. This analysis revealed the following relationships:

$$a_{ka} \approx 0.002 \qquad\qquad (21\text{-}15)$$

$$a_{Va} \approx 0.0045 + 0.0055 \{(k_{1a} - k_{2a})/1.6\}^2 \qquad\qquad (21\text{-}16)$$

Not surprisingly, a_{Va} approaches the value for the individual-species acclimation rate (0.01) when k_{1a} and k_{2a} approach their extremes (0.2 and 1.8). At intermediate values of k_{1a} and k_{2a}, shifts in the community composition seem to slow the apparent acclimation rates of V_{1a} and V_{2a} at the ecosystem level.

With these final corrections, the aggregate model faithfully represents the community biomass and productivity (Fig. 21-1b). However, the estimation of a_{ka} and a_{Va} was clearly a "curve fitting" exercise. Ideally, it would be better to derive relationships analogous to those in Eqns. (21-15) and (21-16) from the species-level equations. However, these formal derivations of ecosystem properties from species properties are often mathematically intractable even for a simple model such as the one presented here. There is usually no alternative to using measurements made directly at the ecosystem level.

CONCLUSIONS

What then has been gained by this lengthy process of developing an aggregate model? Why not simply develop an empirical model directly? First, because of the mechanistic basis for the derivation, the aggregate model is a well-founded hypothesis that can be tested against independent data. As long as the same species are present, the aggregate model would be expected to faithfully represent ecosystem biomass and productivity. The aggregate parameters would, of course, change if a different set of species were represented or if any of the present species were eliminated, but these changes too could be incorporated into an appropriately derived aggregate model.

A second, perhaps more important reason is that an explanatory linkage of ecosystem properties to species properties has been maintained. Although the species are not explicitly represented in the aggregate model, implicitly they are represented in almost every aspect of the model. The basic structure of the aggregate model is very similar to the species model, parameters representing important within-species processes are still in the aggregate model with basically the same meaning, and the important between-species interactions have been synthesized into a few, mechanistically interpretable relationships [Eqns. (21-9)–(21-16)].

The final reason for deriving an aggregate model from the species model is that a great deal can be learned about the manifestation of within- and between-species properties at the ecosystem level. Through the failure of the naive model, we were able to quantify the errors associated with a simple "average" representation of our community of "functionally redundant" species

(Fig. 21-1b). Through the sequential calibration, we identified some new relationships that were either significantly transformed from their species-level representation, such as the S_a-B_{*_a} relationship [Eqn. (21-6) vs. (21-11)], or were not explicity represented at the species level, such as the k_{ia}-E_i relationship [Eqns. (21-9) and (21-10)]. Through the calibration to nonequilibrium data we uncovered an unanticipated interaction between the rate of acclimation of the V_{ia} and the shifts in community structure [Eqn. (21-16)]. Finally, through the derivation of the aggregate model we learned how the interactions among functionally redundant species are manifested at the ecosystem level. Changes in the relative abundance of functionally redundant species is manifested at the ecosystem level [Eqns. (21-13) and (21-14)] in much the same way as within-plant acclimation is manifested at the species level [Eqns. (21-3) and (21-4)].

22

SPECIES COMPENSATION AND COMPLEMENTARITY IN ECOSYSTEM FUNCTION

Thomas M. Frost, Stephen R. Carpenter, Anthony R. Ives, and Timothy K. Kratz

SUMMARY

Functional complementarity occurs when ecosystem processes are maintained at constant levels despite stresses that induce shifts in the populations driving those processes. Understanding when such complementarity occurs depends on an integration of perspectives from population and ecosystem ecology. Here, we examine the extent to which functional complementarity is linked with compensatory dynamics among species that carry out a particular ecosystem function. Our approach combines analyses of long-term zooplankton data from a whole-lake acidification experiment with a theoretical treatment of species compensation. Results from the acidification of Little Rock Lake, WI indicated that the biomass of cladocerans, copepods, rotifers, and total zooplankton remained at high levels despite the loss of component species from each group. Compensatory increases by other taxa were responsible for this complementarity of function. Theoretical considerations indicated that the degree of compensation occurring among species in response to environmental change increased in response to two different factors: the functional similarity of interacting species and the degree to which an environmental change acts nonuniformly on their interactions. Finally, we tested the extent to which compensation in unperturbed systems could predict functional complementarity. We analyzed a 7-year record from the reference basin of Little Rock Lake and found that substantial compensation occurred only in the natural dynamics of rotifers and cladocerans during some seasons. Complementarity in response to acidification was evident, however, among copepods and total zooplankton in addition to rotifers and cladocerans, and could not have been predicted solely on the basis of compensation prior to stress. Taken together, our analyses reveal how functional complementarity is linked to a variety of compensatory interactions among species responding to stress.

INTRODUCTION

Ecosystem processes such as primary production or decomposition are sometimes maintained at near constant levels despite stresses that generate substantial shifts in the species carrying out those processes (Schindler, 1987; Howarth, 1991, Lawton and Brown, 1993; Anderson, Ch. 10). Such resistance to change at the ecosystem level reflects a complementarity of system function; species with similar function replace those lost through stress. Functional complementarity, sometimes termed redundancy, also occurs in situations where invading species cause substantial changes in a community but do not affect the ecosystem processes occurring there (Vitousek, 1990). Function is not always complementary, however. There are numerous cases of substantial shifts in ecosystem processes that can ultimately be linked to changes in the populations of a single species (Vitousek, 1986; Vitousek and Walker, 1989; Carpenter and Kitchell, 1993; Pollock et al., Ch. 12). Determining when functional complementarity occurs and understanding the factors that control it depend on a nexus between population and ecosystem ecology.

How widespread is the phenomenon of functional complementarity? More importantly, what are the circumstances under which complementarity is likely to break down? These are difficult questions to answer. Much of the information on complementarity comes from studies of species invasions (Vitousek, 1986, 1990). Some invasions have caused substantial changes in ecosystem processes but many others appear to have had minimal impact at the ecosystem scale (Vitousek, 1990). The loss of species from ecosystems also produces varied results. During a whole-lake acidification experiment, many phytoplankton species disappeared but total primary production remained unchanged (Schindler, 1987). In contrast, during the same experiment it appeared that the entire process of nitrification was halted below a pH level of 5.7 (Rudd et al., 1988), a change that may have involved the loss of all nitrifying species. These varied results indicate the need for general approaches with which to evaluate both the occurrence of functional complementarity and the factors that control it.

Ideally, investigations of functional complementarity would be based on concomitant measurements of an ecosystem process and the dynamics of the participating species (see Shachak and Jones, Ch. 27). Such detailed and diverse measurements are not commonly available. In this chapter we explore the use of an alternative approach that examines the dynamics of groups of species that are associated with a common ecosystem process. This approach is based on the notion, suggested by Lawton and Brown (1993), that functional complementarity will be linked with the occurrence of compensatory changes among functionally similar species. If a species is lost from an ecosystem, the process carried out by that species could be maintained at a constant level only if similar species increased to compensate for its loss. Conversely, if there were not compensatory increases by other species, the magnitude of the particular ecosystem process would have to decline unless the rare circumstances occurred where the factor that caused the loss of the species itself led to a direct increase in the rate of the ecosystem process.

This chapter explores the importance of compensatory processes in functional complementarity in three sections. We first examine the links between compensation and functional complementarity using data from a whole-lake acidification experiment. We then develop a theoretical treatment of the factors controlling compensation among species. Finally, we test the extent to which compensation among populations under natural conditions can be used to predict functional complementarity in response to stress. Our overall goals are to explore the factors underlying functional complementarity and to examine techniques for predicting the potential for complementarity in unperturbed ecosystems.

COMPENSATORY RESPONSES DURING EXPERIMENTAL ACIDIFICATION

The experimental acidification of Little Rock Lake provides an opportunity to examine functional complementarity in response to three progressively increasing shifts in zooplankton species composition that occurred in response to increasing stages of acidification. The lake's two basins were divided with a curtain in, 1984. After a baseline period, the treatment basin was acidified during three, 2-year stages to target pH levels of 5.6, 5.1, and 4.7 (Watras and Frost, 1989). The reference basin was not manipulated and its pH remained near 6.1 through the experiment. Brezonik et al. (1993) provide a summary of limnological responses during the acidification.

The biomass of all common zooplankton species was measured frequently throughout the experiment (usually at biweekly intervals during ice-free periods and every 5 weeks under ice cover). Techniques used to quantify zooplankton are summarized in Frost and Montz (1988). Data on individual species have been aggregated to examine the dynamics of major zooplankton groups (rotifers, cladocerans, and copepods) and of total zooplankton. Each aggregate group represents an ecosystem process. The biomass of rotifers and cladocerans represents herbivory by small- and large-bodied forms, respectively. Copepods are omnivorous and their biomass represents a combination of herbivory and carnivory. Total zooplankton biomass represents the potential transfer of primary production to fishes. The extent to which compensation could be expected in an aggregration group would depend on the occurrence of numerical compensation in the species that occur in it. A general treatment of the factors underlying compensation within aggregation groups is presented in the next section.

At the species level, there were major declines for some taxa during the earliest acidification stage to pH 5.6 (Frost and Montz, 1988; Sierszen and Frost, 1993). With further acidification to pH 5.1 and 4.7, more species declined (Brezonik et al., 1993). However, not all taxa decreased with acidification; several increased during each treatment stage. The interaction of these changes yielded a community in the treatment basin that bore little resemblance to that in the reference basin. Comparisons of community similarity between the two basins on each sampling date revealed a systematic divergence that began with the start of acidification in 1985 and increased progressively through the second year at pH 4.7 (Fig. 22-1).

The aggregate response of zooplankton, in terms of total biomass, showed a different pattern with acidification. Shifts in total biomass lagged substantially behind the changes in community structure. There was little evidence of a difference between the basins in total biomass until the pH 4.7 period (Fig. 22-1). Thus, the function of the zooplankton community represented by its total biomass was much less sensitive to acidification than community structure.

Comparisons of species responses with the shifts in biomass of rotifers, cladocerans, and copepods indicate the importance of compensatory responses in the functional complementarity of the zooplankton community. For each zooplankton group, a single species can be identified that has increased from a minor fraction to a substantial portion of a group's biomass as acidification progressed (Fig. 22-2). The importance of these three species is further evident when they are compared with the pattern observed in total zooplankton biomass (Fig. 22-2). Clearly, the biomass of the groups could not have been maintained at the levels occurring during the experiment without the increasing species. Smaller-scale, single-species experiments have indicated that the increasing species do not show a strong increase in reproduction or sur-

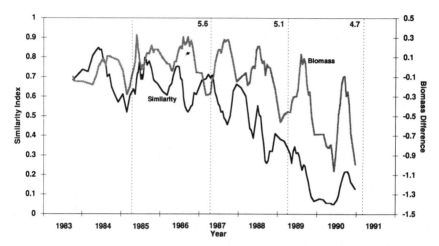

Figure 22-1. Similarity in species composition and difference in total zooplankton biomass between the treatment and reference basins of Little Rock Lake calculated for all sampling dates between, 1983 and, 1991. Both were smoothed using five-point moving averages. Similarity was calculated following Inouye and Tilman (1988). The difference in biomass was calculated following Stewart-Oaten et al. (1986) as log (Treatment Biomass) – log (Reference Biomass). Original biomass units were µg/L.

vivorship under simply the chemical conditions occurring at lower pH levels (Gonzalez and Frost, 1994). Their increase, rather, appears to be a compensatory response, due possibly to a relaxation of interspecific competition for resources, triggered by the loss of other species with acidification.

THEORETICAL PERSPECTIVE ON COMPENSATION

The relationship between compensatory processes and the response of species to environmental changes is not obvious and is potentially very complex. While a thorough exploration is not possible in one brief chapter, we present a simple, general model that illustrates the main connections between compensation, species interactions, and environmental change. Our approach is readily applied to complex data because the parameters can be estimated by multiple regression. The two-species model presented here can be directly extended to more complex ecosystems with multiple components (A.R. Ives, unpublished).

Consider the population dynamics of two species governed by the equations

$$x_1(t + 1) = x_1(t) + b_{1,1}x_1(t) + b_{1,2}x_2(t) + c_1 T(t) + e_1(t)$$
$$x_2(t + 1) = x_2(t) + b_{2,1}x_1(t) + b_{2,2}x_2(t) + c_2 T(t) + e_2(t)$$

$$(22\text{-}1)$$

where $x_i(t)$ denotes the density of species i at time t, and $b_{i,j}$ represents the effect of the density of species j on the population growth rate of species i. The signs of the values of $b_{i,j}$ set the type of interaction between species (e.g., if the species compete, then $b_{1,2}$ and $b_{2,1}$ are negative). $T(t)$ represents an environmental variable such as temperature or pH. The parameter c_i scales the effect of the environmental parameter $T(t)$ on the population growth rate of species i. The term $e_i(t)$ is a random variable that represents variability in the population growth rate of species i that is not included in $T(t)$. Although the strict linearity assumed in the model is inappropriate for most biological systems, suitable transforms may make the linear approximation more realistic. Rather than trying to portray natural systems realistically, however, our goal here is to illustrate factors contributing to compensation.

An example of population trajectories for two competitors generated by Eqn. (22-1) is illustrated in Fig. 22-3A. In this case, the environmental parameter $T(t)$ has a negative effect on both species, but the negative effect on species 1 is greater than that on species 2. After 15 generations, the mean value of $T(t)$ increases, thereby decreasing the population growth rates of both species. The mean population density of species 1, which is affected more by changes in $T(t)$, decreases, while the mean population density of species 2 increases in compensation.

A graphical technique can be used to illustrate two distinct factors that control the extent of compensation between the two species, the similarity between the ecological function of the species and the relative effect of the environmental changes on population growth rates. A critical component in this technique is to define ecological function in terms of the effects of a given species on the population growth rates of all species in the community. Let $\mathbf{B}_1$ and $\mathbf{B}_2$ be the vectors $[\mathbf{b}_{1,1}, \mathbf{b}_{2,1}]$ and $[\mathbf{b}_{1,2}, \mathbf{b}_{2,2}]$, respectively. Thus, $\mathbf{B}_i$ summarizes the density-dependent effect of species i on the population growth rates of both species. Fig. 22-3B depicts these vectors graphically. To make the presentation of the model easier, vectors $\mathbf{B}_1$ and $\mathbf{B}_2$ have been standardized to unit length by rescaling population densities in Eqn. (22-1). Because both species are competitors, $\mathbf{B}_1$ and $\mathbf{B}_2$ are located in the negative quadrant of the graph. Functional similarity between the species is reflected in the angle β between $\mathbf{B}_1$ and $\mathbf{B}_2$; the smaller this angle, the more similar the function of the species measured in terms of population interactions. The effect of environmental factors on the species can be represented in a similar fashion, with $\mathbf{C}$ being a vector $[\mathbf{c}_1, \mathbf{c}_2]$ where $\mathbf{c}_i$ indicates the effect of the environmental factor on the population growth rate of species i. The angle α measures how different the effect of environmental change on population growth rates is relative to the effects of changing populations densities (summarized by $\mathbf{B}_1$ and $\mathbf{B}_2$). Larger values of α represent environmental effects very different from the patterns of density-dependent species interactions.

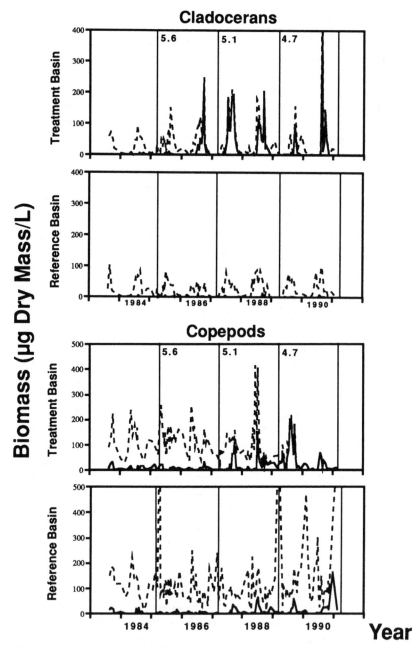

Figure 22-2. Biomass (µg/L of dry mass) of zooplankton groups and increasing species in the treatment and reference basins of Little Rock Lake. *Dotted lines* indicate the biomass of the more aggregated variable. Increasing species were *Daphnia catawba* for cladocerans, *Keratella taurocephala* for rotifers, *Tropocyclops prasinus mexicanus* for copepods, and the sum of these species for total zooplankton.

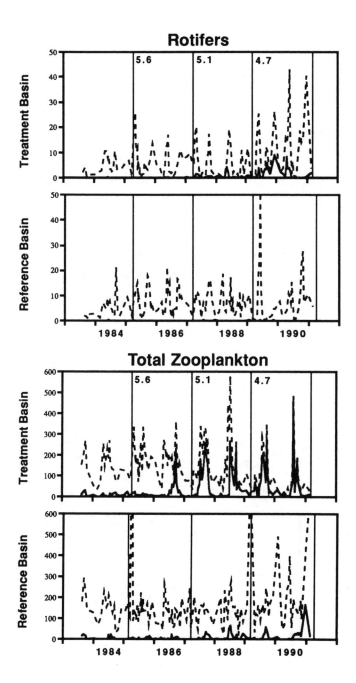

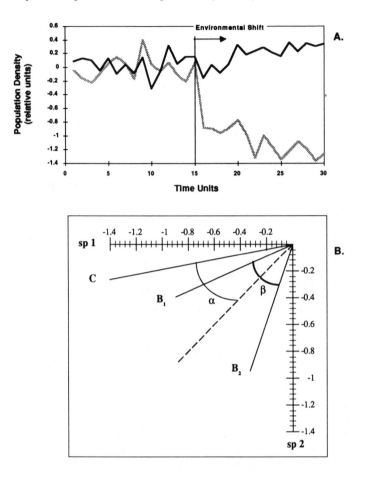

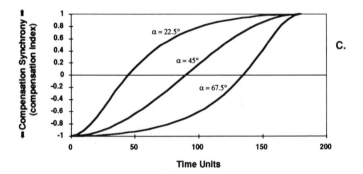

Figure 22-3.

The effects of different values of α and β on the degree of compensation between species 1 and 2 can be quantified using an index of compensation. If compensation occurs, the change in the sum of population densities of species 1 and 2 should be less than the sum of the changes in population densities of both species taken separately. Letting DN_i denote the change in the mean population density of species i caused by changes in the mean value of $T(t)$, an index of compensation can be measured as

$$(DN_1 + DN_2)^2/(DN_1^2 + DN_2^2) - 1 \qquad (22\text{-}2)$$

When there is complete compensation $(DN_1 = -DN_2)$, this expression equals -1. When populations are synchronous $(DN_1 = DN_2)$, it equals 1. The degree of compensation occurring for a set of simulations systematically varying values of α and β is shown in Fig. 22-3C.

Overall, these analyses illustrate how two different factors affect the degree of compensation that occurs among species. Compensation increases as the degree of functional similarity increases between species (β decreases). Furthermore, if the effect of the environmental change on population growth rates is similar to the effects of changing species densities (α is small), then compensation will occur only between very similar species. In contrast, if the effect of the environment is very different from the effects of the species interactions (α is large), the potential for compensation is much greater. Although our example involves two competitors, this general analysis applies to any other type of species interaction and for any number of species (A.R. Ives, unpublished).

This simple analysis highlights three aspects of compensation. First, the function of a species should be defined in terms of its effects on the population growth rates of other species in the community, as opposed to other ways that they are affected by other species. Second, the effect of the environment on species must be measured with respect to the effects on species interactions.

Figure 22-3. (A) Population trajectories for two competing species generated from Eqn. (22-1). Initially, random variables $T(t)$, e_1, and e_2 are governed by normal distributions with means of 0 and standard deviations of 0.1. After 15 generations, the mean of $T(t)$ is increased to 0.5. Values of the other parameters in this example are $b_{1,1} = -0.894$, $b_{2,1} = -0.371$, $b_{1,2} = -0.447$, $b_{2,2} = -0.928$, $c_1 = -1.79$, and $c_2 = -0.0371$. (B) Projected shifts in population growth rates of species 1 and 2 with changes in the densities of species 1 (B_1) and species 2 (B_2), and changes in the environmental parameter T. Beta is the angle between B_1 and B_2, and α is the angle between C and the angle bisecting B_1 and B_2. In this example, vectors $B_1 = [-0.894, -0.371]$, $B_2 = [-0.447, -0.928]$, and $C = [-1.79, -0.0371]$, the lengths of B_1 and B_2 have been fixed to be the same, and the vector $B_1 + B_2$ bisects the angle between B_1 and B_2 (see text for explanation of units for axes). (C) Compensation measured by Eqn. (22-2) using data generated using Eqn. (22-1) graphed as a function of β for three values of α. In this example, $\| B_1 \| = \| B_2 \| = 1$.

The critical measure of the environmental effect is the angle α. Although one might intuitively expect that an environmental stress affecting two species in exactly the same way could not lead to compensation, this is not the case. Even if the vector **C** lay along the 45° line in Fig. 22-3B, indicating identical environmental effects on both species, the angle α could be large, implying strong potential for compensation. Third, the potential for compensation can be analyzed using regression techniques to extract the strengths of species interactions. Although this will likely involve nontrivial statistical problems in application to real data sets, the approach above at least places the analysis of compensation in a data-oriented context.

PREDICTING FUNCTIONAL COMPLEMENTARITY IN UNDISTURBED ECOSYSTEMS

As discussed previously, evaluations of functional complementarity are usually based on responses to stress such as species invasions or environmental contamination. It would be useful to evaluate the potential for complementarity in unstressed ecosystems in order to predict ecosystem responses to stress. Our theoretical analysis indicates that both knowledge of the differential effects of possible stress agents and information on the degree of functional similarity occurring within an unperturbed community might be useful in predicting complementarity.

In many cases, information on the differential effects of a stress may not be obtainable without actually applying that stress to an ecosystem. Indirect, food web effects can predominate in such situations and it is difficult to extrapolate the results from one system to others without very similar community structure (Webster et al., 1992). The degree of similarity occurring among species within a system can be inferred, however, by examining species dynamics under natural conditions. This information could be particularly useful if functional complementarity were more likely among species exhibiting compensatory patterns when no stress was evident. Here we test for such an association using a simple index to assess compensation in unperturbed systems using long-term population data. We apply this analysis to data from Little Rock Lake's reference basin and assess the extent to which the functional complementarity occurring in the treatment basin could have been predicted on the basis on compensation occurring in the reference system.

Previous analyses of compensatory processes have focused on density compensation (e.g., Tonn, 1985). Such measures evaluate the tendency for populations to achieve a consistent biomass in different habitats or years. The index we suggest focuses on variance compensation. Rather than assume that populations should converge on a particular density, this index allows for differences in total biomass generated by natural variability across years or habitats. Evidence for compensation is derived from comparisons of inter-annual variability in population parameters and their aggregates.

The index of variance compensation is based on the relationship between the variance of aggregate variables that sum the abundance of several populations and the variances of the individual populations (Box et al., 1978). For n species

$$\text{Var} \left(\sum_{i=1}^{n} S_i \right) = \left[\sum_{i=1}^{n} (\text{Var } S_i) \right] + \left[2 \sum_{i=1}^{n} \sum_{j=1}^{i-1} \text{Cov}(S_i S_j) \right] \tag{22-3}$$

where S_i is the biomass (or density) of species i, Var is the variance, and Cov is the covariance. It is important to note that this relationship does not hold if data have been transformed (e.g., as a logarithm) prior to aggregation.

When populations vary independently, their covariance is zero. Thus, in the case of independently fluctuating species, the variance of their aggregate should be equal to the summed variance of the component species. When species exhibit compensatory dynamics their covariance is negative, and the variance of the aggregate would be less than the sum of the variances for all of the species. The opposite occurs when species dynamics are synchronized. Thus, simple additivity of variance between component species and their aggregate can be used as an indication of either a lack of species interactions or a lack of similar patterns among component species. The ratio between the variance of an aggregate variable and the sum of the variances of its component species can be used as an index of the relationships occurring among groups of species. Variance ratios substantially less than one indicate compensation whereas those greater than one indicate synchrony. Note that the compensation index presented previously is based on a mean response to an environmental shift and is not directly applicable to nonstressed populations.

Because the reference basin of Little Rock Lake was unmanipulated aside from the installation of a curtain, we can use data collected there between, 1984 and 1990 to reflect natural variability by the zooplankton community. Analyses are presented separately for each season to test for compensation within each period rather than phenomena such as seasonal succession. The data used in our calculations were the mean biomass ($\mu g/L$) recorded for each species during each season of each year. Species were aggregated at two levels: major zooplankton groups (rotifers, cladocerans, and copepods) and total zooplankton.

We tested for the strength of the deviation of the variance ratios from a null model of species independence using a randomization procedure. The biomass of each species was randomized among years prior to calculating aggregate biomasses and variances. One thousand permutations were conducted and the number of values less (in the case of compensation) or greater (in the case of synchrony) than the value determined for the actual data was recorded.

Cladocerans showed the strongest deviations from independence (Fig. 22-4). During fall, the variance exhibited by total cladoceran biomass was almost four times less than would have been expected if species varied independently. This suggests that substantial compensation occurred at that time. In contrast, it appeared that cladoceran species were largely synchronized during winter and summer. During spring, species appeared to vary independently.

Rotifers exhibited strong evidence for compensation during spring and summer and an indication of compensation during winter (Fig. 22-4). They showed no evidence of synchrony during any season. Copepods appeared to be largely independent during all seasons except perhaps fall, when synchrony was suggested (Fig. 22-4). They showed no indication of compensation.

Interestingly, there was no evidence for compensation or synchrony when the dynamics of the entire zooplankton community were aggregated as total biomass. This indicates that nonindependent variability patterns by one major group can be masked by contrasting dynamics exhibited by other groups. Aggregations involving functionally less similar groups appear to have the potential to conceal compensation at lower levels of aggregation.

How well could patterns of compensation in Little Rock's reference basin have predicted the patterns of functional complementarity that were observed with acidification? Not particularly well. It is clear that both cladocerans and rotifers pass through seasonal windows of compensation. Both of these groups exhibited evidence for functional complementarity (Fig. 22-2). At the same time, despite evidence of complementarity for copepods and total zooplankton, variance ratios gave no indication of compensation for either groups. Because several of the species responsible for functional complementarity were rare prior to acidification, a parameter based on surveys that were less dependent on abundant species could prove to be a better predictor of the potential for complementarity.

DISCUSSION

Little Rock Lake's zooplankton community exhibited considerable functional complementarity in its response to acidification. Compensatory processes, in which increases by a few species matched declines by other taxa, were responsible for maintaining zooplankton biomass at a constant level. These results support the suggestion that functional complementarity should generally be linked with compensatory population dynamics. Functional redundancy and species compensation are two sides of the same coin.

A fundamental question is whether compensation in response to environmental stress can be predicted from the pattern of population compensation

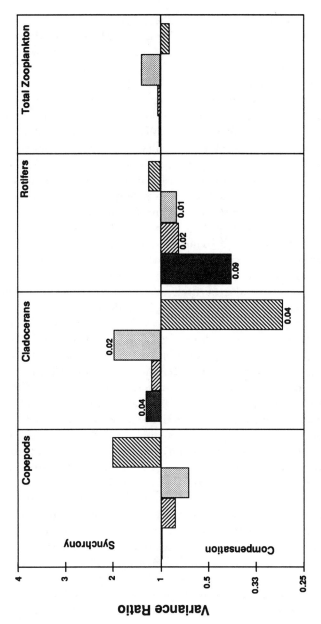

Figure 22–4. Variance ratios for major zooplankton groups and total zooplankton biomass in the reference basin of Little Rock Lake (see text for explanation). Ratios are presented in the sequence winter, spring, summer, and fall for each group. Numbers indicate the proportion of 1,000 randomizations that produced values greater than or equal to those actually observed. Only proportions <0.10 are reported.

before the environmental stress is applied (see Pace et al., Ch. 7). Unfortunately, judging from our results in the section on predicting functional complementarity in undisturbed ecosystems, the answer is not very well. There are several reasons for this. First, critical to the compensation observed during the experimental acidification of Little Rock Lake were rare species that reached unprecedented abundance following acidification. Rare species present a particular problem for predicting responses to environmental stresses because their rarity makes them unlikely subjects for experimentation. Also, in the analysis applied in the section on predicting functional complementarity in undisturbed ecosystems, variance ratios were used to assess compensation in the unstressed system. Because a rare species will contribute little to the total variance of an aggregated group, its potential for compensation will not be apparent. Second, our theoretical analysis demonstrates the importance of the differential effects of the environmental stresses on the population growth rates of species in a community. The pattern of environmental fluctuations that drives the variance/covariance pattern of species abundances before a major environmental stress may be very different from the pattern imposed by the environmental stress itself, making it impossible to predict the latter from the former. Third, the response of population densities to continuous, frequent environmental fluctuations is different from the response to long-term, directional stresses such as acidification. This can be explained using the analogy of different types of perturbation experiments. Bender et al. (1984) distinguish two types of perturbation experiments, PULSE experiments in which perturbations are applied once and then relaxed, and PRESS experiments in which perturbations are applied continuously (see Pace et al., Ch. 7). Continuous environmental fluctuations operate like repeated PULSE experiments, whereas directional environmental stresses operate like PRESS experiments. In the same way that PULSE and PRESS experiments need to be analyzed differently, so do continuous and directional environmental changes. As we show in the theoretical analysis, the critical information needed for predicting compensation in response to long-term environmental stresses is the density-dependent impact of species on population growth rates.

It is unfortunate that compensation in natural populations does not serve as a better predictor of functional complementarity. Such information could be more easily obtained than data on the myriad forms of stress that might act on ecosystems. Ecologists seeking to define situations when ecosystem function breaks down will need to consider the direct actions of a wide range of stressing agents as well as the numerous indirect effects that have the potential to work their way through food webs.

It is important to keep in mind that our evidence for functional complementarity in response to acidification does not suggest that individual species are unimportant in maintaining ecosystem function. For example, although functional complementarity occurred after 2 years of acidification to pH 5.1, it is not certain that it would have been maintained for a more extended period (see Pace et al., Ch. 7). Nor is it clear how Little Rock Lake, or ecosystems in general, will respond to the interactions of more than one stressor. Finally, the compensating species that were responsible for functional complementarity in Little Rock were all rare prior to acidification. Such species might have been considered unimportant if the effects of acidification were not anticipated.

We have not attempted to discuss all aspects of functional complementarity here. Vitousek (1986) reviews several other aspects of complementarity that are beyond the scope of this chapter. In keeping with the spirit of this volume, rather, we have focused on an explicit examination of the links between compensatory population dynamics and the resistance of ecosystem processes to changes with stress. We have shown, both empirically and theoretically, that an understanding of the factors stabilizing ecosystem processes can be gained from an knowledge of the population dynamics of species driving those processes.

ACKNOWLEDGMENTS

This work was supported by the Andrew W. Mellon Fund, the National Science Foundation, the U.S. Environmental Protection Agency, and the Trout Lake Station. We thank Pam Montz for her assistance in providing the zooplankton data and Shelley Arnott, R.P. McIntosh, D.L. Strayer, and Kathy Webster for their helpful suggestions.

23

ELEMENTAL STOICHIOMETRY OF SPECIES IN ECOSYSTEMS

Robert W. Sterner

SUMMARY

One way that species and ecosystems are linked is through elemental stoichiometry. Consumer–resource interactions follow stoichiometry in that mass must balance when resources are turned into consumer biomass plus waste products. It is important to describe the constaints within which elemental concentrations can vary within species in ecosystems. It is shown that the elemental content of species varies genetically and environmentally, but within limits. Certain elements have correlated concentrations in different species. Constraints on elemental content result in interactions among cycles of individual elements. Species stoichiometry partly controls biogeochemical cycling.

Data on elemental content are limited for all groups except perhaps primary producers. Nevertheless, certain general patterns in elemental content can be advanced. Autotrophs exhibit reduced nutrient content at low growth rate. In addition, they show considerable interspecific variation. Animals exhibit limited intraspecific, but large interspecific variation. There is evidence that the elements nitrogen (N) and phosphorus (P) along with certain trace elements vary because organisms differ in the amount of protein synthesis material (nucleic acids, proteins) they contain.

Stoichiometry introduces multiple stable states in a variety of systems, including plant–herbivore systems in both terrestrial and aquatic environments.

INTRODUCTION

To study anything as complex as an ecosystem, simplifications are necessary. A common simplification in ecosystem analysis is to consider organisms to be a single substance such as biomass or energy. This simplification is warranted if the composition of organisms does not affect how they forage, grow, or process materials. However, it is sometimes necessary to consider organism composition. For example, N per unit biomass in terrestrial plants influences herbivory on that plant (Mattson, 1980) and the mineralization of detritus from that plant (Hobbie, 1992). In these cases, secondary production and nutrient cycling are affected by the ratio of N to biomass. At times such as these, models of multiple substances should be more successful at explaining system dynamics than single-substance models (Bolin and Cook, 1983).

Biogeochemical cycles interact in various ways. Interactions among elements can be divided into two broad categories: constraints on composition and regulation of rates. Constraints include such phenomena as physiological limits to the %N in an organism's tissues. Another type of constraint is that element concentrations may be constant at some points in the ecosystem but vary in others. Still another example is that the concentrations of two elements may vary across species but have close correlations. Such correlations mean that the cycles of both elements are tied to each other and to the community-level dynamics of the success of these different species. Regulation of rates includes phenomena such as the rate at which an herbivore grows when feeding on food of high vs. low C/N ratio. This chapter focuses mainly on constraints.

In chemical stoichiometry, the numbers of atoms of each element must be equal on both sides of a reaction. In ecological stoichiometry, one organism consuming another is a complex chemical reaction involving reactants (resources) and products (consumer biomass and wastes). Every consumer–resource pair in an ecosystem has a mass balance constraint similar to a chemical reaction. Stoichiometric principles apply to any substances that are not broken apart during chemical reactions; these include all elements and some biomolecules. This chapter will deal exclusively with elements.

The total elemental composition of organisms reflects some of the primary stoichiometric constraints encountered during their evolution. Most of the elements in the top four rows of the periodic table are essential for one or more organisms (Fig. 23-1). All essential elements are light, perhaps because heavy elements are rarer (Bowen, 1979). The concentrations of elements in biomass and in the earth's crust are correlated (Markert, 1992a), which probably reflects selective pressure for use of more abundant resources. Essential elements come from nearly every column of the periodic table, because organisms require many different chemical properties (Fraústo da Silva and Williams, 1991). Elements can be divided into three functional groups: (1) structural (C, H, O, N, P, S, Si, Ca), (2) electrolytic (K, Na, Ca, Cl, Mg), and (3) enzymatic (V, Cr, Mo, Mn, Fe, Co, Ni, Cu, Zn, B, Sn, Se, F, I, Mg) (Markert, 1992a). The bulk biological elements (Fig. 23-1, shaded) are required by all living things. Some trace elements, in contrast, are required by only some organisms. Essentiality is not a prerequisite to biological cycling: many nonessential elements are found in biomass at concentrations similar to the concentrations of essential trace elements (Markert, 1992a). As analytical techniques have improved, the list of essential elements has lengthened (Markert, 1992b). Stoichiometric constraints on elemental composition should be a factor in nearly all element cycles.

Figure 23-1.   The elements essential for one or more species. (Compiled from Fraústo da Silva and Williams, 1991 and Markert, 1992a.) Shaded cells denote the bulk biological elements that together account for 99.9% of biomass of most organisms. Elements marked with * are required only by animals, and those marked ** are required only by plants. It is uncertain if the elements marked with a question mark are essential.

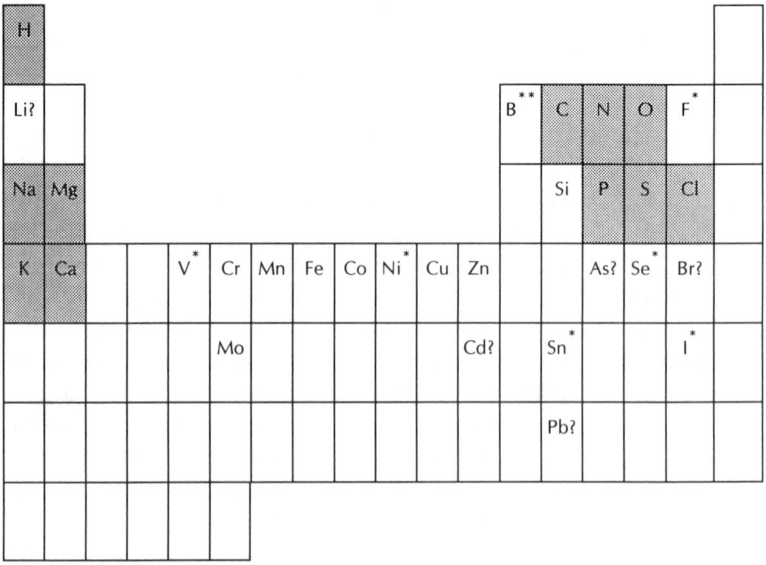

STOCHIOMETRIC CONSTRAINTS ON ORGANISMS

Intra- and Interspecific Patterns in Elemental Content

Organisms do not passively reflect the elemental composition of their resources. Living things cannot be made from arbitrary proportions of elements. Thus, a set of important constraints are biological limits to elemental composition. Organisms hold some elements within narrow bounds, while other elements are more variable. By definition, essential elements have a lower bound on concentration, and all elements have upper bounds, because no living thing can be 100% a single element. A major part of stoichiometric ecosystem studies is understanding constraints and variations in elemental content.

Elemental composition can be expressed either as percent by mass (I use dry mass) or as a ratio of one element to another (I use atomic ratios, not mass ratios). A universal biological constraint is that most of the mass of living cells consists of carbohydrate, lipid, protein, and nucleic acid. All these contain abundant C, H, and O, and the latter two contain significant amounts of N. Thus, protoplasmic growth requires more of these four elements than others. C, H, and N are among the few elements found in higher concentrations in biomass than in the earth's crust, and O has a similar concentration in both (Markert, 1992a). Extracellular products for support and defense, such as shells, tests, carapaces, bones, etc., may contain very high quantities of Ca, P, or Si. These may also lack N almost completely. Composition of extracellular material can differ considerably from organism to organism (Reiners, 1986). For stoichiometric models to advance, we must understand the regulation of elemental content and how elemental content relates to ecological properties such as growth rate.

Much more is known about the elemental composition of autotrophs than other organisms. Autotrophs exhibit much intra- and interspecific variation. Vascular plants have highly divergent elemental content in above- vs. belowground biomass and in woody vs. nonwoody tissue. However, even undifferentiated plant tissues exhibit variation. Elements that make up a large fraction of cell mass, such as C or O, vary less than elements with lower concentrations, such as N, P, or S (see, e.g., Reynolds, 1984, p. 160 or Bowen, 1979, p. 69). An exception to this rule is H, which is usually <10% of mass, and varies within close limits like C. C typically makes up 40–50% of dry mass of most organisms. In contrast, cell contents of N and P in many autotrophs, including phytoplankton, range 5- to 10-fold.

One relationship between growth rate and elemental content was developed by M.R. Droop for algae with growth limited by a single nutrient (Droop, 1974). The inverse of cellular nutrient content and population growth rate can be linearly related, which leads to Droop's internal stores model:

$$\mu = \mu' \left(1 - \frac{1}{Qk_Q} \right) \tag{23-1}$$

where μ = population growth rate (time^{-1}), μ' = theoretical maximal growth rate (never attained), Q = cell quota, or nutrient content of cell (moles/cell), and $k_Q = Q$ when $\mu = 0$ (k_Q is a "minimal cell quota"). In addition, μ_{max} is defined as the maximum *attainable* growth rate. Expressing Q as nutrient/C, one can plot μ/μ_{max} vs. C/element, and a straight line would be a good fit to the Droop internal stores model. Sometimes, the fit is good, but in other cases it is not. For example, in a comparison of two freshwater algae (Fig. 23-2), one is seen to fit the Droop model well, but the other does not. Departures from

Figure 23-2. Elemental stoichiometry of nutrient limitation in two freshwater algae grown under single nutrient limitation, *Cyclotella meneghiniana*, a diatom (diamonds), and *Scenedesmus acutus*, a green (squares). Vertical axes indicate relative growth rate (RGR) = μ/μ_{max}; horizontal axes indicate cellular element ratios. (Top) Algal C/P ratio under P limitation. (Bottom) Algal C/N ratio under N limitation. Straight lines represent least-squares fit to the data, which is the Droop internal stores model of growth [Eqn. (23-1)]. *Scenedesmus* was grown in chemostats (Sterner, 1993). *Cyclotella* was grown in semi-continuous cultures. C and N were determined by CHN analysis and P was determined by persulfate digestion. *Cyclotella* follows the linear internal stores model but *Scenedesmus* does not. At low growth rate, the C/P ratios of these two species diverge.

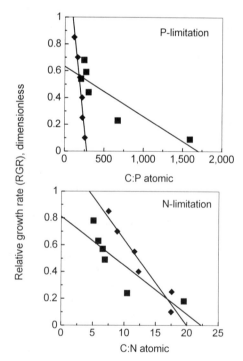

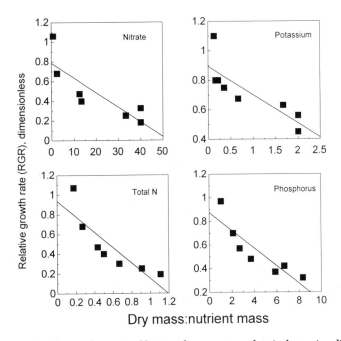

Figure 23-3. Elemental content of lettuce plants grown under single nutrient limitation. (Data from Burns, 1992.) Plot labels indicate the chemical measured in the plants, and the horizontal axes indicate the inverse of the content of that chemical (dry mass/nutrient mass). The vertical axes represent relative growth rate (RGR), μ/μ_{max}. Straight lines represent least-squares fit to the data, which is the Droop internal stores model of growth [Eqn. (23-1)]. All plots show departure from the straight-line fit.

linearity are evident in lettuce as well (Fig. 23-3). In all these cases, though, limitation of the producer's growth by N or P resulted in a considerably reduced amount of that element compared with C. Figs. 23-2 and 23-3 show several major features of species stoichiometry: variation of elemental content between and within species and the constraint on that variation. Elemental content has a lower bound set by the minimal quota, k_Q (x-intercept in Figs. 23-2 and 23-3). The Droop model is one expression of a general phenomenon in primary producers, namely that low growth rate is correlated with low content of the nutrient that limits growth. Thus in a wide diversity of autotrophs studied in controlled conditions, there is a good correlation between the degree of nutrient limitation of growth and elemental content.

In the field, factors such as multiple nutrient limitation or interspecific variability could potentially obfuscate any simple single-species patterns. Nevertheless, a field study of lake communities that included experimental determination of growth limitation showed that the C/P ratio and nutrient limitation were related (Fig. 23-4). In this lake, primary production was limited partly

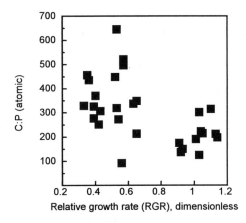

Figure 23-4. Seston C/P ratio (particulate matter between 0.7 and 80 μm) vs. an index of nutrient limitation in Joe Pool Lake, Texas for 1991–1992 (Sterner, 1994). Seston C and P were determined using CHN analysis and persulfate digestion followed by molybdate determination of phosphate, respectively. Nutrient limitation was determined in experimental bioassays where *in situ* phytoplankton were diluted with filtered lake water in the proportion of 0.1 whole water/0.9 filtered water, and some were spiked with N, P, Si, and a mixture of trace metals while other vessels were unspiked. Vessels were incubated in the laboratory at *in situ* light and temperature for 4 days. Nutrient limitation was calculated as the ratio of seston C growth in controls divided by growth in nutrient supplements (1.0 = no limitation, 0.0 = maximum limitation).

by P, and C/P was low when nutrient limitation was weak, but it was variably high under nutrient limitation. The field data are considerably "noisier" than the laboratory data, but the trend is evident. Field C/P ratios vary over a wide range and are partly determined by producer growth rates.

The most rigorous way to study constraints and variation in species stoichoimetry is to subject species to varying nutritional treatments and see how elemental composition varies. A less rigorous method is to survey species in the field for statistical associations between elements (Garten, 1976). The major weakness of survey data is that the role of the environment in these correlations cannot be independently assessed. A survey of 45 elements in standard plant reference material (total of 12 species) yielded the interesting observation that all tight correlations ($|r| > .9$) between pairs of elements across different plant species were positive (Markert, 1993). Such data on many elements also suggest potential physiological bases of constraints. The elements P and N had a correlation of +.84, an association that might come from the involvement of both these elements in protein synthesis. Garten (1978) also argued for the existence of a "nucleic acid-protein" set of elements (including P and N). Survey data thus help delineate some constraints on element composition and further suggest some actual biochemical constraints that cause these interactions.

Given the fact that primary producers show considerable genetic (inter-specific) and environmental (intraspecific) variation in elemental content, the next step is to compare that variation in other members of the ecosystem. Unfortunately, data on other groups are limited. Also, to properly assess constraints, we must have good estimates of the full range of variation, not just means. Unfortunately, most studies downplay or even reject extreme values [e.g., "some of the data have been deliberately rejected if they were collected for (a) diseased or deficient organisms; (b) contaminated organisms, or those growing in abnormal media. . . ; or (c) *accumulator organisms*, which concentrate particular elements in a striking manner," (Bowen, 1979), p. 86, his italics]. Due to a paucity of study coupled with a general failure to purposely produce a maximum range of element concentration, there is much uncertainty about elemental constraints in heterotrophs.

Consider metazoans. Some intraspecific variation is known. The best data on the elemental composition of vertebrates raised on different diets are for agriculturally important species (McDowell, 1992). Dietary P deficiency is known to reduce P in bone (McDowell, 1992). Nevertheless, intraspecific variation in elemental content of animals appears to be much less than in primary producers. For example, when individual cohorts of *Daphnia* were fed a set of three foods with N/P ratios from 22 to 194, the *Daphnia* N/P ratio varied only from 20 to 22 (Sterner et al., 1993). Field samples have revealed only a small amount of variation in elemental content of *Daphnia* (Berberovic, 1990; Sarnelle, 1992), which is correlated with animal size. A Droop plot for an animal using the axes of Figures 23-2 and 23-3 would be close to a vertical line.

On the other hand, interspecific variability of elemental content of metazoans may be as large as in autotrophs. To some extent, this is due to extracellular products; for example, vertebrate bone contains more P by weight even than Ca so that vertebrates should have a higher P content than invertebrates. But, there is a surprising amount of variation in elemental content of even similar invertebrates of close phylogenetic affinity. The freshwater zooplankton contain a group of species of similarly sized Crustacea (copepods and cladocerans) that all inhabit a similar ecological niche. In spite of this ecological similarity, the N/P ratio in crustacean zooplankton is known to vary from 13 to 50 (Andersen and Hessen, 1991) and perhaps more as additional species are studied. Shifts in species composition thus must rearrange biogeochemical pools, which seems to produce signals in nutrient cycling evident at the ecosystem level (Sterner et al., 1992). It is uncertain what the physiological basis of this interspecific variation is. Plant survey data (above) suggested an importance of nucleic acid and protein synthesis in elemental content. *Daphnia*, which have a low N/P ratio (about 15) also have an unusually high RNA content (8% of dry mass) (McKee and Knowles, 1987). Copepods have a high N/P ratio (30–50) and a low RNA content [data from marine species, RNA ranges .03 to .7%, (Båmstedt and Skjoldal, 1980)]. Such a large difference in concentration of RNA, which is very low in N/P (3.75 assuming equimolar nucleotide makeup) and high in P (10% by mass), might account for these differences in the N/P ratio of zooplankton species. Given the above figures, RNA-P could account for about 50% of *Daphnia* P and could explain all of the known differences in P content of these organisms. Organisms with a high RNA content should have higher rates of somatic growth when food is plentiful. On the other hand, the requirement of P in the *Daphnia* diet makes this animal vulnerable to shifts in food quality (Sterner et al., 1993; Sterner, 1993). Stoichiometric constraints based on protein synthesis might thus relate to the ability of the different species to convert food into their own biomass and thus to life history strategies.

Regulation of Elemental Content

There are two principle ways that organisms can produce biomass with an elemental composition different from the available resource spectrum: behavioral foraging choices and physiological control of assimilation and excretion.

Organisms that ingest parcels of solid material make foraging decisions such as when and where to feed, which food packets to ingest and to ignore, etc. If elemental stoichiometry enters these foraging decisions, we should see organisms behaving to ingest food with elemental content balanced for growth and reproduction with obvious implications for patterns of nutrient flux in food webs and ultimately entire ecosystems. Many diverse animals behave in a manner consistent with this prediction. African ungulates (McNaughton, 1988, 1990) and the tropical mammalian folivore, *Kerodon rupestris* (Willig and Lacher, 1991), forage in habitats that improve their intake of mineral elements. Copepods select particles of high N content out of mixtures of cells of high and low content (Butler et al., 1989). Stoichiometry of food choice suggests that organisms will forage in a way to minimize the imbalance between their nutritional requirements and their diet. Such a pressure on food choice is balanced with other constraints on energy intake, time minimization, avoiding predators, and so on, and probably is not always involved in diet choice (Belovsky, 1990).

As discussed above, individual plants of high mineral content within a population should be the ones with the highest growth rates. Selective herbivory on the segment of the population with a high growth rate will lower plant production because the slower growing segment is left unharvested. Such consumers are the opposite of "prudent predators" (Slobodkin, 1968). An equal biomass of competitively repressed individuals might presumably increase their growth rates, but this cannot be instantaneous. Another potential complication is the separation of tissues with high photosynthetic rate from those with a high growth rate. Nevertheless, the stoichiometry of food selection suggests that herbivores have a greater impact on plant growth rate than would be estimated without consideration of stoichiometry.

Physiological adjustments of consumers include regulation of assimilation and excretion. Both are probably involved in most situations. Phosphorus excretion, for example, varies with food P content in vertebrates (Morse et al., 1992) and invertebrates (Olsen et al., 1986). Adjustments of assimilation result in changes in egested material such as feces whereas adjustments to excretion result in changes in such things as urine, milk, or soluble release products. Thus, there are some potentially important nutrient cycling implications at the level of physiological control of elemental content. If species differ in patterns of nutrient intake and excretion, so too will patterns of nutrient cycling.

Ecosystems present organisms with a spectrum of elemental contents that may be improperly balanced for their requirements. Maintaining elemental balance must have some costs, but what does homeostasis really cost an organism? There must be many ways to measure this cost, ranging from thermodynamics to energetics. Animals consuming and assimilating a diet high in C must respire the excess, leading to an excess of energy compared to growth requirements. Such excess energy can be used for enhancing uptake of limiting elements, but growth efficiency on a high C diet is reduced from a more balanced diet. Complete analysis of the cost of homeostasis is impossible here, but one example will show one less than obvious cost. The chuckwalla, an herbivorous desert lizard, consumes fruits that are so high in potassium (K) that the osmotic "load" this animal receives from ingesting a single fruit must be eliminated before it eats again (Smits, 1985), and detoxification of the K^+ can take more than a day (Smits et al., 1986). Thus, homeostasis of K^+ costs this animal an entire day after ingestion of a single meal. Evidently, maintenance of elemental homeostasis has population-level consequences.

THE POPULATION/ECOSYSTEM INTERFACE: STABILITY

There are many ways that the stoichiometry of species links population and ecosystem ecology. To close this chapter, I will deal with only one: multiple stable states. This example will serve to illustrate the importance of many of the patterns discussed above. In community studies, it is theoretically possible to have more than one set of species densities with locally stable equilibria, which is referred to as a "multiple stable state" (Sutherland, 1974). Initial conditions decide which of these equilibria the community approaches with time. Although easily found in theoretical models, multiple stable states in real communities are controversial (Connell and Sousa, 1983; Sutherland, 1990).

Elsewhere, I (Sterner, 1990) solved for the N/P recycled by a homeostatic herbivore as a function of the N/P of that herbivore and the N/P of its food. In this model, herbivore homeostasis introduced an instability into the interaction between herbivore and primary producer. When the producer had a low N/P ratio, the N/P recycled was still lower, and conversely when the N/P in the producer was high, the N/P recycled was even higher. Limitation of algal growth tended toward pure limitation by N or P, and not toward colimitation by the two simultaneously. Homeostasis of N/P in the herbivore introduced an instability to the herbivore–algae interaction. This is a multiple stable state for the community made up by the herbivore, its food species, and the two nutrients N and P. Species equilibrium was not assumed, but strict homeostasis was.

Experimental demonstration of a multiple stable state generated by stoichiometry comes from a *Daphnia*–algae continuous culture system (Sommer, 1992). A set of algal cultures of different dilution rate, D, were first established. At equilibrium $\mu = D$. At low D, algae had high C/P ratios (>900). At high D, algae had lower C/P ratios. *Daphnia* were then inoculated into all cultures. Culture D was maintained after *Daphnia* inoculation. Two distinctly different trajectories were observed: in cultures of high algal C/P ratios, *Daphnia* did not grow, and algal biomass remained high. However, in cultures of low C/P, *Daphnia* grew and reduced algal biomass by a large amount (Fig. 23-5). The system was very sensitive to small changes in D near the transition value. The initial C/P determined which of two equilibria the system approached. In the high-*Daphnia*–low-algae community, algal C/P ratio remained high, but in the low-*Daphnia*–high-algae community, algal C/P ratio was greatly re-

Figure 23-5. Outcomes of long-term culture study (Sommer, 1992) of *Scenedesmus acutus* (measured as POC) and *Daphnia magna* (plotted as numbers per liter). Two distinct communities, determined by dilution rate D, are in evidence. At $D \geq 0.55$, *Daphnia* reached high density, and algal density was reduced. At $D \leq 0.50$, *Daphnia* density remained low and algal density remained high. Points represent means of last 3 days of measurements.

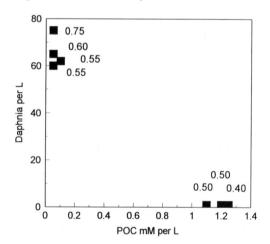

duced. *Daphnia* grazing must have resulted in increased algal growth rates. The first system (high-algae, low-*Daphnia*) was a low turnover system for both algae and consumer and the other system (low-algae, high-*Daphnia*) was a high turnover system for both. A positive feedback chain mediated by stoichiometry resulted in two distinctly different equilibria. At low D, algal C/P was high, which makes them poor-quality food for this herbivore (Sterner, 1993) which has a high P content (see above). In such conditions, the herbivore has low growth, and does not recycle P (Olsen et al., 1986). When grazing rates and P recycling are both low, the result is high algal populations of low growth rate. In contrast, at high D, algal C/P ratio was low, and algal quality as food is higher. Herbivore growth rate and P recycling thus are both higher, creating a high-grazing, high-nutrient cycling environment, which results in low biomass of rapidly growing algae.

Similar feedbacks have been considered in terrestrial plants (see also Wedin, Ch. 24). Hobbie (1992) has contrasted ecosystems of low and high productivity. In low-productivity systems, plants invest heavily in C-based secondary metabolites, which slows herbivory as well as mineralization from litter. Herbivore selectivity acts to remove palatable species that would generate high mineralization rates. Herbivores thus lower mineralization, and productivity remains low. In high-productivity environments, in contrast, grazing increases nutrient availability because plant regrowth contains higher N concentration due to deposition of feces and urine, or due to grazing-induced shifts in plant allocation, leading to lower rates of N immobilization.

For another example of multiple stable states, Wedin and Tilman (1990) have shown that differing species of grasses on initially identical soils result in highly contrasting N mineralization such that species with low tissue N concentration, which are the competitively superior species for N, produce low N mineralization. The presumed stable equilibrium is for the competitively superior plant species to continue to dominate. Species with high tissue N concentration, which are competitively superior species for light, produce high N mineralization, enhancing their chances of competitive success. Thus, litter feedbacks might help promote their dominance in high-productivity environments. The common denominator to stability studies concerning stoichiometry is in the existence of multiple stable states.

CONCLUSIONS AND OUTLOOK

In this chapter, we have seen how elemental contents vary both within and between taxa, but that there are distinct patterns to this variability. When we link species and ecosystems, species stoichiometry comes into play. By incorporating the elemental content of species into community models, we can begin to understand how biogeochemical pools are configured in different food webs. Species stoichiometry introduces dynamics into ecological interactions that are not predicted by other means; for example, herbivore–plant stability is an area where stoichiometry has potentially great influence. One area in need of further investigation is in understanding the biological nature of elemental constraints; knowing these will allow for a greater degree of generalization than might be possible by just accumulating examples. If, as has been argued here, elements involved in protein synthesis are tightly linked in organisms, there are significant biogeochemical implications, and perhaps most excitingly, a link will be drawn, not just between species and ecosystems, but between molecular biology and ecosystems. Further implications would be in the life history strategies of different species, and how this correlates with their role in nutrient processing in ecosystems.

Stoichiometry is not a hypothesis; it is a restatement of the first law of thermodynamics. It is a framework that serves to organize a set of otherwise disparate observations about biological systems. I believe its major contribution will be in suggesting new questions to explore.

ACKNOWLEDGMENTS

I thank an anonymous reviewer for comments. Financial support was provided by NSF Grants BSR-9019722 and DEB-9119781.

24

Species, Nitrogen, and Grassland Dynamics: The Constraints of Stuff

David A. Wedin

SUMMARY

Herbivores, decomposers, and fire are three alternative consumers of primary production in grasslands, each with different requirements for and effects on carbon (C) and nitrogen (N). Differences between plant species in their tissue C and N chemistry can determine which of these three consumer pathways the bulk of primary production will follow in a particular system. This stoichiometric approach is applied to native humid grasslands and their dominant tallgrass species.

INTRODUCTION: THE CONSTRAINTS OF STUFF

Two dominant sets of constraints act on ecological systems: the constraints of evolution and the constraints of, for lack of a better term, stuff. To link species and ecosystems and to integrate the approaches of population/community ecologists and ecosystem ecologists both sets of constraints must be understood and acknowledged (Sterner et al., 1992; Sterner, Ch. 23; DeAngelis, 1992; Ch. 25; Holt, Ch. 26).

The constraints of evolution are phylogeny and natural selection. Because each species has a unique phylogenetic history, only a tiny subset of the possible combinations of physiological and morphological traits currently exist. Given this phylogenetic history, the "choices" open to a species for life history or resource allocation patterns are also limited by natural selection to a reduced number of viable "strategies" (Grime, 1979; Tilman, 1988). A common criticism of ecosystem ecology is that it has generally failed to acknowledge these evolutionary constraints (McIntosh, 1981; Holt, Ch. 26). These critiques often focus on hypothesized emergent properties at the ecosystem level and successional pathways leading to greater ecosystem stability and efficiency, pathways dependent on seemingly altruistic behavior of early successional species (e.g., Odum, 1969). Because of loose evolutionary thinking, or a total lack of it, many evolutionary ecologists have ignored developments in ecosystem ecology .

But organisms are not simply sets of evolved traits. There are constraints associated with the stuff organisms are made of. Organisms are made of organic matter and require reduced C substrates for energy and cell construction. In addition, as stressed by Sterner et al. (1992), and Sterner (Ch. 23), organisms have unique requirements for elements other than C, in other words, unique elemental stoichiometries. Trophic levels, and species within trophic levels, can differ widely in both the elemental contents of their biomass, and in the residence time of various elements in living biomass. The constraints of stuff are the constraints of chemistry, the constraints involved with the transformations of energy and matter, and the constraints that biochemical pathways and biogeochemical cycles place on organisms and ultimately on entire ecosystems. Just as the constraints of evolution have been widely ignored by ecosystem ecologists, the constraints of stuff have been widely ignored by population, community, and evolutionary ecologists.

This chapter discusses the consequences of differences among grass species in their C and N chemistry for major ecosystem processes. Litter quality, forage quality (e.g., percentage crude protein), C/N stoichiometry, tissue C/N ratio, and N-use efficiency are all related ways of expressing the basic C and N chemistry of plants. Species-level differences in tissue chemistry have important consequences for plant community dynamics, plant–herbivore interactions, and the disturbance regimens of grasslands. A species' C/N stoichiometry depends not only on tissue N concentrations, however, but also on patterns of biomass allocation to various tissues (e.g., leaves, stems, and roots) and the lifespans or turnover rates of those tissues. Correlations among these traits across numerous species suggest that evolution has constrained plants to a limited number of viable strategies for dealing with the constraint of nutrient limitation (Grime, 1979; Chapin, 1980, 1987; Tilman, 1988; Tilman and Wedin, 1991).

NITROGEN AND THE DYNAMICS OF
NATIVE HUMID GRASSLANDS

A stereotype of grasses and grasslands is that they are primarily adapted to moisture limitation and high rates of grazing. However, all grasslands are not created equal; major differences in structure and function exist across grassland types. Huntley and Walker (1982) concluded that grasslands and savannas form a continuum from high plant/low animal biomass to low plant/high animal biomass, and that this corresponds to a gradient of increasing soil nutrient availability and decreasing moisture availability. They classified African grasslands and savannas into two types: dystrophic, or nutrient-limited tall grasslands of relatively humid environments, and eutrophic, or moisture-limited short grasslands found in drier environments. This same pattern occurs in native grasslands worldwide. In most humid grasslands, N is the limiting soil nutrient.

Some of the world's most endangered ecosystems are the native humid grasslands. These include the North American tallgrass prairie, the South American pampas, the Russian tallgrass steppe, the South African high veld, and the tussock grasslands of Australia and New Zealand. All of these grasslands have, on average, very low concentrations of available soil N and show large increases in productivity in response to N additions. Dominant grasses in these systems tend to have low tissue N concentrations (i.e., high C/N ratios) in both live and senesced tissues compared to grasses of shorter arid grasslands or managed grasslands. The low tissue N concentrations reflect the species' high N-use efficiency, the ratio of net productivity to N uptake. These dominant tall grasses also maintain large root systems, with over half of annual net primary production allocated belowground, and tend to have a bunch or tussock growth form (Tilman and Wedin, 1991). Many of the native humid grasslands have been largely displaced by woody vegetation or non-native grasses, often Eurasian species, within the last century. For example, overgrazing, anthropogenic N inputs, and the exclusion of fire have all led to the displacement of native tall C_4 grasses by non-native C_3 grasses in North American tallgrass prairie, especially in its eastern distribution (east of the 95th meridian), where it is least moisture-limited.

Systems dominated by strong feedbacks often exhibit highly nonlinear responses once thresholds in the levels of driving variables are crossed (DeAngelis, 1992). I propose that the dynamics of native humid grasslands and savannas, including their widespread and rapid demise, can be understood only when we realize that strong feedbacks involving N link soils, plants, herbivores, and fire. Central to these feedbacks is the C/N ratio chemistry of the dominant grasses of native humid, tallgrass systems (Fig. 24-1). The low tissue N concentrations of these grasses leads to: (1) low forage quality and de-

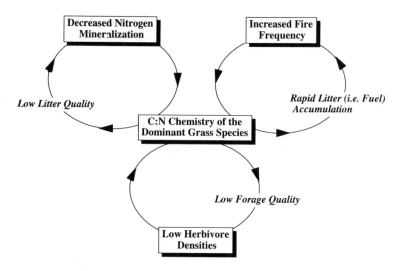

Figure 24-1.   Interacting feedbacks in N-limited native tallgrass ecosystems. See text for details.

creased rates of herbivory, (2) slow litter turnover and increased fuel loads for fires, and (3) high rates of N immobilization in decomposing litter and roots and reduced soil N mineralization and availability. Not only does the N-use physiology of the dominant grasses play a major role in herbivory, fire regimens, and N cycling, these processes interact strongly. Both the complexity of the native humid grasslands and the importance of N were alluded to by Huntley and Walker (1982): "N has been shown in all savannas to be of great significance . . . but despite many thousands of N measurements, in all its forms, an understanding of the N cycle in savanna still eludes us."

The Vegetation–Herbivore Feedback

A common misconception is that the high rates of primary production of native humid grasslands are correlated with high rates of herbivory. In fact, forage quantity and forage quality are strongly inversely correlated in native grasslands as one crosses regional moisture gradients (Bremen and de Wit, 1983). The dominant C_4 grasses of North American tallgrass prairie, African high veld grasslands, Australian tall tussock grasslands, and South American pampas all decrease to 4–5% crude protein (0.6–0.8% N) by mid-growing season and at senescence have dropped to 2–3% crude protein (0.3–0.4% N) (Moore, 1970; Huntley and Walker, 1982; Wedin and Tilman, 1990). Animal scientists give 6% crude protein (0.9% N) as the minimum forage protein content on which ruminant grazers can maintain their body weight (Van Soest, 1982). The high veld grasslands of Africa are called "sour" grasslands because of this unacceptably low forage quality.

The high N-use efficiencies and low tissue protein concentrations of the dominant grasses in these humid grasslands are therefore a major constraint on herbivores. Although this conclusion is best documented for large ruminant grazers, it also appears to hold for other herbivores, including aboveground and belowground invertebrates (Ritchie and Tilman, 1992; Seastedt et al., 1988). Discussing early attempts to graze native Australian tall grasslands, Gardener et al. (1990) concluded that "initially overgrazing of the grasses was not a problem as pasture quality was extremely low in the dry season and cattle tended to die at grazing pressures still safe for the majority of pasture species."

Correlated with the low tissue N concentrations of these grasses is a high concentration of structural compounds that deter grazers, including cellulose, hemicellulose, lignin, and suberin. Ultimately, forage quality is determined by the ratio of structural components such as lignin to metabolic components such as N (Van Soest, 1982). This same ratio is used as an index of litter quality in models of plant decomposition. Although patterns of plant C chemistry are beyond the discussion here, the point is that digestion by a rumen microbial community and decomposition by a soil microbial community are conceptually the same process and are regulated by the same plant chemical parameters (Pastor and Naiman, 1992).

The Vegetation–Fire Feedback

As one moves from semi-arid short grasslands to humid, taller grasslands, fire replaces grazing as the major consumer of aboveground plant biomass in native grasslands: "Fire is an herbivore that does not require protein for growth" (Bell, 1982). Fire is now accepted as critical in determining the distribution of North American tallgrass prairie and other similar systems (Axelrod, 1985; Collins and Wallace, 1990). As discussed in the next section, the high C/N and lignin/N ratios in litter of the dominant grasses in these systems causes slow decomposition, which, together with their high productivity, leads to rapid fuel accumulation. This is half of a positive feedback between these grasses and fire (Fig. 24-1). Fire also favors the dominance of these species. Their phenology, physiology, root-shoot allocation patterns, and competitive strategies are all adapted to fire and to the resulting high light—low N environment (Knapp and Seastedt, 1986; Collins and Wallace, 1990; Tilman and Wedin, 1991; Wedin and Tilman, 1993).

Although both herbivory and fire remove aboveground plant biomass, they have opposite effects on N turnover and availability in grasslands. While grazing accelerates N turnover (McNaughton et al., 1988, Holland et al., 1992), fire volatilizes most of the N in plant litter and reinforces N limitation in the system (Ojima et al., 1994). Thus, the positive feedback between certain groups of grasses and fire as well as fire's direct effect on N cycling are central to understanding the persistence of grasslands in both tropical and temperate humid areas (D'Antonio and Vitousek, 1992).

Litter Chemistry and the Vegetation–Soil N Feedback

The previous sections emphasize the importance of N limitation in native humid grasslands. Our recent studies of long-term grass monocultures suggests that the C/N chemistry of the dominant grasses is both a cause and a consequence of this N-limitation (Wedin and Tilman, 1990; Wedin and Pastor, 1993).

For example, we have followed the C and N dynamics of litterbags containing either aboveground litter or roots of four grass species (D.A. Wedin and J. Pastor, unpublished). *Schizachyrium scoparium* is a dominant C_4 grass of North American tallgrass prairie, whereas *Agropyron repens*, *Poa pratensis*, and *Agrostis scabra* are C_3 species that displace *Schizachyrium* following major soil disturbance (e.g., cultivation) or N fertilization. During the course of decomposition, the broad initial differences among species and tissue types in C/N ratios disappeared (Fig. 24-2). By the time litterbags had reached approximately

Figure 24-2.   Convergence of C/N ratios in decomposing litter and roots for four grass species during a 3-year litterbag study in east-central Minnesota, U.S.A. (Unpublished data of D.A. Wedin and J. Pastor.) Aboveground litter was placed in 1-mm mesh bags on the soil surface, and roots were buried in polyester bags. See Wedin and Tilman (1990) for details of the species and experimental monocultures from which plant materials were taken. Complete species names: *Schizachyrium scoparium, Agropyron repens, Poa pratensis,* and *Agrostis scabra.*

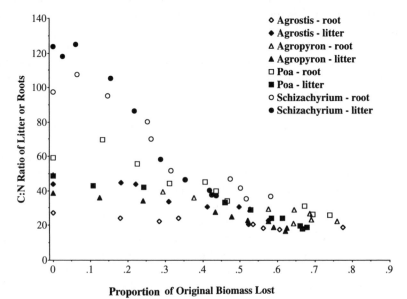

60% mass loss, all tissue types had converged on a similar chemistry with C/N ratios of about 20. Thus, considering the endpoint of grass litter decomposition alone, species differences in litter chemistry appear to be insignificant. Instead, the endpoint of decomposition is presumably regulated by the nutritional requirements and energetics of the microbial decomposer community.

However, the rates of decomposition and N dynamics during decomposition differed sharply among species. Low-quality litter from the prairie grass *Schizachyrium* decomposed slowly (60% still remaining after 2 years) and immobilized large amounts of N. There was no net release of N from *Schizachyrium* within 3 years (see also Pastor et al., 1987). In contrast, higher quality litter from the C_3 grass *Agropyron repens* decomposed faster (40% remaining after 2 years) and showed no net immobilization of N.

The high potential for N immobilization in the litter and root detritus of dominant tallgrass prairie species such as *Schizachyrium* translates into decreased N turnover and reduced N availability in tallgrass prairie. Wedin and Tilman (1990) found that *in situ* N mineralization diverged significantly within 3 years among monocultures of five grass species growing on initially homogeneous soils. Net N mineralization rates decreased sharply in soils under *Schizachyrium* and *Andropogon gerardi*, another tallgrass prairie dominant. In contrast, *Agrostis scabra*, a native early successional C_3 species, had much higher litter quality (C/N ratio = 24 for roots) and significantly increased N mineralization rates. Based on laboratory incubations with soils from these same monocultures, Wedin and Pastor (1993) concluded that the grass species need affect only a small labile pool of soil organic matter (in this case <3% of total soil organic matter) to cause the 10-fold divergence in N mineralization rates observed in the field (Wedin and Tilman, 1990). This suggests that soil N dynamics can rapidly track shifts in the species composition of grassland vegetation, and that feedbacks between plant species and soils may be quite strong (see also Hobbie, 1992).

C/N Stoichiometry and Grassland Consumer–Resource Interactions

The assertion that detrital pathways, trophic dynamics and disturbance regimens interact to determine grassland dynamics suggests that ecologists must dive into a system of overwhelming complexity, especially if one is interested in processes at the species level for one or more trophic levels (McNaughton et al., 1988). Alternatively, a stoichiometric approach to consumer–resource interactions (sensu Sterner, Ch. 23) in grasslands may generate a useful simplification of the system. Herbivores, decomposers, and fire are, in essence, three alternative consumers of primary production in grasslands, each with a unique C/N stoichiometry. The proportion of plant biomass consumed by each varies widely across grassland types and differences among plant species in C/N chemistry may be a major determinant of the relative strengths of these three sinks for primary production.

Herbivores, fire, and microbial decomposers have different minimum N requirements below which consumption of plant biomass will not occur. As alluded to above, fire has no minimum N requirement for its resource (litter). The detrital pathway can also generally consume plant biomass, such as wood, with very low tissue N, often as low as 0.1%. Herbivores, however, have minimum N requirements for their forage, with actual levels depending on body size, physiological status, ruminant vs. nonruminant classification, and other factors. As the mean N concentration of primary production decreases, consumption is therefore shifted to greater detrital and fire consumption.

Secondly, these three consumers differ in the rates at which consumption occurs, organic C–N bonds are broken, and reaction endproducts are released. Fire has the highest rate of consumption and the rate is independent of plant tissue N concentration. Herbivores have the next highest rate of consumption, although the rate at which consumption and digestion occurs may be quite dependent on plant tissue chemistry. The rate at which litter decomposes and mineralizes C and N appears to be highly dependent on tissue chemistry, and is generally quite slow in N-limited systems.

Finally, these three consumers differ dramatically in the final form in which consumed N is released. All three mineralize N in the sense that they break organic C–N bonds and release part or all of the N they consume in nonorganic forms. For both decomposers and herbivores, consumed N in excess of that required for growth is released in forms readily taken up by vegetation (i.e., urea, ammonium, or nitrate). In contrast, fire volatilizes organic N to a poorly defined combination of ammonium, NO_x gases, and particulates. These forms are essentially unavailable, at least in the short term, to terrestrial vegetation.

Thus each of the three consumer–resource interactions discussed here (herbivore–grass, decomposer–grass, and fire–grass) has a unique stoichiometry. Each has different requirements for and effects on C and N. Differences among grass species in C/N ratio can determine both which pathway their above or belowground biomass will take and the rate of consumption and mineralization of N by the consumer. Although this stoichiometric approach ignores vast differences in the ecology of herbivores, fire and decomposers, it may be a powerful simplification based on the underlying chemical and mass balance constraints all three processes face.

CONCLUSIONS

By considering an ecosystem's underlying chemical constraints while resisting the temptation to fragment the study into traditional ecological subdisciplines (e.g., disturbance regimes, trophic dynamics, nutrient cycling, and plant community dynamics), ecologists may find key parameters or processes that can explain much of that system's dynamics. Identifying those key processes also brings clarity to the question "Do species differences matter to overall ecosystem dynamics and stability?" In this case, we can focus on how and why grasses differ in their C and N chemistry.

Because all plants have some plasticity in nutrient and biomass allocation patterns, part of our answer is below the species level. Recent studies have shown that physiological shifts in N-use efficiency and allocation pattern among populations within grass species have led to rapid shifts in soil N availability in response to fire (Ojima et al., 1994) and grazing (Holland et al., 1992). These shifts in N availability in turn affect plant community and trophic dynamics. As stressed by Chapin (1980), plasticity in nutrient use is itself a trait selected for or against under different conditions. More research is needed on inter- vs. intraspecific variation in plant resource use and tissue chemistry. However, it appears that much of the variance is at interspecific or higher levels for natural vegetation. For example, differences in litter chemistry among species are usually larger than differences among individuals or populations within species (e.g., Pastor et al., 1984).

Wedin and Tilman (1990) found that differences in photosynthetic pathway (C_3 vs. C_4) explained much of the divergence in N cycling under the grass species they studied. In most native humid grasslands, the dominant species are C_4 grasses. Because C_4 species have reduced requirements for N-containing photosynthetic enzymes, they generally have lower leaf N concentrations than C_3 species (Field and Mooney, 1986). Thus, the high N-use efficiency of these C_4 grasses is partly explained by their high photosynthetic N-use efficiency (rate of CO_2 uptake per unit leaf N).

The overall C/N stoichiometry or N-use efficiency of a species depends not just on its photosynthetic N-use efficiency, however, but also on its biomass allocation pattern, tissue longevity, and rates of N retranslocation (Aerts, 1989). These traits have evolved under strong selection in plants (Chapin, 1980; Tilman, 1988). For example, among 21 grass species from successional grasslands in Minnesota, all of the early successional species had relatively low litter C/N ratios (i.e., low N-use efficiencies) regardless of photosynthetic pathway (D.A. Wedin, unpublished). Physiological and biochemical constraints alone cannot account for this pattern. There is no constraint of stuff that says old field colonists cannot have high tissue C/N ratios. Rather, evolutionary constraints may dictate an unavoidable tradeoff between the high allocation to reproduction and high growth rates required by successful old field colonists and a high N-use efficiency (Tilman, 1988; Tilman and Wedin, 1991).

An understanding of why grass species differ in C and N chemistry in some cases and not in others requires consideration of both the constraints of evolution and the constraints of biochemistry/biogeochemistry (i.e., stuff). Understanding the consequences of these species differences for grassland dynamics requires a clearer and more integrated picture of the stoichiometry of grass–consumer (herbivore, fire, and decomposer) relations.

ACKNOWLEDGMENTS

This chapter grew out of numerous discussions with collaborators D. Tilman and J. Pastor. Our research was supported by the University of Minnesota and the National Science Foundation (BSR-8811884).

25

Relationships Between the Energetics of Species and Large-Scale Species Richness

Donald L. DeAngelis

SUMMARY

Although energy is frequently mentioned along with a number of other factors affecting ecological communities, it does not occupy the central position that it does in physics or chemistry. In particular, energy concepts at the ecosystem level seem very weakly connected with energy at the species level and with the dynamics of species populations.

This chapter addresses the general question of how important a role energy plays in current ecological thinking. In particular, recent examination of species richness data on a continental scale indicates strong correlations with available energy. An explanation of these patterns is discussed. First, factors such as spatial heterogeneity and disturbances are important at the local level. Second, increased available energy favors the success of specialized strategies. Third, this explanation can be scaled to larger spatial scales to help explain species richness–energy curves.

INTRODUCTION

Energy plays a crucial role in all aspects of ecological research, including the following: biophysical ecology, which is the study of the coupling of individual organisms to their environments through an exchange of energy and matter

(Gates, 1980); food web ecology, where efficiency of energy transmission up the food chain plays a role in food web structure (Lindeman, 1942; Yodzis, 1984); community theory, where relationships between available energy and species richness in ecological communities have been found (Wright, 1983; Currie, 1991); ecosystem theory, where hypotheses for static and dynamic properties of ecosystems such as succession involve energy principles (E.P. Odum, 1969); and evolutionary theory, where the thermodynamics of self-organizing systems can be applied to biological evolution (Morowitz, 1968; Taube, 1985; Brown, Ch. 2).

In physics and chemistry, energy has played a unifying role, explaining most phenomena in terms of energy transformations. In reviewing the early development of energy concepts in ecology, Cherrett (1990) noted that a similar formulation of ecology in energetic terms was sought: "As all life requires energy, it was soon perceived that its supply to, use by, and eventual loss from living systems might provide a unifying resource around which ecosystems could be organized. Perhaps the economy of nature was explicable in terms of this universal currency." Despite this early enthusiasm and despite the fact that ecologists regard energy as a common thread, ecology today is still awaiting a comprehensive theoretical basis to explain patterns at all scales. H.T. Odum (1983) made perhaps the most ambitious attempt to formulate a unified ecological theory based on thermodynamics. However, his systems theory of ecology has not gained broad acceptance across the community of ecologists. Recent critics assessed the prospects of a thermodynamic theory of ecoystems as "dismal" (Mansson and McClade, 1993).

It has been understood for the past 100 years that the flow and degradation of free energy maintains systems far from equilibrium (see also Brown, Ch. 2). Systems as diverse as the convective cycles in the atmosphere and the living systems on the earth, both termed "dissipative systems," are maintained by solar energy. Both matter and energy are integral to these cycles. However, in the convective cycles in the atmosphere energy is fundamental in a way that specific material elements are not. Without energy flow, material cycles could not occur, so it is absolutely indispensable, whereas a wide variety of gaseous elements in various proportions could constitute the atmospheric flow patterns.

The situation is somewhat different in biological and ecological systems. Certain material elements appear to be indispensable to living organisms. Without them, no living system could be maintained, so they are in a sense equivalent in importance to energy. Another difference between physical systems and ecological systems is that the latter grow to bump up against limits on material availability. This does not usually happen in the physical systems that we use to study self-organization. The result is that energy tends to be the limiting factor in the physical systems we are familiar with but chemical elements tend to be limiting or colimiting in ecological systems (Sterner, Ch. 23; Wedin, Ch. 24).

This limitation by material elements (water, carbon, and nutrients) in ecological systems complicates ecological systems and, in particular, makes it very difficult to reduce their dynamics to transformations of a single currency, energy, as is done in physical systems. It is certainly true that energy can be used to "buy" material in many ways (Brown, Ch. 2). With sufficient energy, plants can extend their roots farther to obtain water and nutrients and animals can roam farther to obtain some nutrient that is limiting (e.g., salt for ungulates). However, the "currency exchange rates" between energy and the nutrient can be complex and difficult to calculate. Therefore, a quantitative unifying theory based on thermodynamics may be not be practical in the near future.

Another way of looking for unifying ideas may be to attempt to explain observed phenomenological patterns from a more qualitative perspective. I will give two examples that illustrate this general approach.

OBSERVATIONS REGARDING SPECIES AND ENERGY

Energy and Food Chain Length

Two of the most intriguing problems in ecology involve explaining the number of trophic levels (chains in which species A is eaten by species B is eaten by species C is eaten by species D . . .) in food chains, and the species richness of ecological communities.

A key observation relative to the first problem is that trophic chains are limited in length. The explanation based on energy is that as energy is transferred through each trophic level, only a small fraction is passed on to the next higher level, so that an increasing amount of primary production would be necessary to maintain populations of higher and higher trophic levels (see Lindeman, 1942; Hutchinson, 1959). If such an explanation is valid, then it would indicate that a strong relationship exists between individual energetics and ecosystem energetics, since energy efficiency at the individual species level would limit the trophic structure of ecosystems. Yodzis (1984) reinforced this energetic view by showing from published food webs that a greater proportion of ectothermic than endothermic consumers were supported in these webs. As ectotherms are more efficient at transferring energy than endotherms, this suggests that energy limitation may be important in food chains.

Pimm and Lawton (1977) argued that the energy limitation explanation is insufficient, because they noted that "Primary productivity ranges over at least four orders of magnitude in terrestrial systems, over three in aquatic systems . . . yet the number of trophic levels supported . . . does not seem to be correlated with the productivity of the base of the food chain." They hypothesized instead that the length of food chains is limited by dynamic stability and showed in simple models of species populations in chains that longer chains tended to have longer times of return to steady-state equilibrium following a perturbation than shorter chains. If such long return times posed a threat to long-term survival of populations at the top of the trophic chains, then food chain dynamics rather than energy limitation governs chain length.

In some sense a compromise has emerged in the recognition that energy limitation is not decoupled from food chain stability. It has been shown that the return time to equilibrium is not independent of energy (DeAngelis, 1980). Although this dependence is not enough to affect the patterns Pimm and Lawton (1977) found in commonly studied food webs, it shows that it is not possible to cleanly distinguish the roles of dynamics and energy in food webs. The individual energy efficiencies of organisms in the food web, the energy flow from the base of the food web (an ecosystem-level property), and food chain level dynamic properties all interact to determine food chain length.

Species Richness and Energetics

A central issue in ecology is species richness. Both the reason for differences in species richness and its implications for ecosystem functioning are important. As with trophic chain length, different possible explanations have been hypothesized. Again one can point to phenomenological relationships, such as the so-called species–energy relationship.

Currie (1991) examined the species richness of birds, mammals, amphibians, reptiles, and trees in 336 quadrats covering the North American continent. He found that 80–93% of the variability of the vertebrates could be explained by annual potential evapotranspiration (PET), whereas for trees actual evapotranspiration (AET) was more closely correlated with species richness. Both PET and AET are measures of incident energy per unit area. The tightness of the relationship, showing species richness increasing monotonically with energy, argues for some simple explanation.

One explanation for this pattern was suggested by Connell and Orias (1964), who argued that more productivity should mean a higher total number of individuals of all species. As each species has a higher probability of survival if the numbers of individuals can reach high levels, more species might coexist in systems with higher productivity (Currie, 1991).

This purely energy-based explanation has problems, however. For one thing, the pattern of richness does not always conform to the above pattern in comparable systems. "Under high productivity, as in estuaries and polluted lakes, intense competition by dense populations can keep diversity very low. . . In the cline of diversity on the ocean floor it is the unproductive dark and abyssal depths that have the high diversity, whereas the shallow productive waters have low diversity" (Colinvaux, 1986). Highly productive systems, such as *Spartina* marshes, can have very low species richness (Currie, 1991). Also, Currie points out that there are many data to indicate that species richness peaks as a function of increasing resources rather than continuing to increase. It is apparent that energy availability alone cannot be a total explanation for species richness. Other aspects of the natural history of species must play a role.

Energy at the Individual Level

Can a theory of large-scale patterns of species richness vs. energy in the biosphere be built starting from energetics at the level of individuals? Biophysical ecology (e.g., Gates, 1980) often focuses on the short-term fluxes of energy between organisms and their environment. A broader perspective is necessary for analyzing the origins of such relationships as the species–energy curve, since the persistence of a species is a function of its whole life cycle, not just energy balances at a specific time.

The perspective of the whole life cycle was stated by Solbrig (1981): "The thesis for this book is that organisms can be viewed as systems for the capture, transformation, and transmission of energy, and that survivorship of a lineage depends on the effectiveness with which organisms process energy." An example would be a tree's temporal strategy of allocation of energy to growth and seed production over its life cycle. The study and comparison of life cycles of individuals in given environments may tell us whether a few generalist or many specialist strategies will be most successful over a number of generations.

Wiegert (1988) argued that for living organisms energy is the ultimate limiting factor. This argument requires some detailed explanation, because it is clear that water, nutrients, and various special habitat requirements such as nesting sites are often the proximal limiting factors. However, over long evolutionary time scales, an increase in available energy will open the possibility for organisms to develop mechanisms to conserve or more efficiently acquire any of the other limiting resources. As noted earlier, it may be very difficult to determine the currency exchange ratios between energy and nutrients, but it seems like a safe assumption that additional energy may increase the number of feasible strategies for acquiring limiting nutrients.

Consider the following scenario. In a small lake a single fish species exists, which feeds on both pelagic and benthic prey through its whole life cycle. Because it feeds on both, the fish cannot be optimally adapted to either type of foraging and thus it forages less efficiently than would two fish species, each specialized on one type of prey. There is a cost of specialization, however, because the available food in the pelagic region and the benthic region taken individually may be too limited to support a large enough population for a species to be viable for long. It is possible that there is a threshold in available energy in the lake below which only the one generalist species can persist, but above which two species can persist.

This type of argument is fairly standard in trying to explain species richness, particularly the coexistence of several species in a feeding guild, but I think the opportunity now exists for analyzing this argument in great detail through computer simulations of individual-based models. Either sympatric speciation or invasion and subsequent changes in niche widths may be simulated by modeling many individuals with the genetic variability for a rich repertoire of possible physiological and behavioral attributes in a complex model environment. A simple model of this type has been used by Rice (1984) to examine the possibility of sympatric speciation from a single species along an environmental gradient. This approach could be extended so that several species might be modeled simultaneously, each with a different strategy of exploiting its environment, some being more specialized than others. The success of individuals surviving and passing on their genes would depend on how well they exploit their resources and allocate energy to reproduction. Simulating the system over many generations may lead to a trend toward maintenance of either a diverse community of specialists or only a few generalists, depending in part on the amount of energy available.

Carrying out the modeling program outlined above would require a great deal of detailed information on energetic costs of various strategies of food acquisition and other behavioral and physiological adaptations. In the near future, biophysical ecologists may be able to provide such information only for limited situations. In the absence of such models I will speculate below (on admittedly somewhat uncertain arguments) what might be found.

RELATING INDIVIDUAL SPECIES ENERGETICS TO LARGE-SCALE PATTERNS

Any explanation for species richness will hinge mainly on explaining under what circumstances many specialist life cycle strategies will be superior to a few generalist strategies. The superiority of some strategies over others for a given environmental situation can be determined only by detailed physiological studies of individual organisms under those situations. If a given life cycle strategy can lead to a favorable energy balance for growth and survival to reproduction, then it is a feasible one for survival in the environment.

Before detailed investigations are done, we cannot easily say much more. However, I believe that certain generalizations will emerge. In particular:

1. Some heterogeneity must be present in the environment for specialized strategies to have any possible advantage. This heterogeneity may come from structural heterogeneity, such as topographical and edaphic variation, or may result from a disturbance regimen that continually breaks up the biotic cover of the landscape.

2. A sufficient free energy supply must be available for specialists to persist in this heterogeneous environment. Otherwise only a generalist could persist.

Both the "intermediate disturbance" hypothesis (Connell, 1978; Huston, 1979) and the hypothesis of spatial heterogeneity of limiting nutrients (Tilman, 1982) offer reasonable sources of heterogeneity at the local scale, which can be shown experimentally and through modeling to allow competitors to coexist. I will focus on disturbances, which have the effect of creating spatial heterogeneity. Consider the graph in Fig. 25-1a, in which the frequency or some other measure of the efficacy of disturbances is plotted along the x-axis. The number of coexisting species are plotted on the y-axis in this hypothetical example.

When there are no disturbances, one species of competitor, the species best able to compete under stable conditions for the limiting factor (light or a nutrient), eliminates all the others. In the opposite extreme, when the rate of disturbance is high (e.g., gaps are being opened up), at most only the species that has the highest rate of growth (ability to recover quickly from disturbances) persists. In the intermediate cases a large number of species may coexist.

What does this have to do with the species–energy curve? Let us assume now that there is an increase in available energy. This has the effect of shifting the curve in Fig. 25-1a to the right (see Fig. 25-1b). The reason is that an increase in the energy allows faster growth rates, which means that competitive displacement can occur faster. Therefore a higher disturbance rate is necessary to produce the same species richness as a lower one. It is probably also true that the higher amount of energy allows a higher peak number of species, because there is more energy available to support specialist strategies that would be excluded at lower amounts of energy. We can continue to increase the level of energy, which yields a series of curves as in Fig. 25-1b.

It is true that in most actual experiments of this type, some limiting nutrient is used rather than energy and that in particular experiments energy may not always produce this pattern, as it is not limiting the growth rates. However, my argument is similar to Wiegert's (1988) mentioned earlier that in a broad sense more energy permits more strategies to exist for acquiring nutrients and using them to accumulate biomass faster.

Now let us see what patterns exist for a fixed disturbance frequency and varying levels of energy. If one increases the energy along the three dashed lines shown in Fig. 25-1b, one obtains the three curves shown in Fig. 25-1c.

The above conceptual results are probably valid for small systems. At least they are testable by means of experiment. However, the species–energy curves are derived from observations across large geographical scales. Can the above results be scaled up to large systems? I can only speculate here. Some things should change as we increase spatial scale; the time scale over which one species might competitively displace less competitively competent species should increase and the time scale over which a disturbance regimen can eliminate species susceptible to disturbance should increase. The mere fact of larger scale means that at any given time individuals of even the most endangered

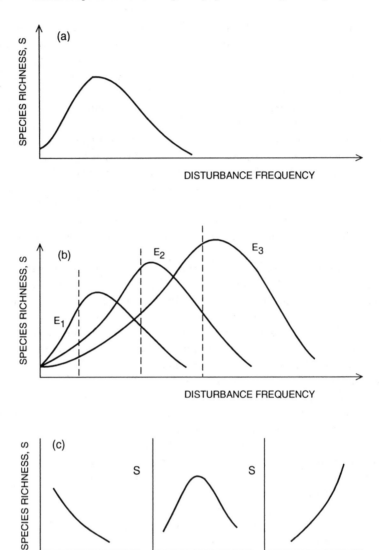

Figure 25-1. (a) Species richness tends to be a maximum at intermediate disturbance frequencies in the environment. (b) Increases in available energy to the ecosystem shift the species vs. disturbance frequency curve to the right, because the competitively dominant species can grow faster and exclude other species at a faster rate. (c) If the disturbance frequency is held constant and energy, E, increased ($E_1 < E_2 < E_3$), three types of patterns of species richness vs. available energy can result.

species are likely to persist somewhere. Therefore, the effect of scale is to change the peaked pattern of the intermediate disturbance curves into more of a plateau shape (Fig. 25-2a). As a consequence, comparisons of different energy levels on a species richness vs. disruption plot for large spatial scale

Figure 25-2.   (a) When the species richness vs. disturbance frequency curve is scaled up from small to large spatial areas, it becomes more plateau-like, because both competitive exclusion and disturbances are less effective at eliminating species at larger spatial scales. (b and c) For a large spatial scale, when available energy, E, is increased the species richness increases monotonically. (The y-axis represents much higher species number than the y-axis in Fig. 25-1.)

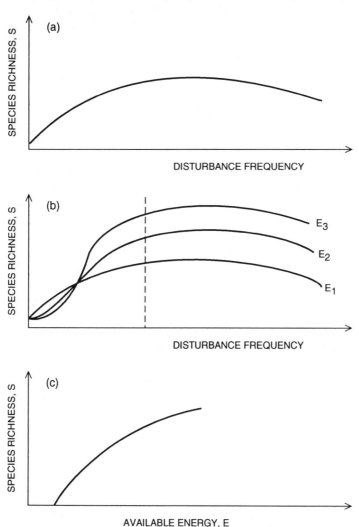

take on the appearance shown in Fig. 25-2b, with the result that we obtain the characteristic species–energy curve shown in Fig. 25-2c.

The above explanation of biotic diversity does not follow from available energy alone, but must rely on dynamics operating at the small scale as well. Factors or mechanisms must be invoked that give specialized strategies a possible advantage. Spatial heterogeneity is one such factor and it can often be achieved through disturbances.

Many disturbances, such as lightning strikes, are exogenous. Others are endogenous, in that they result naturally from the dynamics of the species present. The effects of both have been discussed under the general concept of the mosaic-cycle concept (e.g., Whittaker and Levin, 1977; Remmert, 1991). In nearly every type of ecological system disturbances, including inherent instabilities due to the natural dynamics of the system, create a mosaic pattern where different mosaic patches are the temporary habitat of different species.

CONCLUSIONS

The above explanation starts with species level energetics and with the local-scale dynamics of disturbances and attempts to extrapolate to species patterns on the continental scale. This obviously involves a lot of hand waving and is meant not as a theory but as a perspective from which one might try to launch more specific studies. One thing that this approach must do is account for the apparent exceptions to the general pattern, such as species-rich but energy-poor ocean bottom areas and energy-rich but species-poor *Spartina* marshes. There may be enough leeway in the scheme outlined earlier to do this. For example, the ocean bottom is characterized by a very low disturbance rate, which is consistent with high species richness when the energy input is low (Fig. 25-2b). If a *Spartina* marsh can be thought of as having a low disturbance frequency, a high energy input would imply low species richness (Fig. 25-2b).

This chapter has presented my own highly speculative thoughts based on the more solid work of others. These thoughts lead me to believe that energy at the species and ecosystem level plays a major role in large-scale ecological patterns, but that other dynamic factors such as disturbances are also crucial. Due to limitations of space I have ignored many counterarguments, but I hope that this work provokes a greater interest in searching for general theory that can link all fields of ecology and all spatial scales.

ACKNOWLEDGMENTS

The author thanks M.A. Huston for his useful comments. This research was sponsored by the National Science Foundation's Ecosystem Studies Program under Interagency Agreement No. 40-689-78 with the U.S. Department of Energy under contract DE-AC05-84OR21400 with Martin Marietta Energy Systems, Inc. This is Publication No. 4249 from the Environmental Sciences Division.

26

LINKING SPECIES AND ECOSYSTEMS: WHERE'S DARWIN?

Robert D. Holt

SUMMARY

Ecosystem ecology and evolutionary biology have traditionally been largely separate intellectual endeavors. Yet the organismal traits mechanistically responsible for many ecosystem processes result from evolutionary dynamics; those dynamics, in turn, are constrained by ecosystem processes. An understanding of evolution could enrich ecosystem studies in several ways: (1) it can provide adaptive interpretations of resource acquisition and utilization; (2) it can show how contingency and rarity limit predictability.

INTRODUCTION

Dobzhansky (1973) once famously quipped: "Nothing in biology makes sense except in the light of evolution." It is fair to say that not all ecosystem ecologists agree. Allen and Hoekstra (1992) state that though "ecosystems depend on evolved entities . . . evolution is only tenuously connected to ecosystems." Higashi and Burns (1991) likewise comment, "ecosystem ecology has virtually ignored evolutionary considerations" (see also Grimm, Ch. 1). Indeed, as a skeptical referee of an earlier version of this chapter argued, an understanding of evolution gives no "greater insight [into ecosystems] than knowing what species are present, and what their demographic and ecosystem parameters are."

Given the few, scattered attempts to integrate evolution into ecosystem theory to date, it seems a little premature to assert that evolution should be ignored in ecosystem studies. Other authors have reflected on the need for a cross-fertilization of disciplinary perspectives. For instance, Loehle and Pechmann (1988) outlined several potential implications of evolution for ecosystem theory.

Here I revisit several of the themes of Loehle and Pechmann, and discuss several suggestive examples in which a consideration of evolutionary biology provides essential insights into an ecosystem pattern. I conclude by assessing limitations in evolutionary biology that may in some times and places circumscribe its utility in ecosystem science.

SOME BASICS

It is useful first to remind ourselves of some basic definitions. Ecosystem ecology is the study of fluxes and pools of energy and matter of all sorts (Waring, 1989; Allen and Hoekstra, 1992). Evolutionary biology in its broadest sense is the study of the origins and maintenance of biological diversity, both within lineages (microevolution, e.g., adaptation by natural selection, genetic drift), and among lineages (macroevolution, e.g., speciation, extinction, adaptive diversification, and biogeographical spread). The maintenance of phenotypic variation rests on the within-lineage processes of population genetics (e.g., sex), and the between-lineage processes of community ecology (e.g., mechanisms of local and regional coexistence).

To a reductionist, the biotic component of an ecosystem consists of individuals, which abstractly are arenas for flux and pooling in energy and materials—the fundamental resources for all life. A substantial part [though not all; see Lawton and Jones (Ch. 14) on ecological engineering] of individual organisms' roles in ecosystems is determined by their strategies for resource acquisition, retention (including the avoidance of predation, parasitism, and abiotic mortality agents), and allocation.

But resource strategies are the fruit of evolutionary processes. In some circumstances, particular species have a large effect on ecosystem function (e.g., Huntley, Ch. 8; Pollock et al., Ch. 12; Estes, Ch. 15; Wedin, Ch. 24; D'Antonio and Vitousek, 1992; Carpenter et al., 1993). I argue below that a microevolutionary perspective might sharpen our understanding of the factors determining these species' ecosystem effects.

In other circumstances, individual species effects are harder to discern (e.g., Lawton, 1990a; Lawton and Brown, 1993; Holland, Ch. 13). Micro-evolutionary dynamics may still be pertinent, if suites of species are collectively important and respond similarly in their phenotypic evolution to a common selective regimen (i.e., convergence) (Frost et al., Ch. 22). More generally, one must consider the full distribution among species of those characters pertinent to ecosystem dynamics. The origin of this full phenotypic distribution is explained by a blend of microevolution and macroevolutionary processes.

The traditional separation of ecosystem studies and evolution belies a fundamental dialectic: evolutionary dynamics occur within constraints set by ecosystem processes; ecosystem functions are mediated by individuals, whose traits are due to evolutionary processes. For instance, the relative selective advantage of different resource uptake strategies may be set in part by resource renewal rates (Tilman, 1988) or the total resource pool available (Holt et al., 1994); conversely, the influence of an organism on nutrient dynamics may be determined by the adaptive balance of nutrient leakage vs. storage. Articulating the reciprocal relations of evolutionary and ecosystem dynamics is a very large, and largely uncharted, piece of intellectual territory.

COMPARATIVE ECOSYSTEM STUDIES

Waring (1989) recently stressed the need for a "longer historical perspective and . . . broader geographic scale" in ecosystem studies. Evolutionary perspectives are particularly pertinent when addressing broad, comparative questions (e.g., Holling, 1992).

A tantalizing hint that evolution influences ecosystem fluxes comes from a recent worldwide comparative survey by Milchunas and Lauenroth (1993) on the effect of grazing on aboveground net primary production (ANPP). After ranking the length of the evolutionary association of large grazers and plants, these authors concluded that the percent differences in ANPP between grazed and ungrazed sites declined substantially with an increasingly long shared evolutionary history (outweighing several other factors such as short-term grazing intensity). By contrast, the effect of grazing on plant species composition increased with evolutionary history.

If one knew all the functional relationships and parameter values for species at these sites, one could quantitatively describe the ecosystem effect of grazing without explicit reference to evolution. Such a description would seem to wilfully ignore an important organizing principle that in this case simplifies our understanding of ecosystem patterns: namely, the ecosystem role of large grazers systematically varies as a function of evolutionary history. An ecosystem theory incorporating evolution (via a subtheory for the coevolutionary trajectory of plant–herbivore interactions, as outlined in Milchunas et al., 1988) could be employed outside the domain of nonevolutionary theories, say to predict responses to novel environments.

MODELS OF ADAPTATION AS TOOLS IN ECOSYSTEM SCIENCE

A commonplace of evolutionary biology is that species' traits make *adaptive sense* (Williams, 1992). A detailed understanding of the adaptive nature of resource strategies (acquisition, retention, and allocation)—as governed by constraints, and played out in phylogeny—is a natural bridge linking the perspectives of evolutionary theory and ecosystem science (Loehle and Pechmann, 1988).

Species assemblages in similar physical environments, but on different continents, usually have quite different phylogenetic roots. If these ecosystems have similar structural and functional features, this similarity is likely to be due to evolutionary convergence onto comparable phenotypes from disparate ancestral phenotypes. Such convergence reflects natural selection due to commonalities in both trade-offs and the selective environment.

Tilman (1988) has argued that explicitly considering mechanistic trade-offs provides a powerful tool for interpreting interspecific interactions in communities. Such trade-offs may also be significant in determining the functional role of organisms in ecosystems (e.g., via resource uptake and retention rates).

Analyses of such trade-offs have great promise as an avenue for linking ecosystem, population, and evolutionary perspectives. Consider, for instance, the familiar correlation between vegetation defined in terms of plant life forms and climate (Colinvaux, 1993). Ecophysiological models currently provide excellent descriptors of the climatic ranges of major vegetation types (Woodward, 1987). Such models, I suggest, implicitly reflect the outcome of convergent plant evolution. Biome-level convergent evolution requires a perspective that considers biotic evolution in the context of functional constraints (O'Neill et al., 1986). It is useful to examine one biome descriptor in more detail: vegetation height.

One pattern conspicuous in forested biomes is that at any site canopy height is relatively uniform (in flat terrain), but this average vegetation height varies systematically along major physical gradients (Whittaker, 1973). Because similar patterns arise on different continents with phylogenetically distinct floras, this appears to be convergent evolution at the scale of entire biomes. Aboveground biomass increases with canopy height. Moreover, plant size is a major determinant of plant effects on ecosystems through its control of energy exchange, material fluxes, and responses to disturbance (Chapin, 1993). A quantitative explanation of regional and global variation in vegetation height thus has major implications for terrestrial ecosystem science.

A first step toward such an explanation has been provided by King (1990; see also Givnish, 1988; Friend, 1993), who used a game-theoretical model to predict (with some success) tree height in even-aged monospecific forests. The heart of the model is the trade-off between the competitive advantage of height in competition for light, and the costs of building and maintaining higher stems. One prediction directly pertinent to ecosystem ecology is that the evolutionarily stable optimal height is not the height that maximizes the collective biomass production of a stand. I consider this model to be a nice example of how a consideration of evolutionary dynamics can be used to sharpen our understanding of ecosystem structure and function.

Comparable theoretical insights could be sought whenever a single, focal species (or suite of similar species) has a substantial impact on ecosystem processes via the use, retention, or allocation of resources. For instance, there is increasing evidence that plant species can exert a strong effect on nutrient cycling [see, e.g., Canham and Pacala (Ch. 9); Wedin (Ch. 24)]. Plants in low-nutrient environments allocate disproportionately to roots, have relative low growth rates, use nutrients efficiently, and have high carbon (C)/nitrogen (N) ratios (Hobbie, 1992). Tissues are well protected by secondary metabolites from herbivory, reducing energy flow to higher trophic levels; as an incidental byproduct, these compounds are antimicrobial agents, so litter decomposes gradually. These plant attributes generate positive feedback, accentuating soil nutrient poverty (Hobbie, 1992). The plants occupying these habitats have diverse phylogenetic origins, so this adaptive syndrome once again is an expression of convergent evolution (combined with species sorting). It would be useful to have an evolutionary strategy model (e.g., building on Tilman's ALLOCATE model, Tilman, 1988), comparable in spirit to the plant height model of King (1990), in order to predict quantitatively the combination of plant traits likely to prevail in a given environment. Such a model, to account for existing data, would have to include feedback effects via ecosystem processes.

The ecosystem role of some dominant species may also make them significant components of the selective environment faced by other species. The long-term effect of a focal species [with or without feedback on itself—Lawton and Jones (Ch. 14)] on an ecosystem may be modulated by the evolutionary responses of other species.

This may not seem important if one focuses on a single system over short periods of time, but could be crucial in understanding differences among systems, or one system over long time spans. James Estes (pers. comm.; P.D. Steinberg, J.A. Estes, and F.C. Winter, unpublished) has described a fascinating, plausible example. Sea otter predation is an important determinant of the structure of nearshore marine communities in the North Pacific. Where sea otters are abundant, their preferred prey (invertebrate herbivores, e.g., sea urchins) are scarce, allowing the establishment of luxurious algal beds. Detrital flows from these beds help sustain rich offshore fish populations.

The abiotic environment of the North Pacific is replicated in the South Pacific (e.g., Chilean coasts). But sea otters (and any comparable species) are absent. Herbivore pressure is demonstrably higher in the South Pacific than in the North. This contrast in ecosystems has likely existed as long as sea otters have occupied the North Pacific [one reasonable guess is 10 million years (R. Hoffman, pers. comm.)], generating comparable differences in the magnitude of selection on seaweeds for mechanisms to reduce herbivory by invertebrate grazers.

Estes and his co-workers measured the levels of secondary defensive compounds in algae and found the levels to be much greater in the South Pacific than in the North. He suggests this is an evolutionary response by the algal community to sustained differences in herbivory, and that this evolutionary difference could have substantial consequences for ecosystem processes. In terrestrial ecosystems, as noted above, enhanced allocation of plant resources to secondary compounds can reduce primary productivity; moreover, plant compounds that reduce digestibility often deter decomposition, reducing flows through detrital food chains (Chapin, 1993). One might predict that otherwise similar ecosystems in the South and North Pacific might vary in the same direction, with lower productivity and reduced energy flows via detrital fluxes to offshore environments in the former.

Data are not yet available to test these predictions. But this scenario provides an example of how a consideration of evolutionary biology (namely, cascading evolutionary effects due to the presence or absence of a single, dominant species) can lead to testable ecosystem-level hypotheses.

LIMITS TO THE USE OF EVOLUTIONARY
BIOLOGY IN ECOSYSTEM SCIENCE

I do not want to leave the reader with the impression that evolutionary processes are a necessary ingredient in all ecosystem studies. There are some clear circumstances when evolutionary perspectives would not appear to be very useful. For instance, if a community were comprised of species assembled hodge-podge, without prior contact (as on some islands heavily disturbed by humans), a purely phenetic approach would seem to suffice. As a second example, if an organism is an ecological engineer (sensu Lawton and Jones, Ch. 14) but experiences little or no feedback from its effects on the ecosystem to its own fitness, there would be a decoupling of ecosystem effect from organismal fitness (in contrast to, say, nutrient uptake).

Moreover, there are limitations to our current understanding of evolutionary mechanisms. In particular, even if one subscribes wholeheartedly to the adaptationist program, one cannot ignore the fact that evolution is also highly *contingent*: evolution works in a blindly local sense with the materials at hand (Dawkins, 1987), constrained by a phylogenetic history that itself recursively reflects the past contingency of evolutionary dynamics (for different perspectives on evolutionary contingency, see, e.g., Ulanowicz, 1986; Kauffman, 1993; Brown, 1994b). If particular species play a dominant role in some ecosystem function, the vagaries of dispersal histories will usually restrict these species (except *Homo sapiens*, alas) to particular geographical regions. This introduces a substantial historical contingency into ecosystem processes.

The theme of contingency deserves a much fuller treatment.

THE IMPORTANCE OF RARITY

An important feature of both evolutionary dynamics and community processes is that mean system states may be rather poor predictors of long-term responses to change; long-term responses often involve the magnification of the frequency or abundance of initially rare, seemingly unimportant phenotypes, alleles, or species (S. Pacala, pers. comm.). For instance, Carpenter et al. (1993) note that "surprises are common in whole-lake experiments" because of the proliferation of previously rare or unknown species (see also Frost et al., Ch. 22). An important consequence of evolutionary processes for ecosystems is that evolution sets the bounds of variation, that is, the range of extreme phenotypes available both within- and among-species, and it is these extremes that, in the end, determine the long-term response of the system to an altered environment.

CONCLUSIONS

A deeper understanding of the linkage between species and ecosystems in the future will require recognition that both species and ecosystems have histories, and that these histories reflect a mixture of predictable results from general laws, and the idiosyncratic results of accidents—evolutionary contingencies. It is this blend of order and chance that makes the study of life such an endlessly satisfying endeavor. In this noble enterprise, evolutionary biology and ecosystem science should be mutually reinforcing partners.

ACKNOWLEDGMENTS

I would like to thank Bob Sterner and Mike Vanni for an invigorating conversation held at 9,000 ft. in the Rockies, and an anonymous reviewer for helpful (even if annoying) comments.

27

Ecological Flow Chains and Ecological Systems: Concepts for Linking Species and Ecosystem Perspectives

Moshe Shachak and Clive G. Jones

SUMMARY

One way to link population/community ecology with ecosystem ecology is to interconnect flows of organismal abundance with flows of materials. Using an example of isopod population dynamics, soil erosion, and hydrology in the Negev Desert, Israel, we show how disparate flows of different currencies can be functionally connected. We use two concepts, Ecological Flow Chains and Ecological Systems, to generate a question-driven, multiflow, multicurrency, multiscalar representation of relationships among isopod, soil, and water flows. We provide general definitions and criteria for these two concepts, so that they can be used inductively or deductively to link many different flows of nature.

UNDERSTANDING THE FLOWS OF NATURE

All ecological processes are spatially and temporally dynamic and can be thought of as flows of different currencies (e.g., solar energy, nitrogen, genes, organismal density, species diversity). Tansley's (1935) "basic units of nature" contain many flows that interact (e.g., materials that affect organismal abundance and vice versa). Some ecologists focus on a particular flow of a particular currency (e.g., information, organisms, energy, materials, or structure). This "uniflow–unicurrency" approach leads to detailed understanding of a particular flow that can be readily measured in nature, but tells us little about how flows interact. Other ecologists focus on entire ecosystems (i.e., networks of flows) using a single, universal currency (see Brown, Ch. 2) such as information, materials, or energy (see DeAngelis, Ch. 25). This "multiflow–unicurrency" approach leads to general understanding of the entire ecosystem at the expense of detailed information about particular flows that can be readily measured in nature.

In order to link population/community ecology with ecosystem ecology we need to combine the advantages of both "uniflow–unicurrency" and "multiflow–unicurrency" approaches retaining detailed information about specific, measurable flows of different currencies (e.g., population dynamics vs. materials cycles) while also tractably linking many flows of different currencies (e.g., population dynamics with material flows). Our approach uses two general concepts—**Ecological Flow Chains** and **Ecological Systems**—to generate question-driven representations of nature that have multiflow, multicurrency, and multiscalar properties. Here, we use an example from the Negev Desert, Israel to show how functional interrelationships among a flow of organisms and flows of materials can be inductively described, explained, and scaled using ecological flow chains (EFCs) that are connected to produce an ecological system (ES). We then develop formal definitions of EFCs and ESs, and briefly outline the principles for inductively and deductively applying these concepts.

INDUCTIVE UNDERSTANDING OF SOME FLOWS OF NATURE: AN EXAMPLE

Relationships Among Isopods, Water, and Soil

Studies in the Negev Desert Highlands, Israel (1972–1993) had the goal of understanding interactions among desert isopod population dynamics and hydrology and soil processes. *Hemilepistus reamuri*, the desert isopod, is a detritivorous arthropod with an annual life cycle (Shachak et al., 1979; Linsenmair, 1984). Isopods live as a monogamous family of parents and offspring in a single soil burrow (Shachak, 1980; Linsenmair, 1984). In February, adult isopods (2 cm long, 200 mg FW) leave the burrow in which they hatched and developed, seeking new sites in which to settle (Shachak and Brand, 1991). Isopods emigrate distances of up to 1 km (Shachak and Newton, 1985). About 10% of the emigrating females are successful in locating a new site for a burrow (Shachak and Yair, 1984). Females select a new site in a patch of soil, digging a shallow burrow (ca. 5 cm depth) and accepting a male after a brief courtship (Shachak et al., 1976). Females are gravid by April, and 80–120 offspring hatch in May (Shachak and Newton, 1985; Shachak and Brand, 1988). Parents and young live in the burrow until the following year, communally digging down to the relatively high soil moisture (ca. 6–10% by weight) that persists throughout the 7-month dry season (May–November) at depths of 50–70 cm (Yair and Shachak, 1982, 1987). Only those families that dug burrows at sites with high soil moisture content survive (about 50% of the families of settling females, i.e., 5% of the total emigrating females) (Shachak and Brand, 1988).

The moisture regimen in a small patch of soil (ca. 50 cm diameter) is determined by the amount of rainfall falling on the patch and infiltrating the soil *and* the amount of runoff water entering and infiltrating the same soil patch (Karnieli, 1982). Average annual rainfall during the 5-month wet season (December–April) in this region of the Negev is 90 mm with high interannual variance (CV 40%) (Sharon, 1980). Direct precipitation within a year is usually insufficient to generate the high soil moisture levels required by isopods (Karnieli, 1982). However, this limestone desert highland (200–450 m above sea level) has watersheds (0.02–0.2 km^2) with rock slabs at the top, rocks with soil patches in the midslope, and mostly soil at the bottom. Runoff water from the adjacent upslope rocks that then infiltrates the soil patches is the critical, major source of soil moisture (Yair, 1985). The ratio of local rock area to soil area determines the amount of runoff. At low rock-to-soil ratios (< 0.2) soil moisture content immediately following a high rainfall event (ca. 25 mm) is low (ca. 20% by weight) because water primarily comes from direct rainfall. At high rock-to-soil ratios (1–5) soil moisture content from the same rain event is much higher (up to 37% by weight), with the runoff contributing as much as two to three times more water to a soil patch than direct rainfall (Yair and Danin, 1980; Olsvig-Whittaker et al., 1983). Soil moisture from runoff can persist in the soil for up to 3 years, and the accumulation and persistence of soil moisture from rain (persists for up to 1 year) and runoff are critical in determining isopod settling and establishment (Shachak and Yair, 1984).

The rock-to-soil ratio at a given locale changes over time and is largely determined by local soil erosion (Yair and Shachak, 1987). Undisturbed surface soil is relatively nonerodible because it is colonized by a microphytic crust (cyanobacteria, algae, mosses, and lichens) (Evenari et al., 1982). Isopods are engineers (sensu Lawton and Jones, Ch. 14; Jones et al., 1994). During feeding and burrowing they consume large quantities of soil. The fecal pellets, which look like tiny bricks (3 mm^3), are composed almost entirely of soil particles. Feces are removed from the burrow every morning and are deposited in a circle around the burrow entrance. By October the large pile of feces (ca. 250 g per burrow representing an excavated volume of ca. 250–350 cm^3) shows that a family has successfully survived (Shachak and Brand, 1988). The accumulation of fecal pellets on the soil surface is the major source of erodible soil in a soil patch. During 1973–1990 erodible soil production by *H. reamuri* on a rocky slope in the Negev averaged 170 ± 109 kg ha^{-1} year^{-1}, which represented about 60% of the total erodible soil production in the watershed (Yair and Shachak, 1987).

The fecal pellets on the soil surface are eroded by overland runoff that occurs when runoff exceeds soil infiltration capacity during periodic rainfall events of higher magnitude (Yair and Shachak, 1987). These runoff events occur about 10 times per year on average in the wet season (max. 23 events), and carry the soil from areas with a high rock-to-soil ratio to areas with a low rock-to-soil ratio, that is, from the upper slope to the midslope and from the midslope to the wadi. Runoff is the most important factor determining soil erosion at these scales (Yair et al., 1978). The feces discarded by the isopods are readily disintegrated under the direct impact of raindrops and runoff, and the resulting erodible soil particles are transported by the overland flow of runoff water (Yair et al., 1978).

Isopod formation of erodible soil plays a major role in determining the local rock-to-soil ratio by facilitating soil erosion. By increasing or maintaining a high rock-to-soil ratio the isopods increase the relative contribution of runoff as a soil moisture source, increasing soil moisture content. Because soil moisture content determines site suitability for isopods and isopod density, there is a feedback to isopods and their soil engineering activities. Ecological questions about a particular flow, that is, isopod population dynamics or water or soil flows, or about the interrelationships between population dynamics and material fluxes, can both be addressed from a functional understanding of these interconnections.

Representing and Connecting the Flows of Isopods, Water, and Soil

We can treat isopod population dynamics, hydrology, and soil erosion as flows. Each flow can be represented by an **Ecological Flow Chain** (EFC) (Fig. 27-1). Each flow chain functionally *describes* the flow of one of the three currencies of interest as a connected series of *organizational state changes*. The Isopod Flow Chain (IFC) consists of changes in abundance of potential settlers to settlers and settlers to successful families. Changes from organizational state to organizational state (i.e., potential settlers to settlers, settlers to successful families) describe the flow of isopod numbers, that is, their population dynamics. The Water Flow Chain (WFC) consists of changes in the amount of rainfall to runoff, rainfall to soil moisture, and runoff to soil moisture that describe this particular hydrologic flow. The Soil Flow Chain (SFC) consists of changes in the amount of nonerodible soil to erodible soil and erodible soil to eroded soil that describe this particular soil flow. Each of these EFCs has dynamic behavior (i.e., changes in the organizational states) that are, in part, intrinsic to the flow chain. Thus changes in the number of potential settlers influences the number of settlers which, in turn, influences the number of successful families; erodible soil is formed from nonerodible soil and is necessary for soil erosion; rain is necessary for runoff to occur; and both rain and runoff determine soil moisture content.

The organizational state changes within each EFC are not sufficient to functionally *explain* the flow of interest. Functional explanation requires interconnections among EFCs that represent the *control by an organizational state in one flow chain on an organizational state change in another flow chain*. In this situation there are six major interconnections (Fig. 27-1): (1) soil moisture control on the flow of numbers of potential settlers to settlers (WFC to IFC); (2) soil moisture control on the flow of numbers of settlers to successful families (WFC to IFC); (3) settler control on the flow of nonerodible to erodible soil (IFC to SFC); (4) successful family control on the flow of nonerodible soil to erodible soil (IFC to SFC); (5) runoff control on the flow of erodible soil to eroded soil (WFC to SFC); and (6) eroded soil control on the flow of rain to runoff via modification of the rock-to-soil ratio (SFC to WFC). This control then changes the amounts of rain becoming soil moisture and the amount of runoff becoming soil moisture, and the amount of soil that can hold moisture from either source.

We call these flow chains, together with their controlling interconnections, an **Ecological System** (ES) (Fig. 27-1). The ES is sufficient to both functionally describe and explain this particular set of relationships between isopod population dynamics and hydrologic and soil processes. In effect, we have combined the advantages of the "uniflow–unicurrency" and "multiflow–unicurrency" approaches into a "multiflow–multicurrency" ES. The ES retains the essential details about a flow of one measurable currency (e.g., the flow of numbers of isopods), with its major controls (e.g., soil moisture) while at the same time tractably linking different flows of very different, but measurable currencies (i.e., a flow of numbers with a flow of materials). Such an ES can then be used to understand the effects of organismal population dynamics on material fluxes and vice versa. However, in order to operationally connect population dynamics with material fluxes, we need to include information on the spatial and temporal scales at which the different types of flows represented by the EFCs operate and interconnect in the ES.

The Scales of the Flow Chains and the Ecological System

Isopod, water, and soil flow chains represent different types of flows operating across different spatial and temporal boundaries. Scalar properties can be added to the functionally descriptive properties of each flow chain by including the spatial and temporal extents at which each of the component organizational state changes within an EFC takes place. Scalar boundaries of each EFC are therefore determined by the collective boundaries of the component organizational state changes (Tab. 27-1).

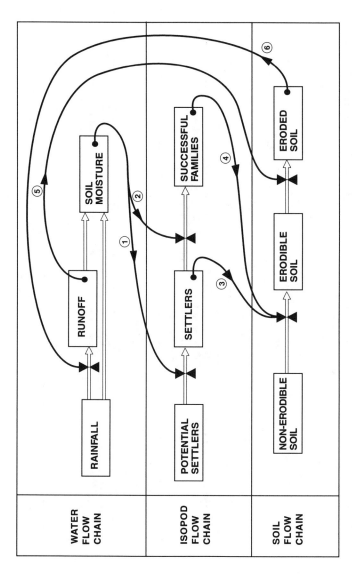

Figure 27-1. The Ecological System representing relationships among isopod population dynamics, soil erosion, and hydrology in the Negev Desert, Israel. Isopod population dynamics (a flow of numbers of organisms), soil erosion, and hydrology (material flows) are each represented as **Ecological Flow Chains** consisting of organizational state changes (open arrows, e.g., rainfall to runoff). The organizational state changes within a flow chain (three changes for water, two for isopods, two for soil) are collectively sufficient to describe the flow of a particular currency (amounts of water, numbers of isopods, amounts of soil). Explaining the flows requires controlling interconnections among flow chains. Here an organizational state within one flow chain controls an organizational state change in a different flow chain (shown as solid arrows with regulatory control points, ►◄). There are six such connections in this ecological system (see text).

Table 27-1. Spatial and temporal scale ranges of each organizational state change within the water, isopod, and soil flow chains in Fig. 27-1; and the rationales determining these scales

Ecological Flow Chain	Organizational State Change	Spatial Scale Range and Rationale	Temporal Scale Range and Rationale
Water flow chain	Rainfall to runoff	*Minimum:* Runoff generation from a patch of soil and its adjacent rock, about 5 m radius	*Minimum:* Runoff event, **a few minutes from a short, low intensity rain event**
		Maximum: Runoff generation over an entire watershed, **0.2 km – 0.2 km²**	*Maximum:* Runoff event, a few hours from a long, high-intensity rain event
	Runoff to soil moisture	*Minimum:* Soil moisture in a soil patch, **a few 100 cm² to a few m²**	*Minimum:* Infiltration of runoff into soil, **a few minutes to a few hours**
		Maximum: Soil moisture in the soil portion of an entire watershed, up to 50% of the watershed area, 0.01–0.1 km²	*Maximum:* Persistence of soil moisture from runoff, **about 2–3 years**
	Rainfall to soil moisture	*Minimum:* Soil moisture in a soil patch, **a few 100 cm² to a few m²**	*Minimum:* Infiltration of rain into soil, **a few minutes to a few hours**
		Maximum: Soil moisture in the soil portion of an entire watershed, up to 50% of the watershed area, 0.01–0.1 km²	*Maximum:* Persistence of soil moisture from rain, **about a year**
Isopod flow chain	Potential settlers to settlers	*Minimum:* Emigration distance, a few m radius	*Minimum:* Emigrating and settling, **a few days**
		Maximum: Emigration distance, **1 km radius**	*Maximum:* Emigrating and settling, a month
	Settlers to successful families	*Minimum:* Family unit of the population in burrow, **0.75 cm radius**	*Minimum:* Raising a family, 6 months
		Maximum: Maximum emigration distance is the population boundary, 1 km radius	*Maximum:* Raising a family, **8 months**
Soil flow chain	Nonerodible soil to erodible soil	*Minimum:* Soil generated from burrow and surrounding feces by settlers and families, about **15 cm radius**	*Minimum:* Erodible soil formation from 1 day's digging by a family, or about 1 month's digging by settlers
		Maximum: Soil generated in the soil portion of an entire watershed, up to 50% watershed area, 0.01–0.1 km²	*Maximum:* Erodible soil formation from digging by a family during the **7-month dry season when there is no runoff**
	Erodible soil to eroded soil	*Minimum:* Erosion in a soil patch, a few 100 cm² to a few m²	*Minimum:* Runoff erosion of soil, a few minutes to a few hours during runoff events
		Maximum: Erosion in the entire watershed, **0.02 km² – 0.2 km²**	*Maximum:* Accumulation of erodible soil during the **7-month dry season between runoff events**

The collective scale boundaries of each flow chain are set by the minimum and maximum spatial and temporal values found across all component state changes, within that flow chain, and are shown in bold. Data on scale ranges come from studies referenced in the text.

Since the functioning of the entire ES is dependent on the connections among EFCs (as well as the EFCs themselves), the scales of these connections must also be identified. There are six connections among the three EFCs (Fig. 27-1), each of which has scalar properties that are determined by the spatial and temporal scales at which an organizational state on one EFC controls an organizational state change on another EFC (Tab. 27-2). The scalar properties of the entire ES are therefore the collective scale boundaries set by both the EFCs themselves *and* their interconnections.

Depicting Scalar Properties of the Ecological System

The spatial and temporal scales of the ES derived from the information in Tabs. 27-1 and 27-2 are shown in Figs. 27-2 and 27-3, respectively. Each organizational state change within an EFC, the boundaries of each EFC, and each connection between an organizational state on one flow chain that controls an organizational state change on a different flow chain, operates at distinctive and different scales. The ES is therefore **multiscalar**.

The maximum spatial scale of the WFC and SFC is the watershed (based on our question of interest we excluded the flow of water and soil out of a watershed via the wadi from consideration). However, the maximum spatial scale of the IFC encompasses a number of watersheds because of isopod emigration and immigration (Fig. 27-2). Because the WFC controls the IFC, water flow events that occur in one watershed can therefore affect isopod flows in another watershed(s) (via emigration). Because the IFC controls a critical portion of the SFC (erodible soil production) isopod flows in many watersheds can affect soil flows in a particular watershed (via immigration). Because the SFC feeds back onto the IFC via the WFC, and the IFC operates across a number of watersheds, both the soil and water flows in one watershed can indirectly influence soil and water flows in other watersheds, via isopod flows. Furthermore, the controls from one flow chain to another overlap more or less continuously across extensive spatial scale ranges. Thus large-scale events can influence small-scale events (e.g., runoff control on settlers to successful families) and small-scale events can influence large-scale events (e.g., settler control on nonerodible to erodible soil).

The maximum time scale of the IFC (which more or less corresponds to the annual life cycle) is slightly longer than that of the SFC. Both the IFC and SFC have a shorter maximum time scale than the WFC (Fig. 27-3). Consequently, the longest term fluctuations in isopod and soil flows will tend to operate over a shorter time frame than the longest term fluctuations in water flows. However, in comparison with the spatial ranges, the temporal scales of controls among flow chains are relatively discontinuous, with a more limited degree of temporal overlap (cf. Fig. 27-2). Consequently, longer term changes in water flow affect shorter term changes in isopod flows; water flow affects soil ero-

Table 27-2. Spatial and temporal scale ranges at which an organizational state within one flow chain controls an organizational state change in another flow chain, for the six controlling interconnections among flow chains (see Fig. 27-1 and text); and the rationales determining these scales

Controls Among Ecological Flow Chains	Spatial Scale Range and Rationale	Temporal Scale Range and Rationale
1. Soil moisture control on potential settlers to settlers	*Minimum*: Soil moisture at a settling site in a patch of soil, a few 100 cm² to a few m². *Maximum*: Soil moisture within the maximum emigration distance of potential settlers, 1 km radius	*Minimum*: Soil moisture accumulation sufficient to make a site for settling, 3 months of wet season in a wet year. *Maximum*: Soil moisture accumulation sufficient to make a site for settling, 5 months of wet season in a dry year
2. Soil moisture control on settlers to successful families	*Minimum*: Soil moisture at a burrow site in a patch of soil, a few 100 cm² to a few m². *Maximum*: Soil moisture within the population boundary set by the maximum emigration distance of settlers, 1 km radius	*Minimum*: Soil moisture accumulation sufficient to raise a family, a wet season of 3–5 months in wet years. *Maximum*: Soil moisture accumulation sufficient to raise a family, three wet seasons of 9–15 months in dry years
3. Settler control on nonerodible soil to erodible soil	*Minimum*: Settler production of erodible soil at settler burrow and adjacent feces from digging, 2 cm radius. *Maximum*: Settler production of erodible soil at settler burrows within the maximum emigration distance of settlers, 1 km radius	*Minimum*: Settler production of a reasonable amount of erodible soil, about 15 days. *Maximum*: Settler erodible soil production while they are settlers, about 1 month
4. Successful family control on nonerodible soil to erodible soil.	*Minimum*: Family production of erodible soil at family burrow and adjacent feces from digging, 15 cm radius. *Maximum*: Family production of erodible soil at family burrows within the population boundary set by the maximum emigration distance, 1 km radius	*Minimum*: Family erodible soil production from 1 day's digging. *Maximum*: Family erodible soil production during duration of family digging, about 7 months
5. Runoff control on erodible to eroded soil	*Minimum*: Runoff into a soil patch with erodible soil from a soil patch and its adjacent rocks, about 5 m radius. *Maximum*: Runoff carrying eroded soil in the entire watershed, 0.02–0.2 km²	*Minimum*: Runoff event, a few minutes to a few hours. *Maximum*: Interval of about 7 months between runoff events in the dry season
6. Eroded soil control on rainfall to runoff	*Minimum*: Erosion conversion of rock to soil ratio at a soil patch and its adjacent rock, about 5 m radius. *Maximum*: Erosion conversion of rock to soil ratio in entire watershed, 0.02–0.2 km²	*Minimum*: Erosion conversion of rock-to-soil ratio, about 1 year, for high ratios, steep slopes, and shallow soil. *Maximum*: Erosion conversion of rock-to-soil ratio, about 10 years, for low ratios, shallow slopes, and deep soil

Data on scale ranges come from studies referenced in the text.

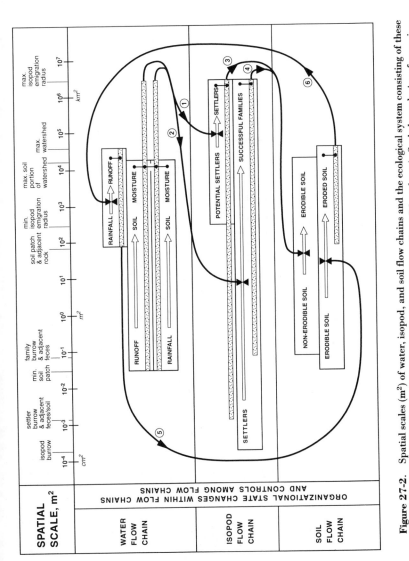

Figure 27-2. Spatial scales (m²) of water, isopod, and soil flow chains and the ecological system consisting of these three flow chains and their six controlling interconnections (1–6; see Fig. 27-1 and text). Scale boundaries of organizational state changes within flow chains are shown as clear boxes containing open arrows that connect one organizational state to another (e.g., rainfall to runoff). Stippled bars represent the scale boundaries of the controls by an organizational state in one flow chain on an organizational state change in another flow chain. The symbol ●— connected to the stippled bar identifies the controlling organizational state. The solid lines and arrows, numbered 1–6, are the controlling connections, and the symbol ➤ denotes the organizational state change in another flow chain that is being controlled.

289

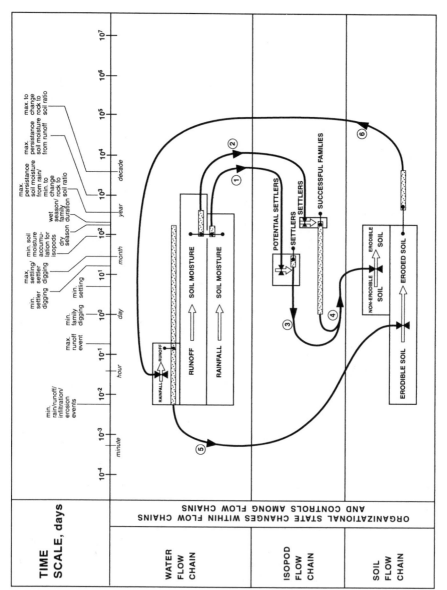

Figure 27-3. Temporal scales (days) of water, isopod, and soil flow chains and the ecological system consisting of these three flow chains and their six controlling interconnections (1–6; see Fig. 27-1 and text). For explanation of conventions and symbols see Fig. 27-2.

sion in the SFC over the same range of time scales; isopod flows affect erodible soil production in the SFC over the same range of time scales; and long-term changes in soil flow affect water flow over much shorter time scales.

It is clear from the multiscalar nature of the ES that a complete, functional understanding of these flows and their interactions cannot be achieved by studies at one, or even a few, spatial and temporal scales. However, the scalar properties of the ES can be explicitly defined and made operational in the field (see Tabs. 27-1 and 27-2).

UNDERSTANDING OTHER FLOWS OF NATURE

The multiflow, multicurrency, multiscalar ES connecting isopod, water, and soil flows (Figs. 27-1 to 27-3) illustrates the application and utility of the EFC and ES concepts for combining key attributes of the "uniflow–unicurrency" and "multiflow–unicurrency" approaches. The ES is functionally descriptive and explanatory. All of the components, flows, and scales of interaction are expressed in real, operational terms that can be directly measured in the field. Consequently, it is possible to develop qualitative and quantitative models of these relationships (see Shachak et al., 1994 for an example).

The isopod, water, and soil flow ES is just one set of interrelationships among flows of nature. However, this example indicates that the EFC and ES concepts should be broadly applicable to any set of functionally interconnected flows of nature. Furthermore, we speculate that the utility of the EFC and ES concepts can go beyond their value in representing a specific example of nature's complexity. The ES example we used has an underlying structure and properties that beg for comparison with other ESs. Such comparisons could provide generalizations about the structural and functional properties of ecological systems. Consequently, in the remainder of this chapter we provide general definitions, criteria, and guidelines for using the EFC and ES concepts to understand other flows of nature.

Definition and Criteria for Ecological Flow Chains

An Ecological Flow Chain is a series of organizational states connecting a measurable flow of materials, energy, structure, information, numbers, or any currency of interest. Changes in organizational states along a flow chain are collectively sufficient to functionally describe the flow of the currency, and are ascertained by measuring specified properties of defined biotic and/or abiotic entities that represent these organizational states. The boundaries of an EFC are determined by the collective spatial and temporal ranges over which the component organizational state changes take place.

The following criteria are used to construct an EFC, based on a question of interest: (1) the general type of flow (e.g., material, energy, structure, information, number of organisms or species). (2) The ecological criterion (sensu Allen and Hoekstra, 1992) that the flow represents (e.g., population = organism flow; community = species flow; ecosystem = material or energy flow; landscape = patch flow). (3) The specific measurable currency of the flow chain (e.g., numbers of a particular organism, number of species, amounts of a particular form of material [nitrogen (N), NH_3^+, soil, water, etc.], energy [solar, metabolizable, kinetic, etc.], information [genes, chemical signals, etc.], or structure [patch number, size, shape, etc.]). The specific measurable currency can be in an aggregated form (e.g., total N or species diversity or soil containing biotic and minerals) or not aggregated (e.g., a particular form of N, or developmental stage of an organism). (4) The type and number of organizational state changes that are necessary to functionally describe the flow (e.g., two in the IFC and SFC, three in the WFC). (5) The specified properties of the defined entities that must be measured to describe each organizational state change (e.g., the number of potential settlers to the number of settlers, the amount of erodible soil to the amount of eroded soil). The entities can be in aggregated form (e.g., soil is a biotic/abiotic aggregate; a successful family is an aggregate of individuals) or not aggregated (e.g., potential settlers). (6) The spatial and temporal boundaries of operation of each organizational state change and hence the collective boundaries of the entire flow chain (see Tab. 27-1).

In principle an EFC can be constructed for any flow. Although we have provided general examples of criteria 1–3, general examples of criteria 4–6 cannot be given without recourse to a specific question about a particular flow (see the three EFCs in our example).

Definition and Criteria for an Ecological System

An Ecological System is a set of at least two Ecological Flow Chains and their controlling interconnections that functionally describe and explain the flow of a measurable currency along at least one of the component EFCs—the focal flow(s) of interest that is selected based on the question being asked. Interconnections among EFCs are necessary to explain the flows. An interconnection represents the control by an organizational state on one flow chain on an organizational state change on a different flow chain. The ES has multiple flows, multiple currencies, and multiscalar properties that are determined by the collective scalar boundaries of each component EFC and the scales at which the controlling interconnections operate.

The following criteria are used to construct an ES, based on a question about the focal flow(s) of interest: (1) the number and types of EFCs necessary to explain the focal flow(s) (e.g., two other flow chains to explain isopod flows); (2) the number of interconnections among flow chains (e.g., six in our example); (3) the specific control points of the interconnections among flow chains (e.g., the organizational state of eroded soil controls the organizational state change from rainfall to runoff); and (4) the specific spatial and temporal scales at which interconnections operate (see Tab. 27-2). In principle an ES can be constructed for any set of interconnected flows of nature, but the ES structure and operation will always be determined by the question(s) asked. In our example, the ES structure was determined by asking how isopod population dynamics affected soil and water material flows and vice versa.

Deductive vs. Inductive Ecological Flow Chains and Ecological Systems

Our example shows that a substantial amount of information is needed to understand the interaction of flows. Most of the required information was at hand before we constructed the ES (i.e., an inductive example). However, the EFC and ES concepts can be used deductively. For example, we could have started by building an SFC- and WFC-based ES that had no IFC. By attempting to explain the SFC solely in terms of the WFC, it would become readily apparent that any agency markedly affecting the production of erodible soil would have to be included (in this case the IFC). Deductive construction of EFCs and their interconnections into an ES, using the criteria, can progressively reveal where information is missing and generates further questions that necessitate development of the ES.

Changes in Ecological System Structure

The structure of the ES entirely depends on the questions being addressed. Any change in focus requires changes in the ES. For example, if we were interested in the flow of soil out of the Negev, the WFC would have to include flash floods and water in wadis (temporary rivers) as organizational states. An interconnection between the WFC and SFC that represents the effects of flash floods moving soil down to the wadis, drainage basins, and the sea (with their spatial boundaries) would be needed. A temporal scale that included time periods with flash floods would be required, and so on.

CONCLUSIONS

Tansley's (1935) "basic units of nature" are complex, dynamic, interacting flows of organisms, materials, energy, information, and structure operating across multiple scales. Gaining tractable, operational, functional, and explanatory understanding of both component flows and entire units of nature is central to goals of enhancing subdisciplinary integration, advancing ecology and its application. Concepts that recognize and effectively deal with the multiflow, multicurrency, multiscalar properties of these units of nature can facilitate these goals. The Ecological Flow Chain and Ecological System concepts can be used to enhance concrete understanding of nature's complexity. The example we chose of interrelationships among isopod, water, and soil flows in the Negev illustrates the approach within the context of linking population and ecosystem ecology (i.e., organismal and material flows). However, we propose that these concepts can be used to understand interrelationships among many different flows linking many subdisciplines in ecology. We hope that our exemplification of the use of these concepts, together with their general definitions, criteria, and guidelines, will encourage ecologists to use these concepts to enhance ecological integration and understanding.

ACKNOWLEDGMENTS

We thank numerous colleagues who have contributed data on isopods, water, and soil; Sharon Okada for help with the graphics; and the Mary Flagler Cary Charitable Trust for financial support. This is a contribution to the program of the Institute of Ecosystem Studies. This article is Publication #179 from the Mitrani Center for Desert Ecology.

CONTEXT

As the previous 27 chapters make abundantly clear, integrating species and ecosystem perspectives presents some exciting intellectual challenges; but it also has profound implications for the interface between ecological science and society. The last four chapters explore some of these issues, in particular the conduct of our science in the discipline of ecology, the education of the next generation of ecologists, and the pressing needs for integration of these perspectives to solve real-world problems.

Biologists charged with the management of environmental resources cannot afford to make fine academic distinctions between populations and ecosystems. Whether we are concerned about impacts of global environmental change (Lubchenco, Ch. 28), lake acidification (Schindler, Ch. 30), or the conservation of spotted owls in the old-growth forests of the Pacific Northwest (Franklin, Ch. 31), we have to be concerned with populations and ecosystems. Nature does not come in two neat box files, one labelled "population problems," the other "ecosystem problems."

Another artificial barrier falls when we integrate population and ecosystem perspectives. There is no distinction between "pure" ecology and "applied" ecology. We need the best science we can muster to solve pressing, real problems; and those same problems pose fundamental challenges, as well as vital insights for "basic" science as is pointed out by Lubchenco (Ch. 28), Schindler (Ch. 30), and Franklin (Ch. 31). The implications for our scientific institutions are profound.

Finally, as educators, we cannot afford to teach another generation of young ecologists that there is population biology and there is ecosystem science and never the twain shall meet. As Slobodkin (Ch. 29) says, whatever we teach students we should not teach them to be like ourselves, or simply to know what we know.

28

THE RELEVANCE OF ECOLOGY: THE SOCIETAL CONTEXT AND DISCIPLINARY IMPLICATIONS OF LINKAGES ACROSS LEVELS OF ECOLOGICAL ORGANIZATION

Jane Lubchenco

SUMMARY

Understanding the linkages between species and ecosystems is of considerable intellectual interest but also has direct relevance to the solution of many environmental problems. Studies of these linkages reflect myriad approaches, no one of which appears to offer a grand synthesis. Rather, the healthy diversity of approaches reflects the complexity of the systems and the variety of interests of investigators. New insights to be gained from linking across levels of ecological organization are urgently needed to help address a wide variety of environmental problems. However, there is a concomitant need to better incorporate what is already known about these linkages into policy and management practices. The next generation of ecologists and ecological societies must have skills, knowledge, attitudes, and structures to meet these needs.

INTRODUCTION

The many excellent chapters in this volume address conceptual frameworks and pragmatic vehicles for linking together population and community ecology with ecosystem ecology. A focus on this linkage is timely and important.

Linking across levels of ecological organization (genetic, individual, population, community, ecosystem, landscape, global) is a formidable challenge. Linkages across these levels are as important as, but arguably more difficult than (though not entirely divorced from), linkages across spatial and temporal scales, both of which have received considerable attention (Steele, 1978; Levin, 1992). Linkages across these three axes (levels, space, and time) require innovative experiments, new theory, long-term data, and creative approaches (Hairston, 1989; Likens, 1989; Real and Brown, 1991; Kareiva et al., 1993).

The varied contributions in this book focus on significant ecological questions about the interface between population/community–ecosystem levels and their linkages. This chapter provides a brief synthesis of these contributions, attempts to set them in a broader context, and considers some of their implications. This contribution notes that changes in society and the environment affect the questions ecologists ask; suggests that these questions, and specifically those at the interface between population/community and ecosystem ecology, are best addressed by a pluralistic approach; and considers how these questions in turn affect the way our discipline is organized and how we teach and communicate the products of our scholarship (Fig. 28-1).

THE CONTEXT OF OUR SCIENCE

As Schneider (1989) points out, we are changing our world faster than our ability to understand the changes. The loss of biodiversity, species extinction, climate change, habitat destruction and alteration, deforestation, desertification, acid precipitation, soil degradation, pollution, eutrophication, and stratospheric ozone depletion are all extremely serious problems that together

Figure 28-1. Some of the relationships between society, ecological research topics, and the communication, dissemination, and utilization of the products of ecological scholarship.

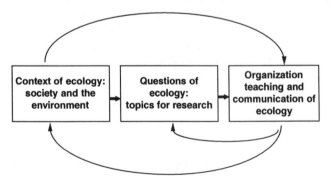

threaten the sustainability of life on earth. The most serious of these are probably those alterations that are irreversible, or involve long time lags or large spatial scales (U.S. EPA, 1990; Vitousek and Lubchenco, 1994). These changes are a direct consequence of the increasing number of people on earth and of the ways and rates at which we use resources and energy. The rates of human population growth and resource and energy use are increasing so fast that the urgency of the resulting problems cannot be understated.

Though clearly interdisciplinary (Myers, 1993), many of these problems are essentially ecological in nature. Therefore, most ecologists both understand the seriousness of the issues and are motivated to help identify and implement solutions. This recognition and motivation for action is exemplified by the Ecological Society of America's Sustainable Biosphere Initiative (SBI; Lubchenco et al., 1991). The SBI set an important precedent in asserting that ecological research priorities must be firmly embedded in the usefulness of the research to solutions of environmental problems. Simultaneously, the SBI articulates the centrality and relevance of basic ecological research to the solutions of critical environmental problems.

The SBI points out that research linking different levels of ecological organization, especially the population/community–ecosystem interface, is particularly important in addressing a wide variety of environmental problems. In a section entitled, "Cross-Cutting Issues in Ecology," the SBI specifies a few general issues that are both common to many specific ecological questions and are "critical to elucidating ecological processes and enhancing the usefulness of ecology for solving practical environmental problems." First among these cross-cutting issues is "Interactions among levels of ecological organization." Specifically, ". . . how phenomena at one level are related to processes operating at other levels . . . must be considered in most ecological investigations. For example, responses at the level of the population, community, and ecosystem must be related to processes at the level of the individual organism, the level at which natural selection acts." (ibid)

Complementing the increased motivation of ecologists to help address environmental problems are increased opportunities for ecologists to contribute to the development of solutions. These increased opportunities result from numerous interacting trends: a growing environmental awareness and ecological literacy among the general public, a rise in the frequency of national and international governmental bodies seeking scientific input, increased sophistication of science journalists, and an improving track record of ecologists providing timely and helpful information (National Research Council, 1990, 1991, 1992; Edmondson, 1991). Thus, the societal and environmental context of our science is changing substantially as the seriousness of environmental problems increases and becomes more widespread. The still quite substantial gap between science and policy, however, presents critical challenges that must be met.

OUR SCIENCE: CHANGES IN THE QUESTIONS WE ASK

These societal and environmental changes have precipitated changes in the kinds of questions ecologists ask. As Jerry Franklin has noted, "The real world has become the cutting edge of our science" (see Ch. 31). The formerly sharp distinctions between "basic" and "applied" ecology are becoming blurred. Many more ecologists than ever before are motivated to address questions that relate to environmental problems.

Concomitant with these changes is the increased realization that much of what was formerly considered esoteric, irrelevant, or basic research is in fact much more relevant than was formerly appreciated. A few examples illustrate this point. Knowledge of plankton population dynamics, species interactions, and nutrient dynamics allowed effective restoration of Lake Washington (Edmondson, 1991). Models derived in part from host–parasite population dynamics shed useful light on human epidemics such as AIDS (Anderson, 1989; Hassell and May, 1989). Understanding the interactions between predator behavior, dispersal ability, and environmental heterogeneity has profound implications for biological control (Kareiva and Odell, 1987). Factors affecting rates of evolution are immediately relevant to evaluating biotic consequences of climate change (Lynch and Lande, 1993). (See Lubchenco et al., 1991 or issues of *Ecological Applications* and *Journal of Applied Ecology* for other examples.) Thus, questions that formerly were pursued for academic interests have now proved to also be important in providing information for managing the environment and its resources.

Increasingly, though, ecologists are motivated to help obtain better answers about certain very specific environmental issues. The need to make policy or management decisions often forces scientists to address difficult problems that might not have been pursued if the need for useful answers didn't force the question. For example, the need to consider ecological consequences of climate change has stimulated a plethora of investigations, most of which require crossing levels of ecological organization (Rosswall et al., 1988; Schlesinger, 1991; Kareiva et al., 1993; Mooney et al., 1993). We can no longer afford the luxury of ignoring particularly difficult problems, if they involve issues that must be addressed.

Some of my recent work, in collaboration with Brian Tissot and John Steinbeck (in preparation), wrestles with linkages across levels of ecological organization and was motivated by the desire to provide better guidance to policy makers. We have focused on the ecological impacts of thermal discharge from the Diablo Canyon Power Plant (operated by Pacific Gas and Electric Company) along the coast of California to understand the possible consequences of increased water temperatures that might be expected under global warming. Our analyses draw on the extensive and well executed sampling conducted for the plant's Thermal Effects Monitoring Program. Based on the knowledge of thermal tolerances of a large number of individual species of seaweeds, invertebrates, and vertebrates, we attempted to predict what changes would occur in the intertidal and shallow subtidal communities of the impacted cove. Our predictions were extrapolations from the physiological level to the community level.

Although the analyses are not yet complete, many of the general trends are obvious. Knowledge of physiological responses of the individual species was insufficient to predict the changes that occurred. Many species known to be tolerant of warmer waters were expected to increase in abundance, but did not do so. Biotic interactions such as competition, predation, and disease had substantial impacts on many of these species, in some cases swamping the direct physiological responses. Thus, it was impossible to predict the community-level responses by simply understanding and summing up the individual responses at the species level.

These results have direct implications for many of the predictions about biotic consequences of climate change, many of which focus solely on direct physiological responses of individual organisms to changes in a single environmental parameter such as temperature or a combination of factors such as temperature and soil moisture. The Diablo Canyon results suggest that one must incorporate multiple levels of ecological organization to understand the changes. Trophic linkages, interaction strength, and engineering links are essential components of the changes. The model we are developing to understand the Diablo Canyon system integrates across physiological, population, community, and ecosystem levels, and in particular synthesizes physiological responses and biotic interactions. Some of the changes observed in Diablo Cove could not have been predicted even with our integrated approach. We fully expect that surprises will be a common feature of perturbed complex systems, even with approaches that integrate across levels.

OUR SCIENCE: CONCLUSIONS FROM A FOCUS ON THE SPECIES–ECOSYSTEM LINK

An understanding of the interface between population/community ecology and ecosystem ecology is particularly relevant to not only the advancement of the discipline of ecology but also to most of the environmental problems facing humanity. As is abundantly clear from this volume, many ecologists have been working at this interface, some for quite some time, but many more now than ever before. What general results have emerged from this body of work?

it contributors to this volume have posed this question in different
s there a general conceptual framework? Are there integrating prin-
niversal currency, useful conversion factors, or general guidelines
for integrating species and ecosystems? Where or when can we aggregate information?

A number of intriguing hypotheses are forwarded in this volume. Brown
(Ch. 2) suggests thinking of organisms and species as complex adapted systems
(CASs) and using reproductive power as a currency; Lawton and Jones (Ch.
14) propose focusing on nontrophic interactions in which organisms modulate
the availability of resources through physical state changes (engineering) as a
conceptual framework; Grimm (Ch. 1) posits thinking in terms of integrating
themes such as patchiness. Rastetter and Shaver (Ch. 21) urges us to think
about the concept of equal limitation; Sterner (Ch. 23) proffers a stoichiometric approach, whereas de Angelis (Ch. 25) focuses on energy, and Shachak
and Jones (Ch. 27) hypothesize ecological flow chains and multiflow, multicurrency, multiscalar ecological systems as a useful approach. This wide range
of ideas in addition to others posited earlier (e.g., Paine's interaction strength
concept;, 1980) represent a wealth of ideas but hardly any consensus.

No single approach or integrating theme or currency emerges as the only
or even the best way to link species and ecosystems. There is of yet no grand
synthesis, no Grand Unifying Theory (or GUT, to borrow a term from physicists) at this interface. Although many physicists hope that a GUT might be
developed that would integrate relativistic theory with quantum theory to give
a coherent picture of the physical world, an ecological GUT appears even more
elusive. [However, as biologists are well aware, a gut is hardly a requirement
for evolutionary success. Entire phyla of metazoa, even some with large bodied individuals (Porifera, Vestimentifera), lack a gut. Other kingdoms lack
guts altogether.]

Regardless of the lack of a GUT, the pluralism of approaches demonstrated
in this volume is healthy. Some approaches are more appropriate or useful for
some questions or systems than for others. Thus, the lack of a GUT should
not be seen as a failure of the science, but rather a recognition of the richness,
the complexity, and the variety of the questions, systems, and taxa involved.
We should revel in and relish this richness.

IMPLICATIONS FOR TEACHING, ORGANIZING, COMMUNICATING, AND USING OUR SCIENCE

There are many implications of both the societal changes and new research
topics, especially those at the population/community–ecosystem interface, for
how we educate students; publish journals; organize meetings, departments,

agencies, and professional societies; construct professional reward systems; write textbooks; communicate to the public or to policy makers; and apply the products of our scholarship (Fig. 28-1). In the brief space allotted, I will focus on a few aspects of two of these areas: training of the next generation of ecologists and the role of professional societies. In the comments below, I include suggestions made by participants at the 1993 Cary Conference, out of which this volume grew and in particular the proscriptions presented during discussions led by Jerry Franklin (see also Ch. 31).

Educating the next generation of ecologists is a formidable challenge (Slobodkin, Ch. 29). Some important considerations for providing the most useful training include not only the obvious provision of expertise and experiences but also modifying current attitudes about research. Cultural changes require a combination of practicing or modelling desired behaviors (as opposed to simply preaching them) but also facilitating and rewarding the changes. Specifically, we should

- Train and reward interdisciplinary research, not only across levels of biology, but also with physical and social scientists;
- Educate students to the need to continue to "grow," to take risks, to stretch in new directions;
- Nurture synthesis and synthesizers;
- Look to the real world for legitimate research topics;
- Look for opportunities to incorporate the process of science into management and policy, for example, by utilizing management as a tool to test hypotheses and generate information which can then be incorporated into revised management strategies;
- Replace "either–or" dichotomies with approaches to determine the relative importance of various factors in determining patterns, coupled with an elucidation of the processes affecting the patterns.

The changing roles of professional societies reflect both societal and disciplinary changes. Often considered simply a vehicle for publishing journals and organizing meetings, professional societies can play a much greater role in facilitating activities of members and providing services to society. During the last ten years, for example, both the Ecological Society of America (ESA) and the British Ecological Society (BES) have played an increasingly stronger role in making the products of their members' scholarship more accessible to decision- and policy-makers, in actively promoting ecological literacy through education and the news media, and in promoting the importance of ecological knowledge and research.

The establishment of ESA's Public Affairs Office in, 1983, for example, facilitated the development and dissemination of a number of policy-relevant

. 28-1), ranging from the ecological consequences of the release of ge-
ineered organisms to a review of the scientific aspects of the Clinton
ɔn's Forest Ecosystem Management Assessment Team report (Marks
_. ɑ1., 1993). The British Ecological Society (BES) publishes a series of summaries
of ecological information on policy-relevant issues (Tab. 28-1). Both societies' reports have drawn on the substantial expertise of their members to articulate the
ecological understanding relevant to the particular policy or management issue.
Rigorous review and approval procedures ensure quality. Credible consensus
statements from respected professional societies such as ESA and BES can be
very effective in presenting the scientific aspects of an issue.

The large number of requests for ecological advice made to ESA's officers,
and the Offices of Public Affairs, and The Sustainable Biosphere Initiative
from U.S. federal agencies and officials, is potent testimony to the newly discovered but largely untapped role for improved ecological input into the U.S.
policy and management arenas. The recent decision by ESA members (August,
1993) to establish a Society Headquarters in Washington, D.C., and to hire
an Executive Director reflects the desire to structure Society offices in a fashion that will allow it to play a more vigorous role in communicating ecological
understanding, in articulating the relevance and importance of ecological research, and in working to improve ecological literacy (Lubchenco et al., 1993).

Table 28-1. Select policy-relevant reports from the Ecological Society of America (ESA) and
the British Ecological Society (BES)

Subject	Citation	Society
1. Ecological consequences of release of genetically engineered organisms	Tiedje et al., 1989	ESA
2. River water quality	Calow et al., 1990	BES
3. Red grouse populations and moorland management	Lawton 1990b	BES
4. Research priorities for ecology	Lubchenco et al., 1991	ESA
5. Criteria for delineation of wetlands	Bedford, et al., 1992	ESA
6. Estuarine barrages	Gray et al., 1992	BES
7. Review of the Clinton Administration's Forest Ecosystem Management Assessment Team report	Marks et al., 1993	ESA
8. Scientific considerations relevant to reauthorization of the Endangered Species Act	Carroll et al., in preparation[1]	ESA
9. Scientific basis of ecosystem management	Christensen et al., in preparation[2]	ESA

[1]C.R. Carroll, C.K. Augspurger, A.P. Dobson, J.F. Franklin, W.V. Reid, C.R. Tracy, D.S. Wilcove, and
J. Wilson.
[2]N.L. Christensen, A.M. Bartuska, J.H. Brown, S.R. Carpenter, M. Collopy, C.M. D'Antonio, R. Francis,
J.A. MacMahon, R.F. Noss, D.J. Parsons, C.H. Peterson, and M.G. Turner.

Changes in the roles of professional societies are often precipitated by the same societal and environmental changes discussed earlier. These new roles are designed to provide members with an effective mechanism for influencing society in general and the state of the environment via changes in environmental management, policy, and citizen awareness (Fig. 28-1).

Thus, societal and environmental changes, ecological research, and the communication–organization–educational aspects of the discipline are intimately connected. Many pressing environmental problems are motivating more ecologists to address different questions, such as the linkages between species and ecosystem. This activity is resulting in a healthy plethora of approaches that enriches the discipline, provides new insight, and results in opportunities for better management and policy decisions. Concomitant changes in training and in communicating and utilizing ecological knowledge feed directly back to society and to the environment.

ACKNOWLEDGMENTS

This contribution was supported by the Pew Charitable Trusts' Scholar in Conservation and the Environment Program.

29

Linking Species and Ecosystems Through Training of Students

Lawrence B. Slobodkin

SUMMARY

Intellectual separation between ecologists concerned with organisms as such and those concerned with ecosystems is neither necessary nor advantageous. Appropriate training at the graduate student level may reduce the problem in the future. If students learn a great deal of biology and natural history, prior to or concurrently with, other ecosystem-related studies they can have greater breadth than their teachers and be more effective in solving the broad array of problems that are likely to be posed for ecology by society. They must also learn intellectual enthusiasm and independence. Several pedagogic techniques are listed.

INTRODUCTION

Historically ecology has been concerned with organism–environment interactions in their full variety and at all levels. However, there is a perception of a possible disjunction within ecology. Is the separateness of studies of species from studies of ecosystems a real empirical dichotomy or might it be merely a problem of the training process of young ecologists? Shortly after the Cary Conference in which this book has its origins there was a symposium sponsored by three ecological associations, which also focused on the interaction of species with ecosystems. There I discussed the problem of drawing inferences on the ecosystem level from information on single species (Slobodkin, 1994b). My assignment here is to discuss the educational problem that may be causing different kinds of ecologists to fail to understand, or even be very much interested in each other.

WHY MUST ECOLOGY REJECT SPECIALIZATION?

Limitation of breadth in ecology may be just a special case of a general problem in scientific education. As information accumulates and techniques become more complex, students tend to become more specialized, generation by generation. For most sciences this is inevitable and perhaps harmless. In some fields there may not be any serious reason to maintain expertise in, or even appreciation of, more than one narrow specialty. This narrowing often occurs without conscious decision, as a consequence of competition for grants and publications. There is even a serious viewpoint that the purpose of graduate education is to bring students to the point of writing their first successful grant request!

I suspect that the phenomenon of students having a narrower specialization than their teachers is particularly true for fields defined by their own internal intellectual structure—for example, some areas of physics and mathematics—or that are defined by use of particular techniques—for example, some areas of molecular biology. It is possible for a mathematician or physicist to reject some empirical or theoretical problem on the grounds of its not being mathematics or physics, although the subject matter seems very close to something that is a part of one or both disciplines. For example, linear hydrodynamic flow has been part of physics and mathematics since the nineteenth century, but turbulent and chaotic flows were relegated to meteorology. They became a serious part of physics and mathematics when a more or less satisfactory theoretical context, the theory of fractals, had been developed for them (Kellert, 1993).

While society permits physics and mathematics to define their own subject matter, not all sciences have this convenience. Geology, chemistry, ecology, and medicine have empirical subject areas (rocks, chemicals, organismic interactions, or illness). Subject matter sciences cannot reject problems relevant to their areas, however difficult or intractable they may prove to be (Slobodkin, 1988). The subject matter of ecology is enormous. The sheer number of species and the multitude of interesting questions that may be asked about each one of them dictates that each ecological question is almost unique. (Think of spotted owls, red tides, landscape beauty, black-footed ferrets, etc).

Uniqueness of ecological questions implies that specific research to answer that question is always required. However, the more we know the *less* question-specific research will probably be needed before generating answers. The situation is somewhat analogous to that in medicine. Diagnostic tests are required to design a treatment program for each case, but an experienced practitioner can design the program more rapidly and more effectively. [But some questions are red herrings in that they are based on illegitimate assumptions, and therefore cannot be answered even in principle (Slobodkin, 1994a).]

Ecology uses information ranging from physiology to evolution to population biology to community and ecosystem dynamics through to geological and atmospheric sciences, soil science, hydrology, and global biogeochemistry. We cannot afford the luxurious simplification that comes with specialization. Seemingly practical solutions to specialized aspects of an ecological problem may prove disastrous if sufficient attention to aspects of the problem as a whole is lacking. Among the most obvious examples are the various deliberate introductions of organisms to serve some particular ecological purpose that then proved to be pests. Australian rabbits, New Zealand blackberries, and Jamaican mongoose come immediately to mind.

There are also cases in which a doctrinaire approach almost guarantees failure to develop understanding. Choosing one easy example from my own work, in attempting to explain what was then known as the Mystery of the Red Tide, researchers of the 1940s and 1950s used phosphorus, nitrogen, and salinity data, if they were biological oceanographers, or used population growth models if they were "theoretical ecologists." Neither of these provided much insight. The "Mystery" vanished in 1952 when elementary hydrodynamics and natural history were included. Of course later workers showed that the explanation of the 1950s was too simple (reviewed in Slobodkin, 1989), but at least it had been made clear that Red Tide organisms, like all other organisms, live in a multilevelled world and can be understood only in terms of their responses on many levels.

In brief, ecology is stuck with, and should revel in, problems in their full complexity. This complexity is, for me, a large part of ecology's intellectual appeal. We must somehow be sure that all young ecologists understand this. If a student loves animal behavior, he or she need not be an expert in biogeochemistry, but must at least understand what the issues of biogeochemistry are. Conversely, if a student is concerned with global warming it should be understood that there is more to worry about than temperature change as such.

Changes in plant growth rate associated with the CO_2 level itself may prove more ecologically significant than the temperature change, and ultimately pathogens and pests may be most significant (P. Vitousek, pers. comm.).

EDUCATING EFFECTIVE ECOLOGISTS.

Are young ecologists being educated in such a way as to understand one or two levels but to be unable to think seriously about all the different levels that are required in ecology? I assume that the capacity to at least appreciate the full spectrum of levels is not only desirable but is necessary in order to be an effective ecologist. How are we to raise the right sort of students, those who can help ecology to solve the very important problems that we hope society will assign to us? Also, how can we continue this process generation after generation. This last point is important because, as time goes on, we must anticipate more, and more difficult problems for ecology, which will require more and more technical specialization, but actual organisms are going to persist in living in a multilevelled world.

Initially this can be considered as a general problem of science education. Not all scientists think of their educational role in the same way. Some believe that the most desirable thing to do with their students is to turn them into replicas of themselves. Others sometimes use their students as technicians on particular projects or as writers of grant proposals. In the first case, the most successful education produces a student whose breadth equals that of the professor. In the second case the student will be somewhat narrower than the professor. But our students must have greater breadth than we do if the science of ecology is to advance.

There is a difference between an academic clone and proper intellectual progeny. The actual work of different generations of scientists will differ because the world will have changed. True progeny can do in the context of their world the equivalent of what the teacher did in an earlier world. Between the time of maturity of a teacher and the time of maturity of his or her students science will have advanced. Intellectual progeny, therefore, need greater breadth and certainly greater sophistication than their teachers.

It is possible to deliberately train students to be of greater breadth than their teachers. Institutions in which this is happening are delightful places and the times when this is occurring are looked back on, at least by students, as "Golden Ages" in a local sense. I have had the good fortune to be associated with three of these. The University of Michigan in the late 1950s and 1960s— when Hairston, Smith, Evans, Dawson, and others were feeling very optimistic about ecology, and Stony Brook in the 1970s and 1980s, when we had the only Ecology and Evolution Department in the world, had this quality. Perhaps because I was a student, the best of these, at least in my memory, was Yale in the late 1940s and 1950s, when G. Evelyn Hutchinson was at the top of his form. The diversity of his students as individuals and as a group was very great indeed, some of them working at each of the levels of organization that are discussed in this volume.

In all three situations, spontaneously generated student seminars were more exciting than formal classes and the faculty felt it was learning more than it was teaching. All three produced students of remarkable breadth, who were quite distinct from each other and from their teachers.

At least to my knowledge, why these golden times occurred has never been carefully analyzed. Perhaps had a formal analysis been attempted on the spot the whole undertaking would not have succeeded, by analogy with the old story about the centipede who became completely tangled by too careful consideration of where to place each leg.

G.E. Hutchinson's methods as a major ecological educator have been analyzed in part. He has written general essays that occasionally touch on how teaching is to be done (Hutchinson, 1979), and he was also the subject of essays by his students (Kohn, 1971; Slobodkin, 1993)

The following are some of the concrete things I know about Hutchinson's teaching method:

Tacked to his wall was a clipping from some magazine which read (in part) "Never try to discourage a student—you are certain to succeed."

It was apparent to his students that he maintained and preserved a catholicity of interests ranging from crystallography to comparative religion. He felt that demonstrating his own intellectual excitement and pleasure was of prime importance in teaching.

Often, when a student developed any sort of independent idea at all, Hutchinson would respond to it by a general comment about its being "interesting" but would accompany this noncommittal response by asking the student if he had read some particularly obscure reference. On being told that the student had not, Hutchinson would pause and then say "Read it. You will find it fascinating." Often the reference was either inaccessible or utterly inapplicable—but Hutchinson would have succeeded in placing the students into the stream of intellectual history, to swim independently into waters that Hutchinson himself did not frequent.

In short, students were brought to think of themselves as independent young intellectuals with considerable promise, rather than as technicians or as parts of a Hutchinson "organization." There were elements of this in every "golden age" that I have heard of. Students trained in this way may not be able to solve problems of differences in level of integration quite automatically. However, they will recognize them and be aware of the need to find solutions.

Conversely, there are ways of almost guaranteeing failure to produce multilevelled students. Teaching of fundamentals can, if done badly, give students the impression that the entire field of ecology consists of only population dynamics or plant–insect interactions or whatever it is that interests the teacher.

Teachers who will produce the kind of ecologists we need must be willing to undertake the rather dangerous business of teaching material beyond their particular research area. In doing this there is a temptation to reify the simplistic. For example, there exist good field ecologists who teach population dynamics as if it consisted of only ramifications of logistic equations (or other mythical figures) and teach energetics as if ecological energetics was simply a subclass of elementary thermodynamics left over from a physical chemistry course. Even at the elementary level it is necessary to be sufficiently tentative in teaching so as to encourage students to learn what you do not know. In our teaching we must be cautious about simplicity but we cannot permit our students to despair because of complexity.

I can imagine some tentative rules for educating ecology students. But before listing them it is vital to bear one substantive point in mind. Love of natural history, anatomy, physiology, and classification and familiarity with natural communities are essential for ecologists. A mass of biological facts and closeness with organisms and their world is prerequisite for any ecologist of quality. Perhaps it is these fundamentals that identify ecological training. Without this background there is extreme danger of focusing on intriguing simulations of ecology without knowing enough to distrust them. Given this background, there are possible pedagogic rules. For example:

1. Because of the close social and historical connection between ecology and various environmental movements of a quasipolitical, quasireligious sort (McIntosh, 1985), it is particularly important to avoid using a lectern as a pulpit. Teaching of sciences must aim at literal truth. We must avoid deliberate inspiration or poetry except in an appropriate context. Ecology at one level is as amoral as mathematics.
2. If simplistic theory is being used for pedagogic purposes make sure that it is not taken for fact.
3. Try to teach practical applications—they will automatically require understanding on all levels of organization.
4. Make clear that we are at an early stage of science so that what you are doing is probably going to be shown false soon.

Perhaps following these rules will produce a sufficiently nonuniform set of intellectual progeny to maintain the intellectual richness and interconnectedness that tomorrow's ecology requires.

In what sense are these suggestions uniquely applicable to ecology? Notice what I have not advocated. I have named no great ecologists of the past as role models, as I might have done if I were discussing education of physicists or mathematicians. I have not suggested that the student be taught to circumvent the "indigestible mass of facts" that has characterized ecology in the past (May, 1975, p. 1), as I might have done in discussing economics or meteorology. I have provided a general list of subjects that are clearly vital to ecologists, although I think I could provide a much more specific list for medical doctors.

Most importantly, I have not listed any particular set of problems that are self-evidently going to grow in importance in ecology as time goes on. I think that lists of this type are counterproductive. Any such list constitutes a simplification of the full range of issues that may concern organisms. Were we to have taught such a list to our students I suspect the next problem down the road would turn out to have not been on our list.

What we want from new ecologists is a high capacity for what has been called deutero-learning—the capacity to learn how to learn (Bateson, 1972). We also want them to be so rich in biological experience that any new experience cannot daunt them, so rich in techniques that no one technique becomes central to their intellectual life, and so practiced in flexibility that their response to a new problem is automatically fresh, new, and realistic.

They are to be trained to face actual, irreducible complexity (Slobodkin, 1992b) and do whatever simplification may be required for themselves and for the purpose of directly answering real questions, rather than follow a fine clear path that may not lead anywhere one wants to be.

ACKNOWLEDGMENTS

This chapter benefited from comments by D. Futuyma and an anonymous referee.

30

LINKING SPECIES AND COMMUNITIES TO ECOSYSTEM MANAGEMENT: A PERSPECTIVE FROM THE EXPERIMENTAL LAKES EXPERIENCE

D.W. Schindler

SUMMARY

The Experimental Lakes Area (ELA) has been the focus of applied ecological studies since, 1968. Both communities and ecosystems have been studied in lakes that were deliberately perturbed with nutrients and strong acids to simulate contemporary lake management problems. Strengths and weaknesses of the approach are examined.

In addition to providing a scientific basis for management of eutrophication and acidification, ELA experiments have yielded several hundred articles in refereed pure science journals.

Following perturbation, it required several years for many features of communities and ecosystems to reach steady state, revealing the pitfalls of using short-term studies to guide applied ecology decisions. Results from the companion Northern Ontario Lake Size Survey (NOLSS) provide important information on spatial (size) scaling for the extrapolation of results from small ecosystems to large ones.

To determine accurately the effects of human perturbation on ecosystems, changes in the baseline state of ecosystems that are believed to be unperturbed must also be considered. Examples of changed "baseline" conditions in lakes of remote areas due to climatic warming, fisheries manipulations, and atmospheric transport of organochlorines are discussed.

Among ecosystem types in a region, lakes may be the best early indicators of human stress, due to the relative ease with which communities and ecosystems can be studied, and the paleoecological records of both aquatic and terrestrial events in lake sediments. The low functional redundancy in the communities of northern lakes allows ecosystem-wide changes to occur, even with slight human perturbation.

INTRODUCTION

Recent discussions with other ecologists reveal that many are still uncomfortable with the notions of extracting "pure" ecology from the study of "applied" problems, and of combining community and ecosystem studies on relatively small budgets. I am convinced that combining all of these elements is necessary in the immediate future, for human stresses are detectable in almost all ecosystems, and budgets for the study of ecology are becoming more and more dedicated to solving practical ecosystem management problems. In this chapter, I review the history of the Experimental Lakes Area (ELA), where all of these elements have been combined for over 25 years. ELA's applied mandate has not hindered the production of good fundamental ecology, and the combination of ecosystem and community approaches has enabled us to decipher the causes of some problems that would have been indecipherable if only community or ecosystem approaches had been used without the other.

Admittedly, it is much easier for limnologists to perform combined community and ecosystem studies, for small lakes are ecosystems where a small research team can examine many features of communities and ecosystem processes, with only a few simple nets, water bottles, and a leaky rowboat. Most freshwater organisms pass through their entire life cycles in less than a year, revealing insights that would take decades of hard work in many other ecosystems. Physical processes also help to keep heterogeneity in small lakes much lower than in terrestrial systems, at least for pelagic systems. As a result, simple sampling designs are feasible.

SOME HISTORICAL REFLECTIONS ON THE CONTRIBUTIONS OF APPLIED ECOLOGY

Applied ecology has contributed greatly to general ecological theory and understanding almost since its inception, as any reader of Charles Elton's (1958) *Ecology of Invasions* or his (1942) *Voles, Mice and Lemmings* must know.

Similarly, E.P. Odum's various treatises on "Ecosystem Stress" (Odum, 1985, 1990) have great value for both pure and applied ecology.

It is certainly no accident that the first international limnological organization was named the International Society for Theoretical and Applied Limnology (SIL). The founders of this society recognized that pure and applied ecologists had much in common, perhaps because human-caused problems were already obvious in aquatic ecosystems in the late nineteenth century. Executives in SIL and recipients of its prestigious Naumann–Thienemann Medal have included nearly equal representatives from pure and applied schools, as well as from community and ecosystem approaches, if we apply these labels as has been customary in the past. Freshwaters continue to be the ecosystems of greatest concern to humans (discussed by la Riviere, 1989; Likens, 1992), and limnologists are fortunate that neither the applied–pure dichotomy nor the community–ecosystem dichotomy has ever been taken very seriously. Certainly, few would dispute the enormous contributions of Vollenweider's (1968, 1976) or Edmondson's (1991) applied studies, or Hutchinson's (1969) reflections on eutrophication to our fundamental knowledge of aquatic ecology. G.E. Hutchinson certainly took the contributions of applied limnology to development of the science of aquatic ecology very seriously (Hutchinson, 1969). In brief, the distinction between applied and pure ecology is not a useful one.

THE EXPERIMENTAL LAKES AREA: EXPERIMENTS IN APPLIED COMMUNITY AND ECOSYSTEM ECOLOGY

The Experimental Lakes Area (ELA) was designed from the start as an applied ecology project. There was neither mandate nor direct funding for doing ecology that did not directly address aquatic management problems. But strangely, more of the science from ELA has found its way into "pure science" journals than into applied ones, although the results have also had major effects on the management of nutrients and acidic deposition in Europe and North America.

ELA studies were designed from the start to have both strong community and ecosystem components. This was due in large part to perceptive hiring by J.R. Vallentyne, the first leader of the limnology section, who strongly believed that both were necessary in order to successfully solve management problems in aquatic ecosystems.

Fashionable as modelling is today, no complicated, overriding supermodels guided ELA'S whole lake experiments. Instead, the most successful experiments were ones designed to strip away as many complicating circumstances as possible, in order to focus clearly on major controversies in freshwater management. For example, whole-lake eutrophication experiments were designed to focus on only three nutrients: phosphorus (P), nitrogen (N) and carbon (C). Similarly, acidification experiments focused only on the fate of acids in lakes, rather than trying to incorporate the complications of lake–catchment interactions. Useful output models from the project have been equally simple. In fact, to call them models would be pretentious, for they are no more than equations. Examples include Schindler et al.'s (1978) model for eutrophication and Kelly et al.'s (1987) for acidification. Both are simply functions of chemical input and water renewal. I'm sure that some would call this simple approach old fashioned.

HOW ALL-INCLUSIVE DO APPLIED STUDIES NEED TO BE?

I have attended a variety of meetings to plan ecosystem or community-scale studies, including manipulations of chemical inputs or predators in lakes to changes in forest practices and densities of mammals in terrestrial experiments. The fear is invariably expressed by workshop participants that funding and expertise are insufficient to allow truly comprehensive long-term studies to be undertaken. As the result, enormous price tags often prevent proposals from being funded, or if they are funded, they are short-lived.

Practical constraints on expertise and funds have always prevented ELA from considering studying all components of communities and ecosystems in its experiments. Instead, it has been necessary in each study to narrow the focus to those components deemed most critical or most likely to be affected by a particular perturbation. For example, instead of replicated treatments with several nutrient combinations including trace elements, we concentrated on P, N and C, in rather simple combinations, and over a rather small, but realistic range of doses. Similarly, in acidification experiments we chose to make separate additions of sulfuric and nitric acids, so that their effects could be isolated. Similarly, we decided not to add aluminum, except for one "acid pulse" experiment (Playle, 1985) or to otherwise try to mimic land–water interactions. We reasoned that these could be well studied at sites where acid deposition was high, and in those areas it would be impossible to separate lake and catchment processes.

Some errors certainly result from such approaches. For example, except for the mesocosm studies of Muller (1980), littoral processes were not studied during the acidification of Lake 223. But once we realized how dramatic the changes to the littoral community were, and how important they were in affecting ecosystem functions, the gap was filled by studies in other lakes (Turner et al., 1991; Kelly et al., 1994). The importance of N/P ratios in addition to P loading was not recognized until the Lake 227 and Lake 226 experiments were both under way, requiring modification of the 227 experiment to test the implications of N/P ratios more fully (Schindler, 1977). More detailed assessments of N/P ratios were also done in mesocosms, although considerable effort was required to properly scale mesocosms to simulate nutrient dynamics in a whole lake (Levine and Schindler, 1992). Comprehensive initial designs would not have prevented these flaws, for the mechanisms involved were not part of the limnological paradigms of the day.

TEMPORAL SCALES

Our simple experimental designs allowed an important advantage: we were able to continue experiments for several years. We were also able to conduct long-term studies of unmanipulated ecosystems, for interpreting the effects of perturbations required an understanding of natural variation. Despite recent attention by ecologists to questions of intercalibration, quality control, replication (or pseudoreplication), and other elements of statistical design, and sophisticated modelling, lags in community and ecosystem response to stress are seldom studied. ELA results show that disregard for time lags can cause serious misinterpretations of the steady-state responses of both communities and ecosystems to changed conditions. Steady-state responses of lakes to changes in chemical inputs require many years. It took 16 years for N concentrations in Lake 227 reach steady-state with inputs from fertilizer, the atmosphere and sediments (Schindler et al., 1987; and D.W. Schindler, D.W. Mayhood, S. Lamontagne, A.J. Paul, P. Leavitt, B. Miskimmin, and B. Parker, unpublished). During this time, the lake became less and less N-limited (S. Guildford and L. Hendzel, unpublished). Fish communities in Lake 226 required 8 years to adapt fully to increased nutrient loading, and another 8 years to fully adapt to the return of nutrient loading to its pristine state (Mills and Chalanchuk, 1987). Lags in community responses to the acidification and deacidification of Lake 223 were even longer, although once suitable conditions returned, ecosystem functions returned with little delay (Schindler et al., 1991). The lack of long-term considerations has led to confusion in the interpretation of "top-down" vs. "bottom-up" effects, for nutrient or fish manipulations in lakes, and especially in mesocosms, are seldom followed for long enough to ensure that responses to either changed chemical inputs or community structure have reached steady state . . . or in fact, to be sure that steady states are reachable at all! If management decisions are to be of long-lasting importance, long-term ecological studies to guide them are essential. Similar conclusions regarding the importance of long-term ecological studies have been voiced by Likens (1992) and Edmondson (1991).

An important consideration for practising ecologists, who must "produce" in order to renew their operating funds, is that even the longest-running experiments at ELA have continued to yield high-quality publications. At present, the Lake 227 experiment is entering its 26th year, spanning three modifications in nutrient loading regimens and one in trophic structure in order to test the steady-state responses of different communities and biogeochemical cycles. The acidification of Lake 223 has continued for 18 years, with both an acidification and a recovery phase. The Lake 226 study spanned 8 years, followed by 8–10 years of recovery studies, and the acidification of Lake 302S is entering its 14th year. The longest of all, the study of natural variation in Lake 239, is in its 27th year. All of the long-term studies continue to yield scientific results and publications at rates greater than newer studies, yet the cost of maintaining each experiment (exclusive of salaries and the maintenance of the ELA field camp) is but a few thousand dollars a year. I believe that a number of experimental areas, where long-term perturbations could be maintained to address a variety of ecological studies, could be a valuable companion to the LTER program, if we are serious about using community and ecosystem ecology as a basis for sound ecosystem management. The advantage of such approaches has already been shown at the Hubbard Brook, Toolik Lake, and Trout Lake sites (Likens, 1992; Peterson et al., 1993; Frost et al., Ch. 22).

SPATIAL SCALES

Consideration of spatial scales is also important in applied ecology, where results from small-scale studies must be extrapolated to much larger systems, and where mistakes can be very costly. As ELA studies became more refined, a companion project was designed for the sole purpose of sharpening our abilities to extrapolate to large lakes: the Northern Ontario Lake Size Study (NOLSS) examines how size affects extrapolation from ELA-size lakes to larger lakes of northern Ontario, including lakes Nipigon and Superior (Fee and Hecky, 1992; Fee et al., 1992).

NOLSS results show that scale has important effects on extrapolation of many results from small freshwater ecosystems to large ones, even when the lakes have similar biotas and geological and climatic settings. For example, the fish of small lakes in northern Ontario had higher concentrations of methyl mercury than larger lakes, because the ratio of methylation to demethylation (M/D) is a function of epilimnion temperature (Bodaly et al., 1993). This is because methylation increases with increasing temperature, whereas demethylation decreases (Ramlal et al., 1993). Because detailed ecological studies, particularly ecosystem experiments, will almost always be carried out on small systems, whereas most management problems focus on large ones, the study of size scales must receive high priority in applied ecology.

THE BASELINE STATE

As mentioned briefly above, knowing the natural variability in unmodified ecosystems is an important reference for applied studies, for without such studies, important and expensive errors can be made in management decisions. For example, acidification appeared to cause phytoplankton production to increase in Lake 223, until it was found that production in all ELA lakes was increasing, as the result of lengthening water renewal times and longer ice-free seasons (Shearer et al., 1987). Acidification appeared to be causing great increases in transparency of experimental lakes, until it was noticed that transparency in Lake 239 was following similar trends, caused by lower DOC yields from catchments and higher rates of DOC removal in lakes (Schindler et al., 1992). These and similar experiences leave me with an uncomfortable feeling that "baseline drift" may affect many of our ecological interpretations. Without long-term measurements, it may be difficult for us to detect whether an ecosystems is still "operating on all its cylinders," or is just a tiny bit out of whack. . . . not enough to see easily, but enough to cause long-term deterioration in community or ecosystem function. Below, I give some additional examples from my own experience in lakes.

LAKES AS CANARIES

Nearly 20 years ago, at dinner during a meeting at the Cornell Ecosystems Center, Herb Bormann and I outlined a rough manuscript entitled "Lakes as Canaries." While the manuscript has gone through two rough draft versions, we have never gotten around to finishing it. I think often of the main point of the manuscript, and believe that it is important enough to mention here. Herb and I believe that lakes have a number of characteristics that make them ideal as early warning systems for ecological damage, in both the lakes themselves and in surrounding catchments. They are therefore analogous to the fabled "coal miners canary" reputed to have warned miners of toxic conditions in early coal mines. In addition to the ease of sampling lacustrine communities and ecosystems mentioned above, lakes maintain a "library" of past community, ecosystem, and watershed events, in the muds at the lake bottom. We are beginning to be able to read the language of these muds, and they do not always have pleasant messages. Paleoecology of lake sediments has been widely used for detecting the changes caused by acid precipitation (e.g., Charles and Smol, 1988); climate (Davis and Botkin, 1985; Ritchie, 1986; Fritz, 1990); fire (Swain, 1973), eutrophication (Swain, 1985); records of contaminant deposition (Muir et al., 1994); and many other problems. In brief, for many broad regional stresses on the environment, we should be able to see signals of problems in lakes earlier than in other ecosystem types.

Had we only looked a few decades earlier in the Adirondacks or eastern Ontario, lake muds would certainly have signalled that acid deposition was slowly changing phytoplankton and zooplankton communities, first by changes in species and then in community diversity. We would have seen that trace metals and organochlorines were accumulating at dangerous levels in arctic regions, despite their remoteness from sources of these substances. It is troubling to find that there has been a continuous, and so far unexplainable, decline in diatom species in lakes of the high arctic since the mid nineteenth century (Douglas, 1993).

As an example of how appearances of pristineness can be deceiving, consider Snowflake Lake, in the alpine zone of Banff National Park. The lake lies 40 km from the nearest road access, in the most remote part of a World Heritage Area, protected by the National Park Service's mandate to maintain ecosystems in a pristine state for the enjoyment of future generations. There is not even a discernible path to the lake. Certainly such a lake must be in pristine condition!

But long-term records for zooplankton and benthos communities and paleoecological studies in Snowflake Lake showed otherwise. Anderson (1972) found that stocking of the lake with rainbow trout, cutthroat trout, and brook trout in the early, 1960s eliminated several important invertebrate species from the food chain of this previously fishless lake, including the original top predator, a large calanoid copepod, *Hesperodiaptomus arcticus*, the only large grazer, *Daphnia pulex*, and the dominant benthic omnivore, *Gammarus lacustris*. The stocked fish did not reproduce, and disappeared from the lake by the late 1970s. Our 1990–1992 surveys of the lake showed that *Daphnia* and *Gammarus* have returned, but *Hesperodiaptomus* had not. In the absence of this voracious predator on rotifers and nauplius larvae, rotifers and cyclopoid copepods are a far greater component of the zooplankton community than they were originally. The size structure of the zooplankton community has shifted to smaller species, and several new species of rotifers are now found in the lake (D.W. Schindler, D.W. Mayhood, S. Lamontagne, A.J. Paul, P. Leavitt, B. Miskimmin, and B. Parker, unpublished). Although there are no older phytoplankton records, paleoecological analyses of pigments in lake sediments indicate a severalfold increase in phytoplankton of most taxonomic groups accompanying fish stocking (P. Leavitt, unpublished data).

Mesocosm experiments with *Hesperodiaptomus* show that its absence, which allows rotifers and small species of crustaceans to flourish, has also caused changes in phytoplankton from small to large species, due to selective cropping of the former (A.J. Paul, unpublished; D.W. Schindler, D.W. Mayhood, S. Lamontagne, A.J. Paul, P. Leavitt, B. Miskimmin, and B. Parker, unpublished). Bioenergetics models indicate that the increase in phytoplankton was probably caused by increased P recycling by stocked fish (P. Leavitt, unpublished). To this day, the pelagic community of Snowflake Lake retains smaller zooplankton species, a higher phytoplankton standing crop, and a less effective top predator (small cyclopoid copepod species) than it did originally. As a result, the transparency of the lake is lower than that of nearby lakes that have unmodified communities.

Reviewers of an earlier draft of this chapter were cynical that such changes were of great significance. After all, Snowflake Lake is still beautiful! Perhaps we should simply forget about managing for pristineness, and concentrate on appearances! I disagree, and shall recount another example of the dangers facing even our most pristine systems, the significance of which few will dispute. Few areas of the northern continents are more remote than the subarctic Yukon. Yet in, 1992, several of the large lakes in remote parts of the Yukon Territories were discovered to contain fish with high concentrations of PCBs, toxaphene, DDT, and other organochlorines. In some cases, including Lake Laberge, made famous by Robert Service's *The Cremation of Sam McGee*, concentrations of organochlorines were similar to those of the lower St. Lawrence Great Lakes, requiring the closure of key aboriginal fisheries for lake trout and burbot. The chemicals appear to have entered via long-range transport in the atmosphere, probably travelling from Eurasia by prevailing wind patterns as is common for pollutants in the Arctic (Barrie et al., 1992). Although concentrations in rain, snow, and surface waters are scarcely detectable, biomagnification of several orders of magnitude by food chains has caused values in fish to be a threat to humans and other consumers of fish (Kidd et al., 1994). Similar problems are occurring throughout the arctic, in both freshwater and marine food chains (Lockhart et al., 1992). As a result, aboriginal people in some remote areas contain higher concentrations of toxicants than if they lived in polluted industrial areas (Kinloch et al., 1992; Lockhart et al., 1992). Even where contaminants are present only at low concentrations, native people are reluctant to eat their traditional foods. Certainly this is disruption of food chains of great concern!

Most disturbing of all, paleoecological analyses of contaminant distributions in lake sediments reveal that these contaminant problems have been occurring for several decades, without our realizing it (Muir et al., 1994).

Clearly, we can no longer trust that ecosystems are fully functional simply because they are in remote locations. Even if ecosystems are still operating normally (at least insofar as we can measure functions with present-day methods), it seems unwise to discard "functional redundancy" that might buffer our ecosystem functions from future stresses. Working in applied ecology one quickly learns that ecologists have little control over the course of development, even when there is strong evidence that ecosystems will be impaired. In brief, I view the study of pristine ecosystems to be an important ingredient in fundamental ecology, in ecosystem management, and in the training of young ecologists. Early detection of change in such systems, whether caused by natural phenomena or man's activities, is an important underpinning to more general ecological studies.

ARE COMMUNITY OR ECOSYSTEM STUDIES MORE USEFUL TO MANAGEMENT?

Odum (1985) outlined several properties of communities and ecosystems that he believed would be changed as the result of anthropogenically imposed stresses. They included both changes in ecosystem function and community structure. However, Odum was forced to draw on studies of a wide variety of ecosystem and community types for examples, because few of the studies included more than one or two of the properties that he wished to examine. Fortuitously, we had measured many of the features hypothesized to be important in single-lake ecosystems that had been either acidified or eutrophied. Many of the properties changed as predicted by Odum. But we also found that many changes are functions of particular ecosystem types or stresses.

The most sensitive species in aquatic communities responded much sooner to acidification and eutrophication than all but one ecosystem function, thus providing the best early-warning indicators of ecosystem stress (Schindler, 1987, 1990a). But important changes in ecosystem functions also occurred in the course of experiments at ELA. The earliest changes were in the role of the atmosphere in relieving N and C deficiencies in P-fertilized lakes (Schindler et al., 1972; Schindler, 1977), and in the metabolism of the littoral algal community during acidification (Turner et al., 1991). In later stages of acidification, changes included the increasing role of sediments in generating acid neutralization (Schindler, 1980, 1986; Kelly et al., 1984; Cook et al., 1986) or consuming oxygen (Schindler, 1990a) and the cessation of nitrification at intermediate pH (Rudd et al., 1988).

But many changes in communities occurred as early as any of these functional changes and usually much earlier, so that they must still be considered to be superior early warning indicators of ecosystem stress. For example, the atmospheric inputs to the aquatic N cycle of Lakes 226 and 227 only became important after invasion of the lakes by heterocystous bluegreen algae, which had a competitive advantage when N/P in nutrients was low (Schindler, 1977; Levine and Schindler, 1992). Filamentous green algae invaded the periphyton before changes in metabolism due to acidification were evident. The changes in the food chain leading to lake trout in Lake 223 occurred before any other ecosystem processes had changed detectably (Schindler et al., 1985, 1990a). But both community and ecosystem studies were necessary to understand the mechanisms causing the changes. For example, *in situ* alkalinity production ceased at pH values of 4.5 due to the development of mats of filamentous green algae in the littoral zone. In thick mats, anoxia occurred above the sediment–water interface in midsummer, causing repositioning of the zone of sulfate reduction from below to above the sediment–water interface. When algal mats decomposed in autumn, exposure of reduced sulfur deposits to oxic waters led to their complete reoxidation, inactivating the major *in situ* mechanism protecting lakes from acidification (Kelly et al., 1994).

In brief, while signs of stress in ecosystems are most evident in sensitive species, both ecosystem and community studies have proven necessary to understand the changes to ecosystems caused by eutrophication and acidification at ELA.

FUNCTIONAL REDUNDANCY

ELA acidification experiments suggest that the components of aquatic communities with low functional redundancy are those most vulnerable to disruption of key food web structures and ecosystem function (Schindler, 1988, 1990a), supporting the arguments made by Frost et al. (Ch. 22). In commenting on these results, Odum (1990) points out that the disappearance of redundancy may signal the impending breakdown of the whole system, and should be cause for alarm. This prediction seems borne out by the Lake 223 study, where several of the organisms most sensitive to acidification were key species in the food chain leading to lake trout, the top pelagic predator, with few functional analogs in the community. Had the lake contained more acid-resistant analogs, the demise of lake trout might not have occurred so early in acidification. One wonders if the reason for the widespread, dramatic effects of acidification on lakes in Norway is that such lakes have very simple food webs, often containing only one abundant benthic crustacean, one cyprinid, and one predatory salmonid. Many of the important species in these simple food chains are very acid-sensitive, such as *Mysis relicta*, *Pontoporeia affinis*, and *Phoxinus phoxinus*.

Several ecosystem functions and about one half of the original species in biotic communities were lost as Lake 223 was acidified from pH values of 6.5–6.8 to 5.0–5.1. Only about one third of the species eliminated by acidification were replaced by more acid-tolerant analogs. Clearly, functional redundancy in this system is not high. Results from acidification studies of Lakes 302S and Little Rock Lake are similar (Schindler et al., 1991; Frost et al., Ch. 22). It seems reasonable to expect that functional redundancy would be lowest in northern, alpine, or other naturally impoverished ecosystems where relatively few species perform key ecosystem functions such as photosynthesis and nutrient cycling, and key positions in the simple food webs are often occupied by single species. With respect to community functioning and structure, this observation is more or less analogous to the "keystone species" concept of Paine (1966). Simple energetic considerations dictate that the most likely trophic level to display low functional redundancy will be that of top predator, where there are seldom more than a few species present. The repercussions of changes in top predators via the "trophic cascade" are well known (e.g., the studies of Paine, 1966; Carpenter et al., 1985; and Power, 1990b; see also Estes, Ch. 15). But in simple ecosystems, diversity is limited by low nutrient inputs, low primary production, harsh environmental conditions, or limited evolutionary time scales, as I have discussed elsewhere (Schindler, 1990b). It follows that in such ecosystems, low functional redundancy can occur at other points in the food chain, as the ELA results show. Functional redundancy will differ among communities and stresses, because different organisms are susceptible to particular perturbations. However, some organisms appear peculiarly vulnerable to a variety of man's stresses. For example, in Precambrian Shield lakes such as those at ELA, *Mysis relicta* is known to be very sensitive to decreasing pH, increasing temperature, decreasing oxygen, increasing trace metals, and increased grazing from fish, making it vulnerable to several of the most common human insults to these systems. Perhaps there are "born loser" species in low-diversity ecosystems that deserve particular attention as we attempt to manage lakes to maintain ecological diversity and ecosystem functioning!

Of course, these hypotheses from ELA experiments need testing in other systems and for other anthropogenic stresses before extrapolating too broadly. The possibility of comparison among ecosystem types is an exciting prospect for the future, of clear value both to the development and application of ecology.

In summary, results from ELA studies show that both community and ecosystem studies are important to manage ecosystems properly, although changes in communities generally occur sooner, providing useful indicators of impending changes in ecosystem functions. They also show that studies designed for applied purposes can yield information that is important to fundamental ecology. Indeed, the need to answer specific questions about how a lake's structure or function might be affected by human perturbation enhances, rather than diminishes, the yield of important ecological information.

THE ROLE OF ECOLOGISTS IN
SUSTAINABLE DEVELOPMENT

Ecologists have an obligation to see that decisions on development that can affect communities or ecosystems are cautiously made. This means that ecologists must be prepared to do more than apply sound ecology, for decisions require that we inevitably must look well beyond even our best ecological studies. It is important that ecologists and their professional societies become involved in debates and decisions on sustainable development that goes well beyond the current *ad hoc* basis. Projects have tended to become larger and larger as the years go by, so that the ecological integrity of entire regions may be jeopardized by a single wrong decision. Unfortunatley, an all too trusting public is confident that we can tell them how to manage their way out of any ecological jackpot that they happen to blunder into while continuing to harvest fossil and renewable resources at dramatically increasing rates. In the past, we seem to have given the public the impression that we know enough to intelligently manage any ecosystem. The reverse is true, for it is difficult to find examples of where we have truly succeeded in our attempts to preserve or manage ecosystems without some losses in biodiversity and perhaps ecosystem function.

In brief, the only sure prescription to ensure a sustainable biosphere is to exploit it very lightly until we understand it better . . . certainly a feat that will require at least several decades or even centuries to accomplish, even with the greatest emphasis on ecology. Perhaps the best index of a well-trained ecologist is that he or she will admit that there are insults to ecosystems that he or she does not know how to predict or mitigate, for which the only sensible course of action is to leave the ecosystems intact, at least for the forseeable future.

ACKNOWLEDGMENTS

Writing of this chapter was supported by NSERC Grant 89673 to D.W. Schindler. Margaret Foxcroft assisted with obtaining references and formatting the manuscript.

31

WHY LINK SPECIES CONSERVATION, ENVIRONMENTAL PROTECTION, AND RESOURCE MANAGEMENT?

Jerry F. Franklin

SUMMARY

Efforts to resolve forest conservation issues in northwestern North America illustrate the necessity to integrate ecological science across multiple organismal, spatial, and temporal scales as do essentially all of the other important environmental issues facing society. Scientific and technical teams are utilizing the best existing information from population, ecosystem, and landscape science to develop and evaluate alternative management approaches. Effective and efficient strategies require simultaneous consideration of many issues, including conservation of species and processes and sustained utilization of resources and interactions between reserved and unreserved portions of the landscape. Such exercises are defining the cutting edge of ecological science. Education of professional ecologists needs increased emphasis on interdisciplinary approaches, synthesis, and intellectual growth.

INTRODUCTION

The answer to the assigned question, "Why link species conservation, environmental protection, and resource management?" seems extraordinarily obvious. In current parlance it is what we call a "no-brainer"! Ultimately, the only way we can accomplish our goals in species conservation, environmental protection, and resource management is by linking them. All three topics are inextricably related and central to the ecological basis for sustaining a habitable planet.

Consider some of the major environmental issues facing the human race: global change including increased levels of CO_2, atmospheric pollutants, and the potential for significant warming; maintenance of biological diversity while providing the food, fiber, energy, mineral, and other resources needed by human society; disposal of waste products from solid wastes to radioactive materials; and assessing and controlling applications of genetic engineering. All of these issues, and many others, are interlinked, and they are directly linked to the greatest environmental issue facing mankind—expanding human populations.

Sustainability—and this litany of related issues and problems—requires the application and integration of science from the molecular to the global level with ecological science one of the central and most relevant elements. It should be obvious to even the most topically chauvinist ecological scientist that there is no effective way to apply our science except to integrate across scales—linking populations, ecosystems, and landscapes. Yet progress toward this objective has been painfully slow across both the scientific and resource management communities.

In this chapter I will try to illustrate the importance of integrating across spatial scales, as well as improving linkages between ecological science and resource management. The controversy over how best to manage the forests of the Pacific Northwest, and the proposed approaches to its resolution, provides my principal example; I will use it to illustrate the importance of linking across multiple scales in understanding and in developing solutions to environmental problems. First, however, I would like to comment briefly on what I view as a major barrier to integration in ecological science—the ideological schisms among ecological scientists.

CULTURAL GAPS IN ECOLOGICAL SCIENCE

Why has it been so difficult to integrate and apply ecological science over a range of scales, assuming that the majority of scientists and resource managers are agreed on the need to do so? Many factors, including the limitations of theory and empirical data bases, especially at larger spatial scales, have doubtless contributed to fragmentation both in the science and in its application. Until relatively recently, there has also been an absence of essential analytic tools and mathematical models. I would like to focus, however, on the negative contributions of traditional educational programs (see also

Lubchenco, Ch. 28, and Slobodkin, Ch. 29) and the professional chauvinism that they have generated.

Linking population and ecosystem approaches to ecology has proved a particularly difficult task for the community of ecologists primarily, I believe, because a cultural gap exists within the profession that reflects differing points of view based, in part, on different training and traditions. To illustrate, let me create *stereotypic* population ecologists and ecosystem scientists.

Population ecologists are theoretical in orientation and, most probably, academics in biology and other basic science departments. Experimentalists, they are constitutionally incapable of accepting as valid any science that is not based on tests of null hypotheses. Reductionists, they believe fundamentally in the predictability of nature if only sufficient data and appropriate models are available.

Ecosystem scientists are empirical and, often, applied in orientation. Typically, they are associated with professional schools—forestry, fisheries, or range science—or have geophysical backgrounds. In addition, they have a holistic outlook and may focus on material and energy cycles, for example. Ecosystem scientists may be less committed to the predictability of nature, often emphasizing probabilities, stochastic processes, and chaos.

The typical dialogue between groups of population ecologists and ecosystem scientists reveals narrow perspectives on both sides, whether or not one relates to my stereotypes. Consider the fruitless debate among population and community ecologists over competition vs. environment as the preponderant factor affecting community structure—as if it had to be one or the other! Of *course* environment can be a powerful factor in structuring biological communities! Or listen to the supposedly serious questions posed by ecosystem scientists as to whether species need to be considered. One look at the giant sequoia (*Sequoiadendron giganteum*) found in the mixed-conifer forests of the Sierra Nevada would quickly put such proposals in perspective. Of *course* individual species can make an immense difference in structure and function of ecosystems!

Many other factors have probably contributed to the rivalry between population and ecosystem scientists, including competition for grant dollars and prestige. Interestingly, my conversations with colleagues suggest that each group believes the other holds the upper hand and is getting more than its share of both. Arrogance is often evident, with population ecologists presenting themselves as the "true" practitioners of ecological science while many ecosystem scientists view themselves as devotees of more relevant science.

Such schisms do not well serve society or ecological science. If we are to address environmental challenges effectively, we must take a broader view. This book will help us to move toward integration by identifying needs and commonalities and building linkages. Professionally generated programs such as the Ecological Society of America's Sustainable Biosphere Initiative (Lubchenco et al., 1991) are also making a major contribution. And I believe that generational change is producing a more integrated body of scientists. Perhaps the old feuds will disappear along with the original protagonists as was the case with the intense "continuum versus discrete community" debate of the 1950s to 1970s.

FOREST CONSERVATION IN THE PACIFIC NORTHWEST

Forest conservation issues in the Pacific Northwest (PNW)—often oversimplified to being about old-growth forests, spotted owls, or, now, salmon—provide an excellent example of the complex environmental challenges facing society and the most relevant scientific cadre, ecologists. Clearly, the knowledge relevant to these issues covers a broad span of ecological science, including population, ecosystem, and landscape ecology.

Late-successional forest and related stream ecosystems—and the associated species and processes—are central elements in the debate (Norse, 1990). During the last century, forest cutting has reduced late-successional forests (which include both mature and old-growth conditions) to about 15% of their original area. Similar drastic changes have occurred in the quality of stream and river habitats. Recent research has identified many hundreds of species and processes that are at least favored by, if not dependent on, the conditions found in late-successional forests and associated aquatic environments (Forest Ecosystem Management Assessment Team, 1993).

Societal concerns about resource management and conservation issues, which include both old-growth forest ecosystems and specific species (and subspecies and genetic stocks), have legal as well as public opinion dimensions that relate to both economic and value judgements. Laws and related regulations—such as the National Forest Management (NFMA), National Environmental Policy, and Endangered Species Acts (ESA)—are creating a central role for scientific information and scientists in defining and resolving these issues. Several vertebrate species—most notably the northern spotted owl (*Strix occidentalis caurina*) and marbled murrelet (*Brachyramphus marmoratus*)—are already listed as threatened with extinction. Just within the range of the northern spotted owl, 314 anadromous salmonid (fish) stocks are identified as being "at risk" (Forest Ecosystem Management Assessment Team, 1993) with many potential candidates for similar listing under the Endangered Species Act. National Forests are required to maintain well-distributed populations of all native vertebrate species, based on the NFMA, which also includes a broad mandate to maintain biodiversity in the National Forests.

As scientists and as a society, we are faced with challenging technical, legal, and social issues that involve all levels of biological organization—races (stocks), subspecies, species, communities, ecosystems, landscapes, and a major geographical region. Let us take one element—the northern spotted owl—and trace some of the relevant scientific elements.

A Strategy for the Northern Spotted Owl

At the outset, any investigation of the ecological significance of the northern spotted owl necessitates taking a population perspective. Information is needed on reproduction, recruitment, adult mortality by age class and sex, and other population parameters. Mathematical models that integrate this information and provide predictions on population trends are also necessary. And, appropriately, the science of population ecology is well represented in legal and management documents such as the Interagency Scientific Committee report (Thomas et al., 1990) and the "Recovery Plan for the Northern Spotted Owl—Final Draft" (U.S. Department of Interior, 1992).

Northern spotted owls are part of a larger biological community, however. They require prey—mammals such as northern flying squirrels (*Glaucomys sabrinus*), redtree voles (*Arborimus longicaudus*), and woodrats (*Neotoma* spp.)—in sufficient quantity to survive and reproduce. There are competing species, such as the barred owl (*Strix varia*), and others that prey on spotted owls, such as the great horned owl (*Bubo virginianus*). Hence, it is necessary to consider community relationships, food webs, and the interacting population dynamics of an array of species. Plants, plant parts, and detritus, as well as invertebrates, enter the picture as the list of relevant vertebrates expands.

The northern spotted owl and associated organisms require habitat so it is necessary to look at the forest. Trees are the primary biological engineers (sensu Lawton and Jones, Ch. 14) producing a rich diversity of structural conditions within the late-successional forests, including a broad array of individual structures. We need to consider trees of many sizes, species and conditions, snags, and down logs (as well as spatial heterogeneity with the forest as a whole), multiple canopy layers, and highly variable understory conditions, which often reflect variable overstory canopy density. This structural complexity provides numerous niches and performs a variety of functions—modified thermal regimes, hiding cover, nesting sites, foraging habitat, etc.—for the northern spotted owl and associated vertebrates.

These forest habitats, and their included structures and processes, are the stuff of ecosystem science! Once again, scientific information is required, this time on the amounts, distribution, and dynamics of the structural components, as well as the functional relationships between the structures and the various organisms. Processes such as primary productivity, decomposition, and other aspects of carbon and nutrient cycling now become relevant. Long-term perspectives, such as successional change in structure, composition, and function and decomposition of large snags and logs, have to be considered along with short-term dynamics.

Still larger spatial perspectives than the stand-level ecosystem must be considered—interactions between patches of different types, the subject of landscape ecology. The size of an existing patch of late-successional forest strongly affects its ability to provide suitable conditions for northern spotted owls and associated organisms. So does its context—the kind of patches that surround it. A small patch of late-successional forest surrounded by recent clearcuts is strongly modified by edge effects; microclimate within a forest patch may be modified for distances of 200 m or more (Franklin, 1992). We should also ask whether there is a sufficient collective area of suitable patches within the home range of the owl to provide the required prey base. The condition of the matrix—the area of the landscape located between habitat reserves—is very important in determining the success of dispersing owls; open conditions can favor predation by great horned owls, for example.

Developing a strategy for the maintenance of the northern spotted owl obviously requires integration and application of scientific information across the full range of organismal, temporal, and spatial scales. The owl management plans developed over the last 5 years clearly reflect this approach, and illustrate a rapid evolution in the sophistication of the analyses. For example, up until 1990 the Forest Service planned to deal with the owl through dispersed "spotted owl management areas" (SOMAs) of 400 to 1500 ha. This was not viewed as a credible plan for maintaining northern spotted owl populations and a scientific team, the Interagency Scientific Committee to Address the Conservation of the Northern Spotted Owl, was convened by the chief of the U.S.D.A. Forest Service, under the leadership of wildlife biologist Dr. Jack W. Thomas. The Interagency Committee's analysis (Thomas et al., 1990) produced a dramatically different strategy, which created a series of large "habitat conservation areas" (HCAs) sufficient to provide habitat for at least 20 breeding pairs of owls spaced at intervals of no more than 18 km. In addition to creating over 3 million ha of forest reserves (HCAs) the strategy also required modified management of the forests on intervening lands (the matrix) in order to improve the potential for successful owl dispersal. This is known as the "50–11–40" rule; it requires 50% of the matrix (nonreserved lands) to be maintained in forests with an average tree diameter of 11 inches at breast height and with a 40% canopy cover.

The development of a matrix prescription was revolutionary with regard to both science and resource management because it recognized that: (1) a corridor-based approach would probably not be very effective at aiding owl dispersal, (2) more favorable overall conditions in the matrix would aid dispersal, and (3) the importance of both the matrix and habitat reserves should be explicitly recognized. Managing for northern spotted owls with an ecosystem perspective has affected management practices over the entire landscape, not just on some reserves.

But Owls Alone Are Not Enough!

As innovative as they were, the conservation strategies developed for the northern spotted owl still focused on a single species; any consideration of old-growth forests, salmonids, or the thousands of other late-successional species was strictly peripheral—and demonstrably inadequate! The HCAs were designed to provide for the maintenance of owl populations but have minimal impact on potential forest harvest. Hence, the habitat reserves, or Designated Conservation Areas (DCAs) as they came to be known in the final northern spotted owl recovery plan (U.S. Department of Interior, 1992), missed hundreds of thousands of hectares of the best remaining old-growth forest as identified by the Scientific Panel on Late Successional Forest Ecosystems (Johnson et al., 1991). A worst-case example was on the Umpqua National Forest in the Oregon Cascade Range, where the DCAs captured <50% of the Most Significant Late-Successional Forest (LS/OG1 of Johnson et al., 1991). The limitations and inefficiencies of addressing the conservation of biological diversity one species at a time were becoming apparent.

Obviously, integrated approaches to studying late-successional forest and stream ecosystems and their constituent species are necessary to avoid conflicting conservation objectives and attain maximum efficiency in designing reserves and prescribing matrix management approaches. Several science-based analyses, developed following the work of the Interagency Scientific Committee (Thomas et al., 1990), represent a rapid evolution in this approach (Johnson et al., 1991; Scientific Assessment Team, 1993; Forest Ecosystem Management Assessment Team, 1993). The analysis conducted for President Clinton (Forest Ecosystem Management Assessment Team, 1993, known as FEMAT) is the most recent and comprehensive effort of this type.

There are many common elements in these integrated approaches. One important feature is the creation of reserve systems that are of maximum collective value in conserving old-growth forests and high-quality stream and fisheries habitat, and in maintaining populations of sensitive or endangered organisms. For example, late-successional reserves are designed to overlap and protect critical watersheds as far as possible; further, reserves are of a size and distribution to meet the criteria established for northern spotted owls. Another common element is extensive riparian protection, even on intermittent and the smallest permanent streams. Yet another important element is the development of management guidelines for the harvested timberlands within the matrix, which provides habitat and other conditions required by a broad array of species and processes, as well as improving patch connectivity. For example, retention of large trees, snags, down logs, and small patches of older forest are required on cutover areas in addition to providing some overall level of forest cover within the matrix as prescribed by the 50–11–40 rule.

These integrated proposals are attempts to move away from species-based and toward ecosystem-based approaches, which is viewed as essential to the development of successful programs for conserving biological diversity (Franklin, 1993). Yet laws such as the ESA and the NFMA require that at least some species receive individual attention; hence a complete shift to an ecosystem approach has not been possible—even had such a shift been viewed as desirable—and is not likely to occur in the near future since changes in current law are unlikely. Hence, individual species and stocks continue to be a concern in the integrated plans even while the problem of assessing how thousands of species will respond to alternative management scenarios is now very clear.

LESSONS FOR ECOLOGICAL
SCIENTISTS AND EDUCATORS

The Pacific Northwestern forest controversy is proving to be a major learning experience in the need to link species conservation, environmental protection, and resource management and how this can be accomplished. Similar lessons are being learned in other forest, steppic, and desert landscapes throughout the world. Species conservation and environmental conservation are spatially pervasive concerns; they cannot be isolated in certain locales within landscapes and regions, nor from the management of natural resources for utilitarian purposes. *Biological diversity is not primarily a "set aside" issue!*

Take-home lessons in the PNW are many. It is clear that we cannot plan one species at a time and that we cannot separate terrestrial from aquatic concerns. Conservation strategies must deal with management practices on the entire matrix; they cannot be based exclusively on the creation and protection of habitat reserves. Creating corridors is just one perspective on the much more important issue of landscape connectivity. These are obvious lessons, perhaps, but their implications have not been fully absorbed by scientists, resource managers, decision-makers, or the general public.

This also illustrates how scientific and policy analyses have become the cutting edge of ecological science. Many (most?) important ecological questions are actually emerging from our efforts to apply existing knowledge in a real world setting, rather than in academic isolation. For example, the hypothesized importance of the matrix as habitat and to provide for landscape connectivity, suggests a whole new array of important questions for scientific study.

We are also learning how science, management, and policy are related in the real world. The theoretical model approximates the following sequence: (1) new knowledge is developed by scientific research; (2) this knowledge is used as the basis for proposing new or altered management practices; (3) the proposed practices undergo experimental testing; and, following modification and perhaps further testing, (4) the new practices are implemented. Major paradigm shifts in resource management almost never follow such a sequence, however. Societal objectives drive paradigm shifts in resource management and, under the best conditions, resource managers have utilized the best existing science to develop management programs that are actually working hypotheses rather than proven concepts. But they are *not* guesses. We know enough to formulate plans or hypotheses that have a high probability of success.

The implementation of dispersed patch clearcutting on the federal timberlands of the PNW in the mid-twentieth century is a perfect example of resource management as a working hypothesis. It reflected societal objectives (as well as the Forest Service's internal objectives) and the best existing scientific information on the ecology of Douglas fir. The system was not tested experimentally prior to being implemented on a massive scale. Alterations in the system came through ad hoc feedbacks since there was no systematized adaptive management procedure.

Once again forest resource management is undergoing a paradigm shift and once again it is being implemented as a working hypothesis. Hopefully, the new paradigm will utilize the best of existing scientific information; this process is being assisted by a social and legal environment that provides science and scientists a central role. And, it is hoped, a systematic program for adaptive management will be implemented along with the new management schemes. It is critical that management programs be constructed and used so as to consciously create scientific knowledge, knowledge that can be used to improve both our understanding and subsequent management of ecosystems.

The lessons for scientists and educators are many. Societal issues of this magnitude require the combined effort of the best of our population, community, ecosystem, and landscape ecologists and the effective collaboration of scientists, resource managers, and interested citizens. The importance of incorporating scientific synthesis and synthesizers in designing the soundest management practices (or working hypotheses) and developing appropriate evaluative and adaptive programs is clear.

Are these linkages, attitudes, and activities adequately reflected in our educational and research programs? They are not, in my opinion. Individual disciplines and organismal levels still dominate our training programs; neither the philosophy nor the practice of interdisciplinary science receives sufficient attention. Similarly, academics are often penalized for interdisciplinary efforts that take them outside traditional departmental boundaries. Yet interdisciplinary approaches need to extend well beyond biology to the physical and social sciences and to resource management professionals.

In addition to changing our educational and reward systems we need to do a much better job of teaching our students about the need to grow intellectually throughout their professional lives and modeling this for them. Most of us do not do this very well. We do not prepare students to let go of old paradigms and embrace new ones. More often, we teach them to aggressively defend the hypotheses that they were taught or developed early in their careers as though they are universal truths for all systems at all times! Aldo Leopold is a good model of an ecologist who grew intellectually; his early writings in support of control of predators and of wildfire (Flader and Callicott, 1991) hardly seem written by the same individual who became the intellectual father of a generation of ecosystem thinkers!

Ecological science has come of age or, at least, has arrived at its age of opportunity. Are ecologists up to the tasks society is laying before us? We certainly will not be if we cannot effectively link species and ecosystems! And if ecologists cannot meet the challenge, society will turn elsewhere for expertise. Consider the decline and potential demise of the forestry profession today. We can and must provide society with the sound ecological knowledge it needs— but the science and its practitioners must undergo some required maturation.

REFERENCES

Chapter numbers or sections in which references are cited are shown in bold face.

ABER, J.D., K.J. NADELHOFFER, P. STEUDLER, and J.M. MELILLO. (1989) Nitrogen saturation in northern forest ecosystems—hypotheses and implications. *Bioscience* 39:378–386. **(9)**

ABRAMS, P.A. (1993) Effects of increased productivity on the abundances of trophic levels. *Am. Nat.* 141:351–371. **(16)**

ADMIRAAL, W., L.A. BOUWMAN, L. HEOKSTRA, and K. ROMEYN. (1985) Qualitative and quantitative interactions between microphytobenthos and herbivorous meiofauna on a brackish intertidal mudflat. *Int. Rev. Gesamt. Hydrobiol.* 68:175–191. **(4)**

ADMIRAAL, W., and H. PELETIER. (1980) Influence of seasonal variations in temperature and light on the growth rate of cultures and natural populations of intertidal diatoms. *Mar. Ecol. Prog. Ser.* 2:35–43. **(4)**

AERTS, R. (1989) Nitrogen use efficiency in relation to nitrogen availability and plant community composition, in *Causes and Consequences of Variation in Growth Rate and Productivity in Higher Plants* (ed. H. Lambers), SPB Academic Publishing, The Hague, The Netherlands, pp. 285–297. **(24)**

ÅGREN, G.I., and E. BOSATTA. (1987) Theoretical analysis of the long-term dynamics of carbon and nitrogen in the soil. *Ecology* 68:1181–1189. **(16)**

ALLAN, J.D., and C.E. GOULDEN. (1980) Some aspects of reproductive variation among freshwater zooplankton, in *Evolution and Ecology of Zooplankton Communities* (ed. W.C. Kerfoot), University Press of New England, Hanover, New Hampshire, pp. 388–410. **(7)**

ALLDREDGE, A.L., and Y. COHEN. (1987) Can microscale chemical patches persist in the sea? Microelectrode study of marine snow, fecal pellets. *Science* 235:689–691. **(5)**

ALLDREDGE, A.L., and C. GOTSCHALK. (1988) In situ settling behavior of marine snow. *Limnol. Oceanogr.* 33:339–351. **(5)**

ALLDREDGE, A.L., and M.W. SILVER. (1988) Characteristics, dynamics and significance of marine snow. *Prog. Oceanogr.* 20:41–82. **(5)**

ALLEN, T.F.H., and T. W. HOEKSTRA. (1992) *Toward a Unified Ecology.* Columbia University Press, New York. **(1; 26; 27)**

ALLEN, T.F.H., and T.B. STARR. (1982) *Hierarchy.* University of Chicago Press, Chicago. **(17; 19)**

ALLEN-MORLEY, C.R., and D.C. COLEMAN. (1989) Resilience of soil biota in various food webs to freezing perturbations. *Ecology* 70:1127–1141. **(16)**

ALLER, R.C. (1982) The effects of macrobenthos on chemical properties of marine sediment and overlying water, in *Animal–Sediment Relations* (eds. P.L. McCall and M.J.S. Tevesz), Plenum, New York, pp. 53–102. **(3; 4)**

ALPINE, A.E., and J.E. CLOERN. (1992) Trophic interactions and direct physical effects control phytoplankton biomass and production in an estuary. *Limnol. Oceanogr.* 37:936–945. **(7)**

ANDERSEN, D.C. (1987a) Below-ground herbivores in natural communities: a review emphasizing fossorial animals. *Q. Rev. Biol.* 62:261–286. **(12)**

ANDERSEN, D.C. (1987b) *Geomys bursarius* burrowing patterns: influence of season and food patch structure. *Ecology* 68:1306–1318. **(12)**

ANDERSEN, T., and D.O. HESSEN. (1991) Carbon, nitrogen, and phosphorus content of freshwater zooplankton. *Limnol. Oceanogr.* 36:807–814. **(23)**

ANDERSEN, V., and P. NIVAL. (1989) Modelling of phytoplankton population dynamics in an enclosed water column *J. Mar. Biol. Assoc. U.K.* 69:625–646. **(18)**

ANDERSON, J.M. (1988a) Invertebrate-mediated transport processes in soils. *Agric. Ecosyst. Environ.* 24:5–19. **(10)**

ANDERSON, J.M. (1988b) Spatiotemporal effects of invertebrates on soil processes. *Biol. Fertil. Soils* 6:216–227. **(10; 16)**

ANDERSON, J.M. (1989) Forest soils as short, dry rivers. *Gesellschaft Ökologie* 17:33–46. **(10)**

ANDERSON, J.M. (1994) Functional attributes of biodiversity in land use systems, in *Soil Resilience and Sustainable Land Use Systems* (eds. D.J. Greenland and I. Szabolcs), Commonwealth Agricultural Bureau International, Wallingford, pp. 267–290. **(10)**

ANDERSON, J.M., and A. MACFADYEN. (1976) *The Role of Terrestrial and Aquatic Organisms in Decomposition Processes.* Blackwell Scientific, London. **(4)**

ANDERSON, R.A. (1992) Diversity of eukaryotic algae. *Biodiversity Conserv.* 1:267–292. **(14)**

ANDERSON, R.M. (1989) Discussion: ecology of pests and pathogens, in *Perspectives in Theoretical Ecology* (eds. J. Roughgarden, R.M. May, and S.A. Levin), Princeton University Press, Princeton, pp. 348–361. **(28)**

ANDERSON, R.S. (1972) Zooplankton composition and change in an alpine lake. *Verh. Internat. Verein. Limnol.* 18:264–268. **(30)**

ANKER, K. (1975) On the influence of sewage pollution on inshore benthic communities in the south of Kiel Bay. 2. Helgolander wiss. *Meersunters* 27:408–438. **(4)**

APPLE, L.L., B.H. SMITH, J.D. DUNDER, and B.W. BAKER. (1984) The use of beavers for riparian/aquatic habitat restoration of cold desert, gully-cut stream systems in southwestern Wyoming, in *Proceedings, American Fisheries Society/Wildlife Society Joint Chapter Meeting*, February 8–10, Logan, Utah, pp. 123–130. **(12)**

ARCHER, S.A., C. SCIFRES, and C.R. BASSHAM. (1988) Autogenic succession in a subtropical savanna: conversion of a grassland to a thorn woodland. *Ecol. Monogr.* 58:111–127. **(20)**

ASPER, V.L. (1987) Measuring the flux and sinking speed of marine snow aggregates. *Deep Sea Res.* 34:1–17. **(5)**

AVNIMELECH, Y., B.W. TROEGER, and L.W. REED. (1982) Mutual flocculation of algae and clay: evidence and implications. *Science* 216:63–65. **(5)**

AXELROD, D.I. (1985) Rise of the grassland biome, central North America. *Bot. Rev.* 51:164–196. (**24**)

BÅMSTEDT, U., and H.R. SKJOLDAL. (1980) RNA concentration of zooplankton: relationship with size and growth. *Limnol. Oceanogr.* 25:304–316. (**23**)

BANTA, G.T. (1992) *Decomposition and Nitrogen Cycling in Coastal Marine Sediments: Controls by Temperature, Organic Matter Inputs and Benthic Macrofauna.* Boston University Marine Program, 1992. Dissertation. (**4**)

BARRIE, L.A., D. GREGOR, B. HARGRAVE, R. LAKE, D. MUIR, R. SHEARER, B. TRACEY, and T. BIDLEMAN. (1992) Arctic contaminants: sources, occurrence and pathways. *Sci. Tot. Environ.* 122:1–74. (**30**)

BARTELL, S.M., W.G. CALE, R.V. O'NEILL, and R.H. GARDNER. (1988) Aggregation error: research objectives and relevant model structure. *Ecol. Modell.* 41:157–168. (**19**)

BASEY, J.M., S.H. JENKINS, and G.C. MILLER. (1990) Food selection by beavers in relation to inducible defenses of *Populus tremuloides. Oikos* 59:57–92. (**12**)

BASNET, K., G.E. LIKENS, F.N. SCATENA, and A.E. LUGO. (1992) Hurricane Hugo: damage to a tropical rain forest in Puerto Rico. *J. Trop. Ecol.* 8:47–55. (**14**)

BATESON, G. (1972) *Steps to an Ecology of Mind.* Chandler Publishing, San Francisco. (**29**)

BEARE, M.H., R.W. PARMELEE, P.F. HENDRIX, W. CHENG, D.C. COLEMAN, and D.A. CROSSLEY. (1992) Microbial and faunal interactions and effects on litter nitrogen and decomposition in agroecosystems. *Ecol. Monogr.* 62:569–591. (**10; 11; 16**)

BEDFORD, B., M. BRINSON, R. SHARITZ, A. VAN DER VALK, and J. ZEDLER. (1992) Evaluation of proposed revisions to the 1989 "Federal Manual for Identifying and Delineating Jurisdictional Wetlands." *Bull. Ecol. Soc. Am.* 73:14–23. (**28**)

BEGON, M., J.L. HARPER, and C.R. TOWNSEND. (1990) *Ecology.* Blackwell, Oxford. (**Issues; 1; 16**)

BEGON, M., and M. MORTIMER. (1986) *Population Ecology: A Unified Study of Animals and Plants.* Sinauer, Sunderland. (**Issues**)

BELL, R.H.V. (1982) The effect of soil nutrient availability on community structure in African savannas, in *Ecology of Tropical Savannas* (Ecological Studies 42) (eds. B.J. Huntley and B.H. Walker), Springer-Verlag, Berlin, pp. 193–216. (**24**)

BELOVSKY, G. (1990) How important are nutrient constraints in optimal foraging models or are spatial/temporal factors more important? in *Behavioural Mechanisms of Food Selection* (ed. R.N. Hughes), Springer-Verlag, Berlin, pp. 255–279. (**23**)

BENDER, E.A., T.J. CASE, and M.E. GILPIN. (1984) Perturbation experiments in community ecology: theory and practice. *Ecology* 65:1–13. (**22**)

BERBEROVIC, R. (1990) Elemental composition of two coexisting *Daphnia* species during the seasonal course of population development in Lake Constance. *Oecologia* 84:340–350. (**23**)

BERDOWSKY, J.J.M., and R. ZEILINGA. (1987) Transition from heathland to grassland: damage effects of the heather beetle. *J. Ecol.* 75:159–175. (**8**)

BERNER, R. (1984) Sedimentary pyrite formation: an update. *Geochim. Cosmochim. Acta* 48: 605–615. (**4**)

BERTNESS, M.D. (1984a) Habitat and community modification by an introduced herbivorous snail. *Ecology* 65:370–381. (**8; 14**)

BERTNESS, M.D. (1984b) Ribbed mussels and *Spartina alterniflora* production in a New England salt marsh. *Ecology* 65:1794–1807. (**14**)

BERTNESS, M.D. (1985) Fiddler crab regulation of *Spartina alterniflora* production on a New England salt marsh. *Ecology* 66:1042–1055. (**14**)

BETTS, J.N. and H.D. HOLLAND. (1991) The oxygen content of ocean bottom waters, the burial efficiency of organic carbon, and the regulation of atmospheric oxygen. *Paleo, Paleo, Paleo* (*Global and Planetary Change Section*) 97:5–18. (**13**)

BIANCHI, T.S., and C.G. JONES. (1991) Density-dependent feedbacks between consumers and their resources, in *Comparative Analyses of Ecosystems: Patterns, Mechanisms, and Theories* (eds. J.J. Cole, G. Lovett, and S. Findlay), Springer-Verlag, New York, pp. 331–340. **(8)**

BLOOM, A.L. (1978) *Geomorphology.* Prentice-Hall, Englewood Cliffs, NJ. **(14)**

BODALY, R.A., J.W.M. RUDD, R.J.P. FUDGE, and C.A. KELLY. (1993) Mercury concentrations in fish related to size of remote Canadian Shield Lakes. *Can. J. Fish. Aquat. Sci.* 50:980–987. **(30)**

BOLIN, B., and R.B. COOK (eds.). (1983) *The Major Biogeochemical Cycles and Their Interactions.* John Wiley & Sons, Chichester, England. **(23)**

BORMANN, F.H., and G.E. LIKENS. (1981) *Pattern and Process in a Forested Ecosystem.* Springer-Verlag, New York. **(9)**

BOSATTA, E., and G.I. ÅGREN. (1991) Dynamics of carbon and nitrogen in the soil: a generic theory. *Am. Nat.* 138:227–245. **(16)**

BOTKIN, D.B. (1992) *Forest Dynamics: An Ecological Model.* Oxford University Press, Oxford. **(9)**

BOULTON, A.J., C.G. PETERSON, N.B. GRIMM, and S.G. FISHER. (1992) Stability of an aquatic macroinvertebrate community in a multi-year hydrologic disturbance regime. *Ecology* 73:2192–2207. **(1)**

BOWEN, H.J.M. (1979) *Environmental Chemistry of the Elements.* Academic Press, London. **(23)**

BOX, G.E.P., W.G. HUNTER, and J.S. HUNTER. (1978) *Statistics for Experimenters.* John Wiley & Sons, New York. **(22)**

BRADLEY M.C., N. PERRIN, and P. CALOW. (1991) Energy allocation in the cladoceran *Daphnia magna* Straus, under starvation and refeeding. *Oecologia* 86:414–418. **(18)**

BREMEN, H., and C.T. DE WIT. (1983) Rangeland productivity and exploitation in the Sahel. *Science* 221:1341–1347. **(24)**

BREZONIK, P.L., J.G. EATON, T.M. FROST, P.J. GARRISON, T.K. KRATZ, C.E. MACH, J.H. McCORMICK, J.A. PERRY, W.A. ROSE, C.J. SAMPSON, B.C.L. SHELLEY, W.A. SWENSON, and K.E. WEBSTER. (1993) Experimental acidification of Little Rock Lake, Wisconsin: Chemical and biological changes over the pH range 6.1 to 4.7. *Can. J. Fish. Aquat. Sci.* 50:1101–1121. **(22)**

BRIAND, F., and COHEN, J.E. (1984) Community food webs have scale- invariant structure. *Nature* 307:264–266. **(17)**

BROECKER, W.S., and T.-H. PENG. (1982) *Tracers in the Sea.* Eldigio Press, Palisades, NJ. **(13)**

BROOKS, P.C., J. WU, and J.A. OCIO. (1991) Soil microbial biomass dynamics following the addition of cereal straw and other substrates to soil, in *The Ecology of Temperate Cereal Fields* (eds. L.G. Firbank, N. Carter, J.F. Darbyshire, and G.R. Potts), Blackwell Scientific Publications, Oxford, pp. 95–111. **(10)**

BROWN, J.H. (1981) Two decades of homage to Santa Rosalia: toward a general theory of diversity. *Am. Zool.* 21:877–888. **(2)**

BROWN, J.H. (1994a) Complex ecological systems, in *Complex Adaptive Systems*, (eds. G.A. Cowan and D. Pines), Santa Fe Institute, Santa Fe, New Mexico (in press). **(2)**

BROWN, J.H. (1994b) *Macroecology.* University of Chicago Press, Chicago. **(26)**

BROWN, J.H., D.W. DAVIDSON, J.C. MUNGER, and R.S. INOUYE. (1986) Experimental community ecology: the desert granivore system, in *Community Ecology* (eds. J. Diamond and T.J. Case), Harper & Row, New York, pp. 41–62. **(7)**

BROWN, J.H., and E.J. HESKE. (1990) Control of a desert-grassland transition by a keystone rodent guild. *Science* 250:1705–1707. **(2; 7)**

BROWN, J.H., P.A. MARQUET, and M.L. TAPER. (1993) Evolution of body size: consequences of an energetic definition of fitness. *Am. Nat.* 142:573–584. **(2)**

BROWN, J.H., O.J. REICHMAN, and D.W. DAVIDSON. (1979) Granivory in desert ecosystems. *Annu. Rev. Ecol. Syst.* 10:201–27. **(15)**

BROWN, V.K., and A.C. GANGE. (1990) Insect herbivory below ground. *Adv. Ecol. Res.* 20:1–58. (8)

BURKHOLDER, J.M. (1992) Phytoplankton and episodic suspended sediment loading: phosphate partitioning and mechanisms for survival. *Limnol. Oceanogr.* 37:974–988. (5)

BURNS, I.G. (1992) Influence of plant nutrient concentration on growth rate: use of a nutrient interruption technique to determine critical concentrations of N, P and K in young plants. *Plant Soil* 142:221–233. (23)

BUTLER, N.M., C.A. SUTTLE, and W.E. NEILL. (1989) Discrimination by freshwater zooplankton between single algal cells differing in nutritional status. *Oecologia* 78:368–372. (23)

CALE, W.G., and P.L. ODELL. (1979) Concerning aggregation in ecosystem modeling, in *Theoretical Systems Analysis* (ed. E. Halfon), Academic Press, New York, pp. 283–298. (19)

CALE, W.G., and P.L. ODELL. (1980) Behavior of aggregate state variables in ecosystem models. *Math. Biosci.* 49:121–137. (19)

CALE, W.G., and R.V. O'NEILL. (1988) Aggregation and consistency problems in theoretical models of exploitative resource competition. *Ecol. Modell.* 40:97–109. (19)

CALE, W.G., R.V. O'NEILL, and R.H. GARDNER. (1983) Aggregation error in nonlinear ecological models. *J. Theor. Biol.* 100:539–550. (19)

CALOW, P., P. ARMITAGE, P. BOON, P. CHAVE, E. COX, A. HILDREW, M. LEARNER, L. MALTBY, G. MORRIS, J. SEAGER, and B. WHITTON. (1990) *River Water Quality.* British Ecological Society, Ecological Issues No. 1, Field Studies Council, London. (28)

CALVERT, S.E., M. BASTIN, and T.F. PEDERSEN. (1992) Lack of evidence for enhanced preservation of organic matter in sediments underlying the oxygen minimum of the Gulf of California. *Geology* 20:757–760. (13)

CAMMEN, L.M. (1980) The significance of microbial carbon in the nutrition of the deposit feeding polychaete *Nereis succinea. Mar. Biol.* 61:9–20. (3)

CANFIELD, D.E. (1989) Sulfate reduction and oxic respiration in marine sediments: implications for organic carbon preservation in euxinic environments. *Deep Sea Res.* 36:121–138. (4)

CANHAM, C.D., A.C. FINZI, S.W. PACALA, and D.H. BURBANK. (1994) Causes and consequences of resource heterogeneity in forests: Interspecific variation in light transmission by canopy trees. *Can. J. For. Res.* 24:337–339. (9)

CANHAM, C.D., and P.L. MARKS. (1985) The response of woody plants to disturbance: patterns of establishment and growth, in *The Ecology of Natural Disturbance and Patch Dynamics* (eds. S.T.A. Pickett and P.S. White), Academic Press, New York, pp. 197–216. (9)

CARGILL, S.M., and R.L. JEFFERIES. (1984) The effects of grazing by lesser snow geese on the vegetation of a sub-arctic salt marsh. *J. Appl. Ecol.* 21:669–686. (8)

CARON, D.A., P.G. DAVIS, L.P. MADIN, and J. McN. SIEBURTH. (1986) Enrichment of microbial populations in macroaggregates (marine snow) from surface waters of the North Atlantic. *J. Mar. Res.* 44:543–565. (5)

CARPENTER, R.C. (1986) Partitioning herbivory and its effects on coral reef algae. *Ecol. Monogr.* 56:345–363. (8)

CARPENTER, S.R. (ed.). (1988) *Complex Interactions in Lake Communities.* Springer-Verlag, New York. (8; 15)

CARPENTER, S.R., T.M. FROST, A.R. IVES, and J.F. KITCHELL. (1994) Complexity, cascades, and compensation in ecosystems, in *Biodiversity: Its Complexity and Role.* National Institute for Environmental Studies, Tsukuba, Japan (in press). (Issues)

CARPENTER, S.R., T.F. FROST, J.F. KITCHELL, and T.K. KRATZ. (1993) Species dynamics and global environmental change: a perspective from ecosystem experiments, in *Biotic Interactions and Global Change.* (eds. P.M. Kareiva, J.G. Kingsolver, and R.B. Huey), Sinauer Associates, Sunderland, Massachusetts, pp. 267–279. (7; 26)

CARPENTER, S.R., T.M. FROST, J.F. KITCHELL, T.K. KRATZ, D.W. SCHINDLER, J. SHEARER, W.G. SPRULES, M.J. VANNI, and A.P. ZIMMERMAN. (1991) Patterns of primary production and herbivory in 25 North American lake ecosystems, in *Comparative Analyses of Ecosystems: Patterns, Mechanisms, and Theories* (eds. J.J. Cole, G. Lovett, and S. Findlay), Springer-Verlag, New York, pp. 67–96. **(8)**

CARPENTER, S.R., and J.F. KITCHELL. (1988) Consumer control of lake productivity. *BioScience* 38:764–769. **(16)**

CARPENTER, S.R., and J.F. KITCHELL (eds.). (1993) *The Trophic Cascade in Lakes.* Cambridge University Press, Cambridge. **(7; 22)**

CARPENTER, S.R., J.F. KITCHELL, and J.R. HODGSON. (1985) Cascading trophic interactions and lake productivity. *BioScience* 35:634–639. **(7; 30)**

CARPENTER, S.R., P.R. LEAVITT, J.J. ELSER, and M.M. ELSER. (1988) Chlorophyll budgets: response to food web manipulations. *Biogeochemistry* 6:79–90. **(7)**

CARPENTER, S.R., and D.M. LODGE. (1986) Effects of submersed macrophytes on ecosystem processes. *Aquat. Bot.* 26:341–370. **(8)**

CARPENTER, S.R., J. MORRICE, P.A. SORANNO, J.J. ELSER, N.A. MACKAY, and A. ST. AMAND. (1993) Primary production and its interactions with nutrient and light transmission, in *The Trophic Cascade in Lakes* (eds. S.R. Carpenter and J.F. Kitchell), Cambridge University Press, Cambridge, pp. 225–251. **(7)**

CATES, R.O., and G.H. ORIANS. (1975) Successional status and the palatability of plants to generalized herbivores. *Ecology* 56:410–418. **(6)**

CHAPIN, F.S. III. (1980) The mineral nutrition of wild plants. *Annu. Rev. Ecol. Syst.* 11:233–260. **(24)**

CHAPIN, F.S. III. (1987) Plant responses to multiple environmental factors. *Bioscience* 37:49–57. **(24)**

CHAPIN, F.S. III. (1993) Functional role of growth forms in ecosystem and global processes, in *Scaling Physiological Processes: Leaf to Globe* (eds. T. Ehleringer and H. Mooney), Academic Press, New York, pp. 287–312. **(26)**

CHAPIN, F.S. III., E.D. SCHULZE, and H.A. MOONEY. (1990) The ecology and economics of storage in plants. *Annu. Rev. Ecol. Syst.* 21:423–447. **(20)**

CHARLES, D.F., and J.P. SMOL. (1988) New methods for using diatoms and chrysophytes to infer past pH of low-alkalinity lakes. *Limnol. Oceanogr.* 33:1451–1472. **(30)**

CHENG, I.-J., J.S. LEVINTON, M.M. MCCARTNEY, D. MARTINEZ, and M.J. WEISSBURG. (1993) A bioassay approach to seasonal variation in the nutritional value of sediment. *Mar. Ecol. Prog. Ser.* 94:275–285. **(3)**

CHERRETT, J.M. (ed.). (1989) *Ecological Concepts: The Contribution of Ecology to the Understanding of the Natural World.* Blackwell Scientific Publications, Oxford. **(17)**

CHERRETT, J.M. (1990) The contribution of ecology to our understanding of the natural world: a review of some key ideas. *Physiol. Ecol. Japan* 27:1–16. **(25)**

CHESSON, P., and N. HUNTLY. (1989) Short-term instabilities and long-term community dynamics. *Trends Ecol. Evol.* 4:293–298. **(8)**

CHESSON, P., and N. HUNTLY. (1993) Hierarchies of temporal variation and species diversity. *Plant Species Biol.* 8:195–206. **(8)**

CHIPMAN, J.S. (1975) Optimal aggregation in large scale econometric models. *Sankhya Ser. C* 37:121–159. **(19)**

CHIPMAN, J.S. (1976) Estimation and aggregation in econometrics: an application of the theory of generalized inverses, in *Generalized Inverses and Applications* (ed. M.Z. Nashed), Academic Press, New York, pp. 549–769. **(19)**

CHRISTENSEN, N.L. (1985) Shrubland fire regimes and their evolutionary consequences, in *The Ecology of Natural Disturbance and Patch Dynamics* (eds. S.T.A. Pickett and P.S. White), Academic Press, Orlando, pp. 85–100. **(14)**

CLARHOLM, M. (1985) Possible roles of roots, bacteria, protozoa and fungi in supplying nitrogen to plants, in *Ecological Interactions in Soil* (eds. A.H. Fitter, D. Atkinson, D.J. Read, and M.B. Usher), Blackwell, Oxford, pp. 355–366. (11; 16)

CLARHOLM, M., B. POPOVIC, T. ROSSWALL, B. SODERSTROM, B. SOHLENIUS, H. STAAF, and A. WIREN. (1981) Biological aspects of nitrogen mineralization in humus from a pine forest podsol incubated under different moisture and temperature conditions. *Oikos* 37:137–145. (11)

CLARK, R.B. (1985) Progress and expectation in biology: comment, in *The Identification of Progress in Learning* (ed. T. Hägerstrand), Cambridge University Press, Cambridge, pp. 61–68. (Issues)

CLEMENTS, R.O., MURRAY, P.J., and STURDY, R.G. (1991) The impact of 20 years' absence of earthworms and three levels of N fertilizer on a grassland soil environment. *Agric. Ecosyst. Environ.* 36:75–85. (10)

CLOERN J.E., and R.T. CHENG. (1981) Simulation model of *Skeletonema costatum* population dynamics in northern San Fransisco Bay, California. *Estuarine Coastal Shelf Sci.* 12:83–100. (18)

COBLENTZ, B.E. (1980) Effects of feral goats on the Santa Catalina Island Ecosystem, in *The California Islands: Proceedings of a Multidisciplinary Symposium* (ed. D.M. Power), Santa Barbara Museum of Natural History, Santa Barbara, pp. 167–170. (15)

COHEN, J.B. (1985) *Revolution in Science*. Belknap Press, Cambridge. (Issues)

COHEN, J.E., R.A. BEAVER, S.H. COUSINS, D.L. DEANGELIS, L. GOLDWASSER, K.L. HEONG, R.D. HOLT, A.J. KOHN, J.H. LAWTON, N.D. MARTINEZ, R. O'MALLEY, L.M. PAGE, B.C. PATTEN, S.L. PIMM, G.A. POLIS, M. REJMÁNEK, T.W. SCHOENER, K. SCHOENLY, W.G. SPRULES, J.M. TEAL, R.E. ULANOWICZ, P.H. WARREN, H.M. WILBUR, and P. YODZIS. (1993) Improving food webs. *Ecology* 74:252–258. (17)

COHEN, J.E., BRIAND, F., and NEWMAN, C.M. (1990) *Community Food Webs: Data and Theory*. Springer-Verlag, Berlin. (17)

COLINVAUX, P. (1986) *Ecology*. John Wiley & Sons, New York. (25)

COLINVAUX, P. (1993) *Ecology*, Second Edition. John Wiley & Sons, New York. (26)

COLLINS, S.L., and L.L. WALLACE (eds.). (1990) *Fire in North American Tallgrass Prairie*. University of Oklahoma Press, Norman, Oklahoma. (24)

CONNELL, J.H. (1978) Diversity in tropical rain forests and coral reefs. *Science* 199:1302–1310. (25)

CONNELL, J.H., and E. ORIAS. (1964) The ecological regulation of species diversity. *Am. Nat.* 98:399–414. (25)

CONNELL, J.H., and W.P. SOUSA. (1983) On the evidence needed to judge ecological stability or persistence. *Am. Nat.* 121:789–824. (23)

COOK, R.B., C.A. KELLY, D.W. SCHINDLER, and M.A. TURNER. (1986) Mechanisms of hydrogen ion neutralization in an experimentally acidified lake. *Limnol. Oceanogr.* 31:134–148. (30)

COOPER, S.D., S.J. WALDE, and B.L. PECKARSKY. (1990) Prey exchange rates and the impact of predators on prey populations in streams. *Ecology* 71:1503–1514. (1)

COÛTEAUX, M-M., M. MOUSSEAU, M-L. CÉLÉRIER, and P. BOTTNER. (1991) Increased atmospheric CO_2 and litter quality: decomposition of sweet chestnut leaf litter with animal food webs of different complexities. *Oikos* 61:54–64. (16)

COWAN, G.A., and D. PINES (EDS). (1994) *Complex Adaptive Systems*. Santa Fe Institute, Santa Fe, New Mexico (in press). (2)

COX, G.W., and C.G. GAKAHU. (1985) Mima mound microtopography and vegetation pattern in Kenyan savannas. *J. Trop. Ecol.* 1:23–36. (14)

CURRIE, D.J. (1991) Energy and large-scale patterns of animal- and plant-species richness. *Am. Nat.* 137:27–49. (25)

DAHM, C.N., and J.R. SEDELL. (1986) The role of beaver on nutrient cycling in streams. *J. Colo.-Wyo. Acad. Sci.* 18:32. (12)

DAILY, G.C., P.R. EHRLICH, and N.M. HADDAD. (1993) Double keystone bird in a keystone species complex. *Proc. Natl. Acad. Sci.* 90:592–594. (14)

D'ANTONIO, C.M., and P.M. VITOUSEK. (1992) Biological invasions by exotic grasses, the grass/fire cycle, and global change. *Annu. Rev. Ecol. Syst.* 23:63–88. (1; 24; 26)

DARLINGTON, J.P.E.C. (1982) The underground passages and storage pits used in foraging by a nest of the termite *Macrotermes michaelseni* in Kajaido. *Kenya J. Zool.* 198:237–247. (10)

DARWIN, C. (1881) *The Formation of Vegetable Mould, Through the Action of Worms.* John Murray, London. (3)

DAVIS, M.B., and D.B. BOTKIN. (1985) Sensitivity of cool-temperate forests and their fossil pollen record to rapid temperature change. *Quat. Res.* 23:327–340. (30)

DAWKINS, R. (1982) *The Extended Phenotype.* Oxford University Press, Oxford. (14)

DAWKINS, R. (1987) *The Blind Watchmaker.* Norton, New York. (26)

DAYTON, P.K. (1985) Ecology of kelp communities. *Annu. Rev. Ecol. Syst.* 16:215–245. (8)

DE VLEESCHAUWER, D., and R. LAL. (1981) Properties of worm casts under secondary tropical forest regrowth. *Soil Sci.* 132:175–181. (10)

DEANGELIS, D.L. (1980) Energy flow, nutrient cycling, and ecosystem resilience. *Ecology* 61:764–771. (25)

DEANGELIS, D.L. (1992) *Dynamics of Nutrient Cycling and Food Webs.* Chapman & Hall, London. (16; 24)

DEANGELIS D.L., S.M. BARTELL, and A.L. BRENKERT. (1989) Effects of nutrient recycling and food-chain length on resilience. *Am. Nat.* 134:778–805. (18)

DEANGELIS, D.L., R.A. GOLDSTEIN, and R.V. O'NEILL. (1975) A model for trophic interaction. *Ecology* 56:881–892. (16)

DEWITT, T.H., and J.S. LEVINTON. (1985) Disturbance, emigration, and refugia: how the mud snail *Ilyanassa obsoleta* (Say) affects the habitat distribution of an epifaunal amphipod, *Microdeutopus gryllotalpa* (Costa). *J. Exp. Mar. Biol. Ecol.* 92:97–113. (3)

DOBZHANSKY, T. (1973) Nothing in biology makes sense except in the light of evolution. *Am. Biol. Teach.* 35:125–129. (26)

DOUGLAS, M.S.V. (1993) *Diatom Ecology and Paleolimnology of High Arctic Ponds.* Queen's University, Kingston, 1993. Dissertation. (30)

DOWNING, J.A. (1986) Spatial heterogeneity: evolved behaviour or mathematical artifact. *Nature* 323:255–257. (7)

DROOP, M.R. (1974) The nutrient status of algal cells in continuous culture. *Mar. Biol. Assoc. U.K.* 54:825–855. (23)

DUBLIN, H.T., A.R.E. SINCLAIR, and J. MCGLADE. (1990) Elephants and fire as causes of multiple stable states in the Serengeti-Mara woodlands. *J. Anim. Ecol.* 59:1147–1164. (14)

DUGGINS, D.O., C.A. SIMENSTAD, and J.A. ESTES. (1989) Magnification of secondary production by kelp detritus in coastal marine ecosystems. *Science* 245:170–73. (15)

DUNBAR, R.B., and W.H. BERGER. (1981) Fecal pellet flux to modern bottom sediment of Santa Barbara Basin (California) based on sediment trapping. *Geol. Soc. Am. Bull.* Part 1, 92: 212–218. (14)

ECKMAN, J.E., A.R.M. NOWELL, and P.A. JUMARS. (1981) Sediment destabilization by animal tubes. *J. Mar. Res.* 39:361–374. (3)

EDMONDSON, W.T. (1991) *The Uses of Ecology: Lake Washington and Beyond.* University of Washington Press, Seattle. (28; 30)

EDWARDS A., and F. SHARPLES. (1986) *Scottish Sea-lochs: A Catalogue.* Nature Conservancy Council, York. (18)

EDWARDS, C.A., B.R. STINNER, D. STINNER, and S. RABBITIN (eds.). (1988) *Biological Interactions in Soil.* Elsevier Press, New York. (10)

EHRLICH, P.R. (1986) *The Machinery of Nature.* Simon and Schuster, New York. (Issues)

EHRLICH, P.R., and E.O. WILSON. (1991) Biodiversity studies: science and policy. *Science* 253:758–762. **(1)**

ELKINS, N.Z., G.V. SABOL, T.J. WARD, and W.G. WHITFORD. (1986) The influence of subterranean termites on the hydrological characteristics of a Chiuhuahuan desert ecosystem. *Oecologia* 68:521–528. **(10)**

ELLIOTT, P.W., D. KNIGHT, and J.M. ANDERSON. (1991). Variables controlling denitrification from earthworm casts and soil in permanent pastures. *Biol. Fertil. Soils* 11:24–29. **(10)**

ELMORE W., and R.L. BESCHTA. (1987) Riparian areas: perceptions in management. *Rangelands* 9:260–265. **(12)**

ELTON, C.S. (1942) *Voles, Mice and Lemmings: Problems in Population Dynamics.* Oxford University Press, Oxford. **(30)**

ELTON, C.S. (1958) *The Ecology of Invasions by Animals and Plants.* Methuen, London. **(30)**

ERICSON, L., T. ELMQVIST, and K. DANELL. (1992) Age structure of boreal willows and fluctuations in herbivore populations. *Proc. Roy. Soc. Edinburgh* 98B:75–89. **(8)**

ESTES, J.A., and D.O. DUGGINS. (1994) Sea otters and kelp forests in Alaska: generality and variation in a community ecological paradigm. *Ecol. Monogr.* (in press). **(15)**

ESTES, J.A., and J.F. PALMISANO. (1974) Sea-otters: their role in structuring nearshore communities. *Science* 185:1058–1060. **(14)**

ESTES, J.A., and P.D. STEINBERG. (1988) Predation, herbivory, and kelp evolution. *Paleobiology* 14:19–36. **(15)**

EVENARI, M.L., L. SHANAN, and N.H. TADMOR. (1982) *The Negev: The Challenge of a Desert.* Harvard University Press, Cambridge. **(27)**

FABER, J., and H.A. VERHOEF. (1991) Functional differences between closely-related soil arthropods with respect to decomposition processes in the presence and absence of pine tree roots. *Soil Biol. Biochem.* 23:15–23. **(16)**

FARRELL, T.M. (1991) Models and mechanisms of succession: an example from a rocky intertidal community. *Ecol. Monogr.* 61:95–113. **(6)**

FEE, E.J., and R.E. HECKY. (1992) Introduction to the Northwest Ontario lake size series (NOLSS). *Can. J. Fish. Aquat. Sci.* 49:2434–2444. **(30)**

FEE, E.J., J.A. SHEARER, E.R. BEBRUYN, and E.U. SCHINDLER. (1992) Effects of lake size on phytoplankton photosynthesis. *Can. J. Fish. Aquat. Sci.* 49:2445–2459. **(30)**

FENCHEL, T. (1987) *Ecology—Potential and Limitations.* Ecology Institute, Oldendorf/Luhe. **(Issues)**

FENCHEL, T. (1992) What can ecologists learn from microbes: life beneath a square centimetre of sediment surface. *Funct. Ecol.* 6:499–507. **(5)**

FIELD, C.B. (1983) Allocating nitrogen for the maximization of carbon gain: leaf age as a control over the allocation program. *Oecologia* 56:341–347. **(20)**

FIELD, C.B., and H.A. MOONEY. (1986) The photosynthesis–nitrogen relationship in wild plants, in *On the Economy of Form and Function* (ed. T.J. Givinish), Cambrige University Press, Cambridge, pp. 25–55. **(20; 24)**

FINZI, A.C., C.D. CANHAM, J.A. SILANDER, and S.W. PACALA. (1993) Resource heterogeneity in forests: effects of canopy tree species on patterns of nitrogen mineralization in a northeast temperate forest. *Bull. Ecol. Soc. Am.* 74(Suppl.):233. **(9)**

FISH, D. (1983). Phytotelmata: flora and fauna, in *Phytotelmata: Terrestrial Plants as Hosts for Aquatic Insect Communities* (eds. J.H. Frank and P.L. Lounibos), Plexus, Medford, pp. 1–27. **(14)**

FISHER, S.G. (1990) Recovery processes in lotic ecosystems: limits of successional theory. *Environ. Manag.* 14:725–736. **(1)**

FISHER, S.G. (1994) Scale, pattern, and process in freshwater ecosystems: some unifying thoughts, in *Aquatic Ecology: Scale Pattern and Process* (eds. P.S. Giller, A.G. Hildrew, and D.G. Raffaeli), Blackwell Scientific, Oxford, pp. 575–591. **(1)**

FISHER, S.G., and L.J. GRAY. (1983) Secondary production and organic matter processing by collector macroinvertebrates in a desert stream. *Ecology* 64:1217–1224. (1)

FISHER, S.G., and N.B. GRIMM. (1991) Streams and disturbance: are cross-ecosystem comparisons useful? in *Comparative Analyses of Ecosystems: Patterns, Mechanisms and Theories* (eds. J.J. Cole, G.M. Lovett, and S.E.G. Findlay), Springer-Verlag, New York, pp.196–221. (1)

FLADER, S.L., and J.B. CALLICOTT (eds.). (1991) *The River of the Mother of God and Other Essays by Aldo Leopold*. University of Wisconsin Press, Madison. (31)

FOREMAN, K.H. (1989) *Regulation of Benthic Microalgal and Meiofaunal Productivity and Standing Stock in a Salt Marsh Ecosystem: The Relative Importance of Resources and Predation.* Boston University Marine Program, Boston, 1989. Dissertation. (4)

FOREST ECOSYSTEM MANAGEMENT TEAM. (1993) *Forest Ecosystem Management: An Ecological Economic, and Social Assessment.* USDA Forest Service, USDI Bureau of Land Management, Fish and Wildlife Service, and National Park Service, Environmental Protection Agency and USDC National Marine Fisheries Service, Portland. (31)

FOSTER, J., and M.S. GAINES. (1991) The effects of a successional habitat mosaic on a small mammal community. *Ecology* 72:1358–1373. (8)

FOSTER, M.F., and D.R. SCHIEL. (1988) Kelp communities and sea otters: keystone species or just another brick in the wall? in *The Community Ecology of Sea Otters*. (eds. G.R. VanBlaricom and J.A. Estes), Springer-Verlag, Berlin, pp. 92–115. (15)

FOWLER, S.W., and G.A. KNAUER. (1986) Role of large particles in the transport of elements and organic compounds through the oceanic water column. *Prog. Oceanogr.* 16:147–194. (14)

FRANK, D.A., and S.J. McNAUGHTON. (1992) Ecology of plants, large mammalian herbivores, and drought in Yellowstone National Park. *Ecology* 73:2043–2058. (8)

FRANKLIN, J.F. (1992) Scientific basis for new perspectives in forests and streams, in *New Perspectives in Watershed Management* (ed. R.J. Naiman), Springer-Verlag, New York, pp. 25–72. (31)

FRANKLIN, J.F. (1993) Preserving biodiversity: species, ecosystems, or landscapes? *Ecol. Appl.* 3:202–205. (31)

FRAÚSTO DA SILVA, J.J.R., and R.J.P. WILLIAMS. (1991) *The Biological Chemistry of the Elements: The Inorganic Chemistry of Life*. Clarendon Press, Oxford. (23)

FRETWELL, S.D. (1977) The regulation of plant communities by food chains exploiting them. *Perspect. Biol. Med.* 20:169–185. (6)

FRETWELL, S.D. (1987) Food chain dynamics: the central theory of ecology? *Oikos* 50:291–301. (15)

FRIEND, A.D. (1993) The prediction and physiological significance of tree height, in *Vegetation Dynamics and Global Change* (eds. A.M. Solomon and H.H. Shugart), Chapman & Hall, London, pp. 101–115. (26)

FRITZ, S.C. (1990) Twentieth-century salinity and water-level fluctuations in Devils Lake, North Dakota: test of a diatom-based transfer function. *Limnol. Oceanogr.* 35:1771–1781. (30)

FROST, T.M., and P.K. MONTZ. (1988) Early zooplankton response to experimental acidification in Little Rock Lake, Wisconsin, USA. *Verh. Int. Verein. Theoret. Limnol.* 23:2279–2285. (22)

GARDENER, G.J., J.G. McIVOR, and J. WILLIAMS. (1990) Dry tropical rangelands: solving one problem and creating another, in *Australian Ecosystems: 200 Years of Utilization, Degradation and Reconstruction* (eds. D.A. Saunders, A.J.M. Hopkins, and R.A. How), Surrey Beatty and Sons (Ecological Society of Australia), Chipping Norton, pp. 279–286. (24)

GARDNER, R.H., W.G. CALE, and R.V. O'NEILL. (1982) Robust analysis of aggregation error. *Ecology* 63:1982. (19)

GARDNER, R.H., B.T. MILNE, M.G. TURNER, and R.V. O'NEILL. (1987) Neutral models for the analysis of broad-scale landscape patterns. *Landscape Ecol.* 1:19–28. (19)

GARDNER, R.H., and R.V. O'NEILL. (1991) Pattern, process and predictability: the use of neutral models for landscape analysis, in *Quantitative Methods in Landscape Ecology* (eds. M.G. Turner and R.H. Gardner), Springer-Verlag, New York, pp. 289–308. (19)

GARTEN, C.T., JR. (1976) Correlations between concentrations of elements in plants. *Nature* 261:686–688. (23)

GARTEN, C.T., JR. (1978) Multivariate perspectives on the ecology of plant mineral element composition. *Am. Nat.* 112:533–544. (23)

GATES, D.M. (1980) *Biophysical Energy*. Springer-Verlag, New York. (25)

GIBLIN, A.E., K.J. NADELHOFFER, G.R. SHAVER, J.A. LAUNDRE, and A.J. MCKERROW. (1991) Biogeochemical diversity along a riverside toposequence in arctic Alaska. *Ecol. Monogr.* 61:415–435. (1)

GINZBURG, L.R., and H.R. AKÇAKAYA. (1992) Consequences of ratio- dependent predation for steady-state properties of ecosystems. *Ecology* 73:1536–1543. (15; 16)

GIVNISH, T.J. (1988) Adaptation to sun and shade: a whole plant perspective. *Aust. J. Plant Physiol.* 15:63–92. (26)

GOLDWASSER, L., and J. ROUGHGARDEN. (1993) Construction of a large Caribbean food web. *Ecology* 74:1216–1233. (17)

GOLEBIOWSKA, J., and L. RYSZKOWSKI. (1977) Energy and carbon fluxes in soil compartments of agroecosystems. *Ecol. Bull.* 25:274–283. (11)

GONZALEZ, M.J., and T.M. FROST. (1994) Comparisons of laboratory bioassays and a whole-lake experiment: rotifer responses to experimental acidification. *Ecol. Appl.* 4:69–80. (22)

GOULD, S.J., and R.L. LEWONTIN. (1979) The spandrels of San Marcos and the Panglossian paradigm: a critique of the adaptationist programme. *Proc. R. Soc. London Ser B.* 205: 581–598. (15)

GRAY, A., L. CAMPBELL, N. CLARK, G. DABORN, J. GOSS-CUSTARD, M. HILL, S. LOCKWOOD, S. MC-GRORTY, R. MITCHELL, S. MUIRHEAD, J. PETHICK, P. RADFORD, R. UNCLES, and J. WEST. (1992) *The Ecological Impact of Estuarine Barrages*. British Ecological Society, Ecological Issues No. 3, Field Studies Council, London. (28)

GREENE R.S.B. (1992) Soil physical properties of three geomorphic zones in a semi-arid mulga woodland. *Aust. J. Soil Res.* 30:55–69. (10)

GRIME, J.P. (1979) *Plant Strategies and Vegetation Processes*. J. Wiley & Sons, Chichester, England. (24)

GRIMM, N.B. (1994) Disturbance, succession, and ecosystem processes in streams: a case study from the desert, in *Aquatic Ecology: Scale, Pattern and Process* (eds. P.S. Giller, A.G. Hildrew, and D.G. Raffaeli), Blackwell Scientific, Oxford, pp. 93–112. (1)

GRINNELL, J. (1923) The burrowing rodents of California as agents in soil formation. *J. Mammal.* 4:137–149. (12)

GROSSART, H-P., and M. SIMON. (1993) Limnetic macroscopic organic aggregates (lake snow): occurrence, characteristics, and microbial dynamics in Lake Constance. *Limnol. Oceanogr.* 38:532–546. (5)

GUNTERT, M., D.B. HAY, and R.P. BALDA. (1986) Communal roosting in the Pygmy Nuthatch: a winter survival strategy. *Acta Congr. Int. Ornithol.* 19:1964–1972. (12)

GURNEY W.S.C., E. MCCAULEY, R.M. NISBET, and W.W. MURDOCH. (1990) The physiological ecology of *Daphnia*: a dynamic model of growth and reproduction. *Ecology* 71:716–732. (18)

HAGEN, J.B. (1992) *An Entangled Bank: The Origins of Ecosystem Ecology*. Rutgers University Press, New Brunswick. (1)

HAIRSTON, N.G., SR. (1989) *Ecological Experiments: Purpose, design, and execution*. Cambridge University Press, Cambridge. (28)

HAIRSTON, N.G., SR., F.E. SMITH and L.B. SLOBODKIN. (1960) Community structure, population control, and competition. *Am. Nat.* 106:249–257. (6)

HALL, C.A.S., and DAY, J.W. (1977) *Ecosystem Modeling in Theory and Practice*. John Wiley & Sons, New York. (**17**)

HANSELL, M.H. (1993) The ecological impact of animal nests and burrows. *Funct. Ecol.* 7:5–12. (**14**)

HARGRAVE, B.T. (1970) The effect of a deposit-feeding amphipod on the metabolism of benthic microflora. *Limnol. Oceanogr.* 15:24–30. (**4**)

HARPER, J.L. (1977) *The Population Biology of Plants*. Academic Press, London. (**10**)

HARRISON, P.G. (1977) Decomposition of macrophyte detritus in seawater: effects of grazing by amphipods. *Oikos* 28:165–169. (**4**)

HARROLD, C., and D.C. REED. (1985) Food availability, sea urchin grazing, and kelp forest community structure. *Ecology* 66:1160–1169. (**3**)

HASSELL, M.P., and R.M. MAY. (1989) The population biology of host-parasite and host-parasitoid associations, in *Perspectives in Theoretical Ecology* (eds. J. Roughgarden, R.M. May, and S.A. Levin), Princeton University Press, Princeton, pp. 319–347. (**28**)

HATTON, J.C., and N.O.E. SMART. (1984) The effect of long-term exclusion of large herbivores on soil nutrient status in Murchison Falls National Park, Uganda. *Afr. J. Ecol.* 22:23–30. (**12**)

HAY, M.E. (1984) Predictable spatial escapes from herbivory: how do these effect the evolution of herbivore resistance in tropical marine communities? *Oecologia* 64:396–407. (**15**)

HE, X., J.F. KITCHELL, J.R. HODGSON, R.A. WRIGHT, P.A. SORANNO, D.M. LODGE, P.A. COCHRAN, D.A. BENKOWSKI, and N.W. BOUWES. (1993) Roles of fish predation: pisciory and planktivory, in *The Trophic Cascade in Lakes* (eds. S.R. Carpenter and J.F. Kitchell), Cambridge University Press, Cambridge, pp. 85–102. (**7**)

HENRIKSEN, K., J. HANSEN, and T. BLACKBURN. (1980) The influence of benthic infauna on exchange rates of inorganic nitrogen between sediment and water. *Ophelia Suppl.* 1:249–256. (**4**)

HENDRIX, P.F., D.A. CROSSLEY, JR., D.C. COLEMAN, R.W. PARMELEE, and M.H. BEARE. (1987) Carbon dynamics in soil microbes and fauna in conventional and no-tillage agroecosystems. *Intecol. Bull.* 15:59–63. (**11**)

HENDRIX, P.F., R.W. PARMELEE, D.A. CROSSLEY, JR., D.C. COLEMAN, E.P. ODUM, and P.M. GROFFMAN. (1986) Detritus food webs in conventional and no-tillage agroecosystems. *BioScience* 36:374–380. (**11**)

HIGASHI, M., and T.P. BURNS. 1991. Enrichment of ecosystem theory, in *Theoretical Studies of Ecosystems: The Network Perspective* (eds. M. Higashi and T.P. Burns), Cambridge University Press, Cambridge, pp. 1–38. (**26**)

HILL, J.D., and S.W. PACALA. (1992) The effects of hayscented fern on forest community structure and dynamics. *Bull. Ecol. Soc. Am.* 73(Suppl.):207. (**9**)

HILL, P.S. (1992) Reconciling aggregation theory with observed vertical fluxes following phytoplankton blooms. *J. Geophys. Res.* 97C:2295–2308. (**5**)

HOBBIE, S.E. (1992) Effects of plant species on nutrient cycling. *Trends Ecol. Evol.* 7:336–339. (**23; 24; 26**)

HOBBS, R.J., and H.A. MOONEY. (1991) Effects of rainfall variability and gopher disturbance on serpentine annual grassland dynamics. *Ecology* 72:59–68. (**8**)

HOLE, F.D. (1981) Effects of animals on soil. *Geoderma* 25:75–112. (**8**)

HOLLAND, E.A., and D.C. COLEMAN. (1987) Litter placement effects on microbial and organic matter dynamics in an agroecosystem. *Ecology* 68:425–433. (**11**)

HOLLAND, E.A., and J.K. DETLING. (1990) Plant response to herbivory and belowground nitrogen cycling. *Ecology* 71:1040–1049. (**12**)

HOLLAND, E.A., W.J. PARTON, J.K. DETLING, and D.L. COPPOCK. (1992) Physiological responses of plant populations to herbivory and their consequences for ecosystem nutrient flow. *Am. Nat.* 140:685–706. (**8; 20; 24**)

HOLLAND, H.D. (1978) *The Chemistry of the Atmosphere and Oceans*. John Wiley & Sons, New York. (**13**)

HOLLAND, H.D. (1991) The mechanisms that control the carbon dioxide and oxygen content of the atmosphere, in *Scientists on Gaia* (eds. S.H. Schneider and P.J. Boston), The M.I.T. Press, Cambridge, pp. 174–179. **(13)**

HOLLAND, H.D. (1994) Early Proterozoic atmospheric change in *Early Life on Earth* (ed. S. Bengtson), Columbia University Press, New York, pp. 237–244. **(13)**

HOLLING, C.S. (1992) Cross-scale morphology, geometry, and dynamics of ecosystems. *Ecol. Monogr.* 62:447–502. **(14; 26)**

HOLT, R.D., J. GROVER, and D. TILMAN. (1994) Simple rules for dominance in systems with mixed exploitative and apparent competition. *Am. Nat.* (in press). **(26)**

HORSLEY, S.B. (1986) Evaluation of hayscented fern interference with black cherry. *Am. J. Bot.* 73:668–669. **(9)**

HOSKIN, C.M., J.K. REED, and D.H. MOOK. (1986) Production and off-bank transport of carbonate sediment, Black Rock, southwest Little Bahama Bank. *Mar. Geol.* 73:125–144. **(14)**

HOUSE, G.J., and R.W. PARMELEE. (1985) Comparison of soil arthropods and earthworms from conventional and no-tillage agroecosystems. *Soil Till. Res.* 5:351–360. **(11)**

HOWARTH, R.W. (1991) Comparative responses of aquatic ecosystems to toxic chemical stress, in *Comparative analyses of Ecosystems: Patterns, Mechanisms, and Theories* (eds. J.J. Cole, G. Lovett, and S. Findlay), Springer-Verlag, New York, pp. 169–195. **(7; 22)**

HUNT, H.W., D.C. COLEMAN, E.R. INGHAM, R.E. INGHAM, E.T. ELLIOTT, J.C. MOORE, S.L. ROSE, C.P.P. REID, and C.R. MORLEY. (1987) The detrital food web in a short-grass prairie. *Biol. Fertil. Soils* 3:57–68. **(10)**

HUNTER, M.D., and P.W. PRICE. (1992) Playing chutes and ladders: heterogeneity and the relative roles of bottom-up and top-down forces in natural communities. *Ecology* 73:724–732. **(15)**

HUNTLEY, B.J., and B.H. WALKER. (1982) *Ecology of Tropical Savannas (Ecological Studies 42)*. Springer-Verlag, Berlin. **(24)**

HUNTLY, N.J. (1987) Effects of refuging consumers (Pikas: *Ochotona princeps*) on subalpine meadow vegetation. *Ecology* 68:274–283. **(8)**

HUNTLY, N.J. (1991) Herbivores and the dynamics of communities and ecosystems. *Annu. Rev. Ecol. Syst.* 22:477–504. **(6; 8)**

HUNTLY, N.J., and R.S. INOUYE. (1988) Pocket gophers and ecosystems: patterns and mechanisms. *BioScience* 38:786–793. **(8; 12; 14)**

HUSTON, M.A. (1979) A general hypothesis for species diversity. *Am. Nat.* 113:81–101. **(25)**

HUSTON M.A. (1992) Individual based forest succession models and the theory of plant competition, in *Individual-Based Models and Approaches in Ecology* (eds. D.L. DeAngelis and L.J. Gross), Chapman & Hall, New York, pp. 408–420. **(18)**

HUSTON, M.A., D.L. DEANGELIS, and W. POST. (1988) New computer models unify ecological theory. *BioScience* 38:682–691. **(Issues)**

HUTCHINSON, G.E. (1957) Concluding remarks. *Cold Spring Harbor Symposia on Quantitative Biology* 22:415–427. **(2)**

HUTCHINSON, G.E. (1959) Homage to Santa Rosalia or why are there so many kinds of animals. *Am. Nat.* 93:145–159. **(1; 25)**

HUTCHINSON, G.E. (1969) Eutrophication, past and present, in *Eutrophication: Causes Consequences, Correctives*, U.S. National Academy of Sciences, Washington, DC, pp. 17–26. **(30)**

HUTCHINSON, G.E. (1979) *The Kindly Fruits of the Earth*. Yale University Press, New Haven. **(29)**

INGHAM, E.R., D.C. COLEMAN, and J.C. MOORE. (1989) An analysis of food-web structure and function in a shortgrass prairie, a mountain meadow, and a lodgepole pine forest. *Biol. Fertil. Soils* 8:29–37. **(8; 16)**

INGHAM, E.R., J.A. TROFYMOW, R.N. AMES, H.W. HUNT, C.R. MORLEY, J.C. MOORE, and D.C. COLEMAN. (1986) Trophic interactions and nitrogen cycling in a semi-arid grassland soil. II. System responses to the removal of different groups of soil microbes or fauna. *J. Appl. Ecol.* 23:615–630. **(10)**

INOUYE, R.S., N.J. HUNTLY, and G.D. TILMAN. (1987b) Responses of *Microtus pennsylvanicus* to fertilization of plants with various nutrients, with particular emphasis on sodium and nitrogen concentrations in plant tissues. *Holarctic Ecol.* 10:110–113. (**8**)

INOUYE, R.S., N.J. HUNTLY, G.D. TILMAN, and J.R. TESTER. (1987a) Pocket gophers (*Geomys bursarius*), vegetation, and soil nitrogen along a successional sere in east-central Minnesota. *Oecologia* 72:178–184. (**8**)

INOUYE, R.S., and D. TILMAN. (1988) Convergence and divergence of old-field plant communities along experimental nitrogen gradients. *Ecology* 69:995–1004. (**22**)

IRONS, J.G. III, J.P. BRYANT, and M.W. OSWOOD. (1991) Effects of moose browsing on decomposition rates of birch leaf litter in a subarctic stream. *Can. J. Fish. Aquat. Sci.* 48:442–444. (**8**)

IVES, R.I. (1942) The beaver–meadow complex. *J. Geomorphol.* 5:191–203. (**12**)

IWASA, Y., V. ANDREASEN, and S. LEVIN. (1987) Aggregation in model ecosystems I. Perfect aggregation. *Ecol. Modell.* 37:287–302. (**19**)

IWASA, Y., S. LEVIN, and V. ANDREASEN. (1989) Aggregation in model ecosystems. II. Approximate aggregation. *IMA J. Math. Appl. Med. Biol.* 6:1–23. (**19**)

JANEAU, J.L., and C. VALENTIN. (1987) Relations entre des termitières *Trinervitermes* spp. et la surface du sol: réorganizations, ruissellment et érosion. *Rev. Ecol. Biol. Sol* 24:637–647. (**10**)

JANZEN, D.H. (1971) Seed predation by animals. *Annu. Rev. Ecol. Syst.* 2:465–492. (**15**)

JEFFERIES, M.J., and J.H. LAWTON. (1984) Enemy free space and the structure of ecological communities. *Biol. J. Linn. Soc.* 23:269–286. (**8**)

JEFFERIES, R.L. (1988) Vegetation mosaics, plant-animal interactions and resources for plant growth, in *Plant Evolutionary Biology*, (eds. L.D. Gottlieb and S.K. Jain), Chapman & Hall, London, pp. 340–361. (**12**)

JENKINS, S.H., and P.E. BUSHER. (1979) *Castor canadensis. Mamm. Species* 120:1–9. (**12**)

JOHNSON, K.N., J.F. FRANKLIN, J.W. THOMAS, and J. GORDON. (1991) *Alternatives for Management of Late-Successional Forests of the Pacific Northwest.* Report to Agriculture and Merchant Marine and Fisheries Committee of the U.S. House of Representatives. Oregon State University, Corvallis. (**31**)

JOHNSTON, C.A., and R.J. NAIMAN. (1990a) Aquatic patch creation in relation to beaver population trends. *Ecology* 71:1617–1621. (**12**)

JOHNSTON, C.A., and R.J. NAIMAN. (1990b) Browse selection by beaver: effects on riparian forest composition. *Can. J. For. Res.* 20:1036–1043. (**12**)

JOHNSTON, C.A., and R.J. NAIMAN. (1990c) The use of a geographic information system to analyze long-term landscape alteration by beaver. *Landscape Ecol.* 4:5–19. (**12**)

JOHNSTON, C.A., J. PASTOR, and R.J. NAIMAN. (1993) Effects of beaver and moose on boreal forest landscapes, in *Landscape Ecology and Geographical Information Systems* (eds. S.H. Cousins, R. Haines-Young, and D. Green), Taylor and Francis, London, pp. 237–254. (**12**)

JONES, C.G., J.H. LAWTON, and M. SHACHAK. (1994) Organisms as ecosystem engineers. *Oikos* 69:373–386. (**8; 14; 17**)

JONES, C.G., and M. SHACHAK. (1990) Fertilization of the desert soil by rock-eating snails. *Nature* 346:839–841. (**14**)

JONES, J.A. (1990) Termites, soil fertility and carbon cycling in dry tropical Africa: a hypothesis. *J. Trop. Ecol.* 6:291–305. (**10**)

JONES, R.L., and H.C. HANSON. (1985) *Mineral Licks, Geophagy, and Biogeochemistry of North American Ungulates.* Iowa State University Press, Ames, Iowa. (**8**)

KALISZ, P.J., and E.L. STONE. (1984) Soil mixing by scarab beetles and pocket gophers in north-central Florida. *Soil Sci. Soc. Am. J.* 48:169–172. (**8**)

KAREIVA, P.M., J.G. KINGSOLVER, and R.B. HUEY (eds.). (1993) *Biotic Interactions and Global Change.* Sinauer Associates, Sunderland. (**Issues; 28**)

KAREIVA, P.M., and G.M. ODELL. (1987) Swarms of predators exhibit "preytaxis" if individual predators use area-restricted search. *Am. Nat.* 130:233–270. **(28)**

KARNIELI, A. (1982) *Spatial variation of soil moisture regime over arid hillslopes.* Hebrew University, Jerusalem, 1982. Dissertation. **(27)**

KASTING, J.F. (1987) Theoretical constraints on oxygen and carbon dioxide concentrations in the Precambrian atmosphere. *Precambrian Res.* 34:205–229. **(13)**

KAUFFMAN, S.A. (1993) *The Origins of Order.* Oxford University Press, New York. **(2; 26)**

KELLERT, S.H. (1993) *In the Wake of Chaos.* The University of Chicago Press, Chicago. **(29)**

KELLY, C.A., J.A. AMARAL, M.A. TURNER, J.W.M. RUDD, D.W. SCHINDLER, and M.P. STAINTON. (1994) Disruption of sulfur cycling and acid neutralization in lakes at low pH. *Biogeochemistry* 24: (in press). **(30)**

KELLY, C.A., J.W.M. RUDD, A. FURUTANI, and D.W. SCHINDLER. (1984) Effects of lake acidification on rates of organic matter decomposition in sediments. *Limnol. Oceanogr.* 29:687–694 **(30)**

KELLY, C.A., J.W.M. RUDD, R.H. HESSLEIN, D.W. SCHINDLER, P.J. DILLON, C.T. DRISCOLL, S.A. GHERINI, and R.E. HECKY. (1987) Prediction of biological acid neutralization in acid-sensitive lakes. *Biogeochemistry* 3:129–140. **(30)**

KENYON, K.W. (1969) The sea otter in the eastern Pacific Ocean. *N. Am. Fauna* 68:1–352. **(15)**

KERFOOT, C.W., and A. SIH. (1987) *Predation: Direct and Indirect Impacts on Aquatic Communities.* University Press of New England, Hanover, New Hampshire. **(15)**

KIDD, K.A., D.W. SCHINDLER, R.H. HESSLEIN, and D.C.G. MUIR. (1994) Correlation between stable nitrogen isotope ratios and concentrations of organochlorines in biota from a freshwater food web. *Sci. Tot. Environ.* 124: (in press). **(30)**

KINDSCHI, G.A. (1988) Effect of intermittent feeding on growth of rainbow trout, *Salmo gairdneri* Richardson. *Aquacult. Fish. Manage.* 19:213–215. **(18)**

KING, D.A. (1990) The adaptive significance of tree height. *Am. Nat.* 135:809–828. **(26)**

KING, A.W., A.R. JOHNSON, and R.V. O'NEILL. (1991) Transmutation and functional representation of heterogeneous landscapes. *Landscape Ecol.* 5:239–253. **(19)**

KING, A.W., R.V. O'NEILL, and D.L. DEANGELIS. (1989) Using ecosystem models to predict regional CO_2 exchange between the atmosphere and the terrestrial biosphere. *Global Biogeochem. Cycles* 3:337–361. **(19)**

KINLOCH, D., H. KUHNLEIN, and D.C.G. MUIR. (1992) Inuit foods and diet: a preliminary assessment of benefits and risks. *Sci. Tot. Environ.* 122:247–278. **(30)**

KITCHELL, J.F. (1992) *Food Web Management.* Springer-Verlag, New York. **(15)**

KITCHING, R.L. (1983) Community structure in water-filled treeholes in Europe and Australia—comparisons and speculations, in *Phytotelmata: Terrestrial Plants as Hosts for Aquatic Insect Communities* (eds. J.K. Frank and L.P. Lounibos), Plexus, Medford, pp. 205–222. **(14)**

KITTREDGE, J. (1948) *Forest Influences: The Effects of Woody Vegetation on Climate, Water, and Soil.* McGraw-Hill, New York. **(9)**

KLADIVCO, E.J., A.D. MACKAY, and J.M. BRADFORD. (1986) Earthworms as a factor in the reduction of soil crusting. *Soil Sci. Soc. Am. J.* 50:191–196. **(10)**

KLEIN, D.A., W.C. METZGER, B.A. FREDERICK, and E.F. REDENTE. (1986). Environmental stress—functional diversity relationships in semi-arid terrestrial microbial communities, in *Microbial Communities in Soil* (eds. V. Jensen, A. Kjoller, and L.H. Sorensen), Elsevier, Amsterdam, pp. 105–114. **(10)**

KLEIN, D.R. (1968) The introduction, increase and crash of reindeer on St. Matthew Island. *J. Wildl. Manag.* 32:350–67. **(15)**

KNAPP, A.K., and T.R. SEASTEDT. (1986) Detritus accumulation limits productivity of tallgrass prairie. *BioScience* 36:662–668. **(24)**

KOHN, A.J. (1971) Some components of the Hutchinson legend. *Limnol. Oceanogr.* 16:157–176. **(29)**

KOOIJMAN, S.A.L.M. (1986) Population dynamics on the basis of energy budgets, in *The Dynamics of Physiologically Structured Populations* (eds. J.A.J. Metz and O. Diekmann), Springer-Verlag, Berlin, pp. 266–297. **(18)**

KOTLER, B.K., and R.D. HOLT. (1989) Predation and competition: the interaction of two types of species interactions. *Oikos* 54:256–260. **(8)**

KREBS, C.J. (1985) *Ecology. The Experimental Analysis of Distribution and Abundance*, Third Edition. Harper & Row, New York. **(Issues; 1; 14)**

KRISTENSEN, E., F. ANDERSEN, and T. BLACKBURN. (1992) Effects of benthic macrofauna and temperature on degradation of macroalgal detritus: the fate of organic carbon. *Limnol. Oceanogr.* 37:1404–1419. **(4)**

KRISTENSEN, E., and T. BLACKBURN. (1987) The fate of organic carbon and nitrogen in experimental marine sediment systems: influence of bioturbation and anoxia. *J. Mar. Res.* 45:231–257. **(4)**

KRISTENSEN, E., M.H. JENSEN, and R.C. ALLER. (1991) Direct measurement of dissolved inorganic nitrogen exchange and denitrification in individual polychaete (*Nereis virens*) burrows. *J. Mar. Res.* 49:681–716. **(4)**

KRUMBEIN, W.E., and B.D. DYER. (1985) This planet is alive—weathering and biology, a multifaceted problem, in *The Chemistry of Weathering* (ed. J.I. Drever), Kluwer-Ac, Dordrecht, pp. 143–160. **(14)**

KVITEK, R.G., J.S. OLIVER, A.R. DEGANGE, and B.S. ANDERSON. (1992) Changes in Alaskan soft-bottom prey communities along a gradient in sea otter predation. *Ecology* 73:413–428. **(15)**

LAL, R. (1987) *Tropical Ecology and Physical Edaphology*. John Wiley & Sons, Chichester. **(10)**

LA RIVIERE, M. (1989) Threats to the world's water. *Sci. Am.* 261:80–94. **(30)**

LAVELLE, P., and A. MARTIN. (1992) Small-scale and large-scale effects of endogeic earthworms on soil organic matter dynamics in soils of the humid tropics. *Soil Biol. Biochem.* 24:1491–1498. **(10)**

LAWTON, J. H. (1989) Food webs, in *Ecological Concepts: The Contribution of Ecology to the Understanding of the Natural World* (ed. J.M. Cherrett), Blackwell Scientific Publications, Oxford, pp. 43–78. **(17)**

LAWTON, J.H. (1990a) Biological control of plants: a review of generalizations, rules, and principles using insects as agents, in *Alternatives to the Chemical Control of Weeds*, FRI Bulletin 155, Rotorua, pp. 3–17. **(26)**

LAWTON, J.H. (ed.). (1990b) *Red Grouse Populations and Moorland Management*. British Ecological Society, Ecological Issues No. 2, Field Studies Council, London. **(28)**

LAWTON, J.H., and V.K. BROWN. (1993) Redundancy in Ecosystems, in *Biodiversity and Ecosystem Function* (eds. E.-D. Schulze and H.A. Mooney), Springer-Verlag, Berlin, pp. 255–270. **(1; 22; 26)**

LEE, K.E. (1985) *Earthworms: Their Ecology and Relationships with Land Use*. Academic Press, Sydney. **(10; 11)**

LEONARD, M.A., and J.M. ANDERSON. (1991a) Growth dynamics of Collembola (*Folsomia candida*) and a fungus (*Mucor plumbeus*) in relation to nitrogen availability in spatially simple and complex laboratory systems. *Pedobiologia* 35:163–173. **(10)**

LEONARD, M.A., and J.M. ANDERSON. (1991b) Grazing interactions between a collembolan and fungi in a leaf litter matrix. *Pedobiologia* 35:239–246. **(10)**

LEVIN, S.A. (1992) The problem of pattern and scale in ecology. *Ecology* 73:1943–1967. **(28)**

LEVINE, J.S. (1991) The biosphere as a driver for global atmospheric change, in *Scientists on Gaia* (eds. S.H. Schneider and P.J. Boston), The M.I.T. Press, Cambridge, pp. 353–361. **(13)**

LEVINE, S.N., and D.W. SCHINDLER. (1992) Modification of the N:P ratio in lakes by in situ processes. *Limnol. Oceanogr.* 37:917–935. **(30)**

LEVINTON, J.S. (1977) The ecology of deposit-feeding communities: Quisset Harbor, Massachusetts, in *Ecology of Marine Benthos* (ed. B.C. Coull), University of South Carolina, Columbia, p. 191–228. **(3)**

LEVINTON, J.S. (1980) Particle feeding by deposit feeders: models, data, and a prospectus, in *Marine Benthic Dynamics* (eds. K.R. Tenore and B.C. Coull), University of South Carolina, Columbia, p. 423–429. **(3)**

LEVINTON, J.S., and T.S. BIANCHI. (1981) Nutrition and food limitation of deposit-feeders. I. The role of microbes in the growth of mud snails. *J. Mar. Res.* 39:531–545. **(3)**

LEVINTON, J.S., and G.R. LOPEZ. (1977) A model of renewable resources and limitation of deposit-feeding benthic populations. *Oecologia* 31:177–190. **(3)**

LEVINTON, J.S., and M. MCCARTNEY. (1991) The use of photosynthetic pigments in sediments as a tracer for sources and fates of macrophyte organic matter. *Mar. Ecol. Prog. Ser.* 78:87–96. **(3)**

LEVINTON, J.S., and S. STEWART. (1982) Marine succession: the effect of two deposit-feeding gastropod species on the population growth of *Paranais litoralis* Müller 1784 (Oligochaeta). *J. Exp. Mar. Biol. Ecol.* 59:231–241. **(3)**

LEVINTON, J.S., and S. STEWART. (1988) Effects of sediment organics, detrital input, and temperature on demography, production, and body size of a deposit feeder. *Mar. Ecol. Prog. Ser.* 49:259–266. **(3)**

LEVINTON, J.S., S. STEWART, and T.H. DEWITT. (1985) Field and laboratory experiments on interference between *Hydrobia totteni* and *Ilyanassa obsoleta* (Gastropoda) and its possible relation to seasonal shifts in vertical mudflat zonation. *Mar. Ecol. Prog. Ser.* 22:53–58. **(3)**

LEWIN, R. (1992) *Complexity.* Macmillan, New York. **(2)**

LIKENS, G.E. (ed.) (1989) *Long-term Studies in Ecology.* Springer-Verlag, New York. **(8; 28)**

LIKENS, G.E. (1991) Human-accelerated environmental change. *BioScience* 41:310. **(Issues)**

LIKENS, G.E. (1992) *The Ecosystem Approach: Its Use and Abuse.* Ecology Institute, Oldendorf/Luhe, Germany. **(Forward; Issues; 1; 2; 17; 30)**

LINDEMAN, R.L. (1942) The trophic-dynamic aspect of ecology. *Ecology* 23:399–418. **(25)**

LINSENMAIR, K.E. (1984) Comparative studies on the social behaviour of the desert isopod *Hemilepistus reaumuri* and of a *Porcellio* species. *Symp. Zool. Soc. Lond.* 53:423–453. **(27)**

LOCKHART, W.L., R. WAGEMANN, B. TRACEY, D. SUTHERLAND, and D.J. THOMAS. (1992) Presence and implications of chemical contaminants in the freshwaters of the Canadian arctic. *Sci. Tot. Environ.* 122:165–243. **(30)**

LODGE, D.M. (1991) Herbivory on freshwater macrophytes. *Aquat. Bot.* 41:195–224. **(8)**

LOEHLE, C., and J.H.K. PECHMANN. (1988) Evolution: the missing ingredient in systems ecology. *Am. Nat.* 132:884–899. **(26)**

LOGAN, B.E., and A.L. ALLDREDGE. (1989) Potential for increased nutrient uptake by flocculating diatoms. *Mar. Biol.* 101:443–450. **(5)**

LOGAN, B.E., and J.W. DETTMER. (1990) Increased mass transfer to microorganisms with fluid motion. *Biotechnol. Bioeng.* 35:1135–1144. **(5)**

LOGAN, B.E., and D.B. WILKINSON. (1990) Fractal geometry of marine snow and other biological aggregates. *Limnol. Oceanogr.* 35:130–136. **(5)**

LOPEZ, G.R., and J.S. LEVINTON. (1987) Ecology of deposit-feeding animals in marine sediments. *Q. Rev. Biol.* 62:235–260. **(3; 14)**

LOTKA, A.J. (1922) Contribution to the energetics of evolution. *Proc. Natl. Acad. Sci.* 8:147–155. **(2)**

LOTKA, A.J. (1925) *Elements of Physical Biology.* Williams & Wilkins, Baltimore. **(Issues; 2)**

LOUDA, S.M., K.H. KEELER, and R.D. HOLT. (1990) Herbivore influences on plant performance and competitive interactions, in *Perspectives on Plant Competition* (eds. J.B. Grace and D. Tilman), Academic Press, San Diego, pp. 414–444. **(8)**

LUBCHENCO, J. (1986) Relative importance of competition and predation: early colonization by seaweeds in New England, in *Community Ecology* (eds. J. Diamond and T.J. Case), Harper & Row, New York, pp. 537–555. **(6)**

LUBCHENCO, J., K. GROSS, R. PEET, L. PITELKA, B. PATTEN, R. PULLIAM, J. FRANKLIN, J. MEYER, and T. JANETOS. (1993) Open letter to ESA members: a proposal to establish ESA Headquarters. *Bull. Ecol. Soc. Am.* 74:98–106. **(28)**

LUBCHENCO, J., A.M. OLSON, L.B. BRUBAKER, S.R. CARPENTER, M.M. HOLLAND, S.P. HUBBELL, S.A. LEVIN, J.A. MACMAHON, P.A. MATSON, J.M. MELILLO, H.A. MOONEY, C.H. PETERSON, H.R. PULLIAM, L.A. REAL, P.J. REGAL, and P.G. RISSER. (1991) The Sustainable Biosphere Initiative: an ecological research agenda. *Ecology* 72:371–412. **(Issues; 28; 31)**

LYNCH, M., and R. LANDE. (1993) Evolution and extinction in response to environmental change, in *Biotic Interactions and Global Change* (eds. P.M. Kareiva, J.G. Kingsolver, and R.B. Huey), Sinauer Associates, Sunderland, pp. 234–250. **(28)**

LYNCH, M., L.J. WEIDER, and W. LAMPERT. (1986) Measurement of carbon balance in *Daphnia*. *Limnol. Oceanogr.* 31:17–33. **(18)**

MACARTHUR, R.H. (1972) Strong, or weak interactions? in *Growth by Intussusception* (ed. E.S. Deevy), Archon Books, Hamden, pp. 179–188. **(15)**

MACKAY N. (1992) Evaluating the size effects of lampreys and their prey: an application of an individual based model, in *Individual-based Models and Approaches in Ecology* (eds. D.L. DeAngelis and L.J. Gross), Chapman & Hall, New York, pp. 278–294. **(18)**

MACMAHON, J.A., D.A. PHILLIPS, J.V. ROBINSON, and D.J. SCHIMPF. (1978) Levels of organization: an organism-centered approach. *BioScience* 11:700–704. **(Issues)**

MANKAU, R., and S.K. MANKAU. (1963) The role of mycophagous nematodes in the soil. I. The relationships of *Aphelenchus avenae* to phytopathogenic soil fungi, in *Soil Organisms* (eds. J. Doeksen and J. van der Drift), North-Holland, Amsterdam, pp. 271–280. **(11)**

MANSSON, B.A., and J.M. MCGLADE. (1993) Ecology, thermodynamics and H.T. Odum's conjectures. *Oecologia* 93:582–596. **(2; 25)**

MARKERT, B. (1992a) Presence and significance of naturally occurring chemical elements of the periodic system in the plant organism and consequences for future investigations on inorganic environmental chemistry in ecosystems. *Vegetatio* 103:1–30. **(23)**

MARKERT, B. (1992b) Multi-element analysis in plant materials—analytical tools and biological questions, in *Biogeochemistry of Trace Metals* (ed. D. C. Adriano), Lewis Publishers, Boca Raton, pp. 401–428. **(23)**

MARKERT, B. (1993) Interelement correlations detectable in plant samples based on data from reference materials and highly accurate research samples. *Fresenius' J. Anal. Chem.* 345:318–322. **(23)**

MARKS, P., J.D. ALLAN, C. CANHAM, K. VAN CLEVE, A. COVICH, and F. JAMES. (1993) *Scientific Peer Review of the Ecological Aspects of Forest Ecosystem Management: An Ecological, Economic and Social Assessment*. A report of the Ecological Society of America, Washington, DC. **(28)**

MARSH, A.G., and K.R. TENORE. (1990) The role of nutrition in regulating the population dynamics of opportunistic, surface deposit feeders in a mesohaline community. *Limnol. Oceanogr.* 35:710–724. **(3)**

MARTIN, J.H., G.A. KNAUER, D.M. KARL, and W.W. BROENKOW. (1987) VERTEX: carbon cycling in the northeast Pacific. *Deep Sea Res.* 34:267–286. **(5)**

MARTINEZ, N.D. (1988) *Artifacts or Attributes? The Effects of Resolution on the Food Web of Little Rock Lake, Wisconsin*. University of Wisconsin, Madison, 1988. Dissertation. **(17)**

MARTINEZ, N.D. (1991a) Artifacts or attributes? Effects of resolution on the Little Rock Lake food web. *Ecol. Monogr.* 61:367–392. (17)

MARTINEZ, N.D. (1991b) *Effects of scale on food web structure.* University of California, Berkeley, 1991. Dissertation. (17)

MARTINEZ, N.D. (1992) Constant connectance in community food webs. *Am. Nat.* 139:1208–1218. (17)

MARTINEZ, N.D. (1993a) Effects of resolution on food web structure. *Oikos* 66:403–412. (17)

MARTINEZ, N.D. (1993b) Effect of scale on food web structure. *Science* 260:242–243. (17)

MARTINEZ, N.D. (1994) Scale-dependent constraints on food web structure. *Am. Nat.* (in press). (17)

MASER, C., S.P. CLINE, K. CROMACK, JR., J.M. TRAPPE, and E. HANSEN. (1988) What we know about large trees that fall to the forest floor, in *From the Forest to the Sea: A Story of Fallen Trees* (eds. C. Maser, R.F. Tarrant, J.M. Trappe, and J. Franklin), Gen. Tech. Rep. PNW-GTR-229, USDA Forest Service, Pacific Northwest Research Station, Portland Oregon, pp. 25–46. (12)

MATSON, P.A., and M.D. HUNTER (eds.). (1992) The relative contributions of top-down and bottom-up forces in population and community ecology. *Ecology* 73:723–765. (8)

MATTSON, W.J., JR. (1980) Herbivory in relation to plant nitrogen content. *Annu. Rev. Ecol. Syst.* 11:119–161. (23)

MAY R.M. (1973) Mass and energy flow in closed ecosystems: a comment. *J. Theor. Biol.* 39:155–163. (18)

MAY, R.M. (1974) *Stability and Complexity in Model Ecosystems.* Princeton University Press, Princeton. (17)

MAY, R.M. (1975) *Theoretical Ecology: Principles and Applications.* W.B. Saunders, Philadelphia. (29)

MCARDLE, B.H., K.J. GASTON, and J.H. LAWTON. (1990) Variation in the size of animal populations: patterns, problems and artefacts. *J. Anim. Ecol.* 59:439–454. (7)

MCCAULEY E., W.M. MURDOCH, R.M. NISBET, and W.S.C. GURNEY. (1990) The physiological ecology of *Daphnia*: development of a model of growth and reproduction. *Ecology* 71:703–715. (18)

MCCLAUGHERTY, C.A., J. PASTOR, J.D. ABER, and J.M. MELILLO. (1985) Forest litter decomposition in relation to soil nitrogen dynamics and litter quality. *Ecology* 66:266–275. (9)

MCDOWELL, D.M., and R.J. NAIMAN. (1986) Structure and function of a benthic invertebrate stream community as influenced by beaver (*Castor canadensis*). *Oecologia* 68:481–489. (12)

MCDOWELL, L.R. (1992) *Minerals in Animal and Human Nutrition.* Academic Press, San Diego. (23)

MCINTOSH, R.P. (1981) Succession and ecological theory, in *Forest Succession: Concepts and Application* (eds. D.C. West, H.H. Shugart, and D.B. Botkin), Springer-Verlag, Berlin, pp. 10–23. (24)

MCINTOSH, R.P. (1985) *The Background of Ecology: Concept and Theory.* Cambridge University Press, New York. (Issues; 29)

MCKEE, M., and C.O. KNOWLES. (1987) Levels of protein, RNA, DNA, glycogen and lipids during growth and development of *Daphnia magna* Straus (Crustacea: Cladocera). *Freshwater Biol.* 18:341–351. (23)

MCKELVEY, R.W., M.C. DENNINGTON and D. MOSSOP. (1983) The status and distribution of trumpeter swans (*Cygnus buccinator*)in the Yukon. *Arctic.* 36:76–81. (12)

MCNAUGHTON, S.J. (1984) Grazing lawns: animals in herds, plant form, and coevolution. *Am. Nat.* 124:863–886. (6)

MCNAUGHTON, S.J. (1985) Ecology of a grazing ecosystem: the Serengeti. *Ecol. Monogr.* 55:259–294. (8)

MCNAUGHTON, S.J. (1988) Mineral nutrition and spatial concentrations of African ungulates. *Nature* 334:343–345. (8; 23)

MCNAUGHTON, S.J. (1990) Mineral nutrition and seasonal movements of African migratory ungulates. *Nature* 345:613–615. (8; 23)

MCNAUGHTON, S.J., and N.J. GEORGIADIS. (1986) Ecology of African grazing and browsing mammals. *Annu. Rev. Ecol. Syst.* 17:39–65. (12)

MCNAUGHTON, S.J., R.W. REUSS, and S.W. SEAGLE. (1988) Large mammals and process dynamics in African ecosystems. *BioScience* 38:794–800. (12; 24)

MEADOWS, P.S. (1991) The environmental impact of burrows and burrowing animals— conclusions and a model. *Symp. Zool. Soc. London* 63:327–338. (14)

MEADOWS, P.S., and A. MEADOWS. (eds.). (1991) *The Environmental Impact of Burrowing Animals and Animal Burrows.* Clarendon Press, Oxford. (14)

MEDIN, D.E., and W.P. CLARY. (1991) Small mammals of a beaver pond ecosystem and adjacent riparian habitat in Idaho. *Research Paper INT-445.* USDA Forest Service, Intermountain Research Station, Ogden Utah. (12)

MELILLO, J.M., J.D. ABER, A.E. LINKINS, A. RICCA, B. FRY, and K.J. NADELHOFFER. (1989) Carbon and nitrogen dynamics along the decay continuum: plant litter to soil organic matter. *Plant Soil* 115:189–198. (9)

MELILLO, J.M., J.D. ABER, and J.F. MURATORE. (1982) Nitrogen and lignin control of hardwood leaf litter decomposition dynamics. *Ecology* 63:621–626. (9)

MELILLO, J.M., A.D. MCGUIRE, D.W. KICKLIGHTER, B. MOORE, III, C.J. VOROSMARTY, and A.L. SCHLOSS. (1993) Global climate change and terrestrial net primary production. *Nature* 363:234–240. (13)

MELILLO J.M., R.J. NAIMAN, J.D. ABER, and A.E. LINKINS. (1984) Factors controlling mass loss and nitrogen dynamics of plant litter decaying in northern streams. *Bull. Mar. Sci.* 35:341–356. (20)

MENGE, B.A. (1976) Organization of the New England rocky intertidal community: role of predation, competition, and environmental heterogeneity. *Ecol. Monogr.* 46:355–393. (6)

MENGE, B.A., J. LUBCHENCO, and L.R. ASHKENAS. (1985) Diversity, heterogeneity, and consumer pressure in a tropical rocky intertidal community. *Oecologia* 65:394–405. (8)

MENGE, B.A., and J.P. SUTHERLAND. (1976) Species diversity gradients: synthesis of the roles of predation, competition, and temporal heterogeneity. *Am. Nat.* 110:351–369. (6)

METZ J.A.J., and O. DIEKMANN. (1980) *The Dynamics of Physiologically Structured Populations.* Springer-Verlag, Berlin. (18)

MIGLAVS, I., and M. JOBLING. (1989a) Effects of feeding regime on food consumption, growth rates and tissue nucleic acids in juvenile Arctic charr, *Salvelinus alpinus*, with particular reference to compensatory growth. *J. Fish Biol.* 34:947–957. (18)

MIGLAVS, I., and M. JOBLING. (1989b) The effects of feeding regime on proximate body composition and patterns of energy deposition in juvenile Arctic Charr, *Salvelinus alpinus. J. Fish Biol.* 35:1–11. (18)

MILCHUNAS, D.G., and W.K. LAUENROTH. (1993) Quantitative effects of grazing on vegetation and soils over a global range of environments. *Ecol. Monogr.* 63:327–366. (26)

MILCHUNAS, D.G., O.E. SALA, and W.K. LAUENROTH. (1988) A generalized model of the effects of grazing by large herbivores on grassland community structure. *Am. Nat.* 132:87–106. (26)

MILLS, K.H., and S.M. CHALANCHUK. (1987) Population dynamics of lake whitefish (*Coregonus clupeaformis*) during and after the fertilization of Lake 226, the Experimental Lakes Area. *Can. J. Fish. Aquat. Sci.* 44 (Suppl. 1):55–63. (30)

MILLS, L.S., M.E. SOULÉ, and D.F. DOAK. (1993) The keystone- species concept in ecology and conservation. *BioScience* 43:219– 224. (6; 15)

MOHLER, C.L., P.L. MARKS, and D.G. SPRUGEL. (1978) Stand structure and allometry of trees during self-thinning of pure stands. *J. Ecol.* 66:599–614. (9)

MOONEY, H.A., E.R. FUENTES, and B.I. KRONBERG (eds.). (1993) *Earth System Responses to Global Change*. Academic Press, San Diego. (28)

MOORE, J.C., and H.W. HUNT. (1988) Resource compartmentation and the stability of real ecosystems. *Nature* 333:261–263. (16)

MOORE, J.C., D.E. WALTER, and H.W. HUNT. (1988) Arthropod regulation of micro- and mesobiota in belowground detrital food webs. *Annu. Rev. Entomol.* 33:419–439. (16)

MOORE, R.M. (1970) *Australian Grasslands*. Australian National University Press, Canberra. (24)

MORGAN, L.H. (1868) *The American Beaver and His Works*. J.B. Lippincott, Philadelphia. (12)

MOROWITZ, H.J. (1968) *Energy Flow in Biology*. Academic Press, New York. (25)

MORSE, D., H.H. HEAD, C.J. WILCOX, H.H. VANHORN, C.D. HISSEM, and B. HARRIS. (1992) Effects of concentration of dietary phosphorus on amount and route of excretion. *J. Dairy Sci.* 75:3039–3049. (23)

MUIR, D.C.G., N.P. GRIFT, W.L. LOCKHART, G. BRUNSKILL, P. WILKINSON, and B. BILLECK. (1994) Historical profiles of semivolatile organochlorines in arctic lake sediments: support for the "cold-condensation" hypothesis? *Sci. Tot. Environ.* 124: (in press). (30)

MULLER, P. (1980) Effects of artificial acidification on the growth of epilithiphyton. *Can. J. Fish. Aquat. Sci.* 37:355–363. (30)

MURDOCH W.W., E. MCCAULEY, R.M. NISBET, W.S.C. GURNEY, and A.M. DEROOS. (1992) Individual based models: combining testability and generality, in *Individual-based Models and Approaches in Ecology* (eds. D.L. DeAngelis and L.J. Gross), Chapman & Hall, New York, pp. 18–35. (18)

MYERS, N. (1993) The question of linkages in environment and development. *BioScience* 43:302–310.(28)

NAIMAN, R.J., C.A. JOHNSTON, and J.C. KELLEY. (1988) Alteration of North American streams by beaver. *BioScience* 38:753–762. (12; 14)

NAIMAN, R.J., T. MANNING, and C.A. JOHNSTON. (1991) Beaver population fluctuations and tropospheric methane emissions in boreal wetlands. *Biogeochemistry* 12:1–15. (12)

NAIMAN, R.J., J.M. MELILLO, and J.E. HOBBIE. (1986) Ecosystem alteration of boreal forest streams by beaver (*Castor canadensis*). *Ecology* 67:1254–1269. (12)

NAIMAN, R.J., G. PINAY, C.A. JOHNSTON, and J. PASTOR. (1994) Beaver influences on the long term biogeochemical characteristics of boreal forest drainage networks. *Ecology* 75:905–921. (12)

NATIONAL RESEARCH COUNCIL. (1990) *Decline of the Sea Turtles*. National Academy Press, Washington, DC. (28)

NATIONAL RESEARCH COUNCIL. (1991) *Restoration of Aquatic Ecosystems: Science, Technology, and Public Policy*. National Academy Press, Washington, DC.(28)

NATIONAL RESEARCH COUNCIL. (1992) *Science and the National Parks*. National Academy Press, Washington, DC.(28)

NIELSEN, L.P., P. CHRISTENSEN, N. REVESBECH, and J. SORENSEN. (1990) Denitrification and photosynthesis in stream sediment studied with microsensor and whole core techniques. *Limnol. Oceanogr.* 35:1135–1144. (4)

NISBET R.M., and W.S.C. GURNEY. (1976) Model of material cycling in a closed ecosystem. *Nature* 264:633–634. (18)

NISBET R.M., E. MCCAULEY, A.M. DE ROOS, W.W. MURDOCH, and W.S.C. GURNEY. (1991) Population dynamics and element recycling in an aquatic plant herbivore system. *Theor. Pop. Biol.* 40:125–147. (18)

NORSE, E.A. (1990) *Ancient Forests of the Pacific Northwest*. Island Press, Washington. (31)

NORTHCOTE, T.G. (1988) Fish in the structure and function of freshwater ecosystems: a "top-down" view. *Can. J. Fish. Aquat. Sci.* 45:361–379. (7)

Noy-Meir, I. (1988) Dominant grasses replaced by ruderal forbs in a vole year in undergrazed mediterranean grasslands in Israel. *J. Biogeogr.* 15:579–587. **(8)**

O'Brien, W.J. (1979) The predator–prey interaction of planktivorous fish and zooplankton. *Am. Sci.* 67:572–581. **(7)**

Odum, E.P. (1969) The strategy of ecosystem development. *Science* 164:262–270. **(24; 25)**

Odum, E.P. (1971) *Fundamentals of Ecology.* W.B. Saunders, Philadelphia. **(Issues)**

Odum, E.P. (1985) Trends expected in stressed ecosystems. *BioScience* 35:419–422. **(30)**

Odum, E.P. (1990) Field experimental tests of ecosystem-level hypotheses. *Trends Ecol. Evol.* 5:204–205. **(30)**

Odum, H.T. (1983) *Systems Ecology.* John Wiley & Sons, New York. **(2; 25)**

Odum, H.T. (1988) Self organization, transformity, and information. *Science* 242:1132–1139. **(2)**

Odum, H.T., and R.C. Pinkerton. (1955) Times speed regulator: the optimum efficiency for maximum power output in physical and biological systems. *Am. Sci.* 43:331–43. **(2)**

Ohtsuka, S., and N. Kubo. (1991) Larvaceans and their houses as important food for some pelagic copepods. *Proc. 4th Intn. Conf. on Copepoda. Bull. Plankt. Soc. Japan* Spec. Vol:535–551. **(5)**

Ojima, D.S., D.S. Schimel, W.J. Parton, and C.E. Owensby. (1994) Long- and short-term effects of fire on nitrogen cycling in tallgrass prairie. *Biogeochemistry.* 24:67–84 **(24)**

Oksanen, L., S.D. Fretwell, J. Arruda, and P. Niemelä. (1981) Exploitation ecosystems in gradients of primary productivity. *Am. Nat.* 118:240–261. **(8; 16)**

Olsen, Y., A. Jensen, H. Reinertsen, K.Y. Børsheim, M. Heldal, and A. Langeland. (1986) Dependence of the rate of release of phosphorus by zooplankton on the P:C ratio in the food supply, as calculated by a recycling model. *Limnol. Oceanogr.* 31:34–44. **(23)**

Olsvig-Whittaker L., M. Shachak, and A. Yair. (1983) Vegetation patterns related to environmental factors in a Negev desert watershed. *Vegetatio* 54:153–165. **(27)**

O'Neill, R.V. (1973) Error analysis of ecological models, in *Radionuclides in Ecosystems* (ed. D.J. Nelson), CONF-710501, National Technical Information Service, Springfield, Virginia, pp. 898–908. **(19)**

O'Neill, R.V. (1979) Transmutation across hierarchical levels, in *Systems Analaysis of Ecosystems* (eds. G.S. Innis and R.V. O'Neill), International Cooperative Publishing House, Fairlands, pp. 59–78. **(19)**

O'Neill, R.V., D.L. DeAngelis, J.B. Waide, and T.F.H. Allen. (1986) *A Hierarchical Concept of Ecosystems.* Princeton University Press, Princeton, New Jersey. **(Issues; 17; 19; 26)**

O'Neill, R.V., and B.W. Rust. (1979) Aggregation error in ecological models. *Ecol. Modell.* 7:91–105. **(19)**

Owen-Smith, R.N. (1988) *Megaherbivores. The Influence of Very Large Body Size on Ecology.* Cambridge University Press, London. **(15)**

Pacala, S.W., C.D. Canham, and J.A. Silander. (1993) Forest models defined by field measurements: I. The design of a northeastern forest simulator. *Can. J. For. Res.* 23:1980–1988. **(9)**

Paine, R.T. (1966) Food web complexity and species diversity. *Am. Nat.* 100:65–75. **(30)**

Paine, R.T. (1969) A note on trophic complexity and community stability. *Am. Nat.* 103:91–93. **(1; 14)**

Paine, R.T. (1974) Intertidal community structure. Experimental studies on the relationship between a dominant competitor and its principal predator. *Oecologia* 15:93–120. **(15)**

Paine, R.T. (1980) Food webs: linkage, interaction strength and community infrastructure. *J. Anim. Ecol.* 49:667–685. **(1; 15; 17; 28)**

Paine, R.T. (1983) Intertidal food webs: does connectance describe their essence? in *Current Trends in Food Web Theory* (eds. D.L. DeAngelis, W.M. Post, and G. Suguihara), ORNL-5983, Oak Ridge National Laboratory, Oak Ridge, pp. 11–16. **(17)**

PAINE, R.T. (1988) Food webs: road maps of interactions or grist for theoretical development? *Ecology* 69:1648–1654. (17)

PALMER, A.R. (1979) Fish predation and the evolution of gastropod shell structure: experimental and geographic evidence. *Evolution* 33:697–713. (15)

PALOHEIMO, J.E., S.J. CRABTREE, and W.D. TAYLOR. (1982) Growth model of *Daphnia*. *Can. J. Fish. Aquat. Sci.* 39:598–606. (18)

PARKER, L.W., P.F. SANTOS, J. PHILLIPS, and W.G. WHITFORD. (1984) Carbon and nitrogen dynamics during the decomposition of litter and roots of a Chihuahuan desert annual, *Lepidium lasiocarpum. Ecol. Monogr.* 54:339–360. (11)

PARKER, M.S., and M.E. POWER. (1993) Algal-mediated differences in aquatic insect emergence and the effect on a terrestrial predator. *Bull. N. Am. Benthol. Soc.* 10:171. (6)

PARMELEE, R.W., and D.G. ALSTON. (1986) Nematode trophic structure in conventional and no-tillage agroecosystems. *J. Nematol.* 18:403–407. (11)

PARMELEE, R.W., M.H. BEARE, and J.M. BLAIR. (1989) Decomposition and nitrogen dynamics of surface weed residues in no-tillage agroecosystems under drought conditions: influence of resource quality on the decomposer community. *Soil Biol. Biochem.* 21:97–103. (11)

PARMELEE, R.W., M.H. BEARE, W. CHENG, P.F. HENDRIX, S.J. RIDER, D.A. CROSSLEY, JR., and D.C. COLEMAN. (1990) Earthworms and enchytraeids in conventional and no-tillage agroecosystems: a biocide approach to assess their role in organic matter breakdown. *Biol. Fertil. Soils* 10:1–10. (11)

PARMELEE, R.W., and D.A. CROSSLEY, JR. (1988) Earthworm production and role in the nitrogen cycle of a no-tillage agroecosystem on the Georgia Piedmont. *Pedobiologia* 32:353–361. (11)

PARMELEE, R.W., J.G. EHRENFELD, and R.L. TATE III. (1993a) Effects of pine roots on microorganisms, fauna, and nitrogen availability in two soil horizons of a coniferous forest spodosol. *Biol. Fertil. Soils* 15:113–119. (11)

PARMELEE, R.W., R.S. WENTSEL, C.T. PHILLIPS, M. SIMINI, and R.T. CHECKAI. (1993b) Soil microcosm for testing the effects of chemical pollutants on soil fauna communities and trophic structure. *Environ. Toxicol. Chem.* 12:1477–1486. (11)

PARTON, W.J, D.S. SCHIMEL, C.V. COLE, and D.S. OJIMA. (1987) Analysis of factors controlling soil organic matter levels in Great Plains grasslands. *Soil Sci. Am. J.* 51:1173–1179. (20)

PARTON, W.J., D.S. SCHIMEL, D.S. OJIMA, and C.V. COLE. (1994) A general model for soil organic matter dynamics: sensitivity to litter chemistry, texture and management. *Soil Sci. Soc. Am. J.* (in press). (20)

PARTON, W.J., J.W.B. STEWART, and C.V. COLE. (1988) Dynamics of C, N, P, and S in grassland soils: a model. *Biogeochemistry* 5:109–131. (9)

PASTOR, J., J.D. ABER, C.A. MCCLAUGHERTY, and J.M MELILLO. (1984) Aboveground production and N and P dynamics along a nitrogen mineralization gradient on Blackhawk Island, Wisconsin. *Ecology* 65:256–268. (24)

PASTOR, J., B. DEWEY, R.J. NAIMAN, P.F. MACINNES, and Y. COHEN. (1993) Moose browsing and soil fertility in the boreal forests of Isle Royale National Park. *Ecology* 74:467–480. (8)

PASTOR, J., and R.J. NAIMAN. (1992) Selective foraging and ecosystem processes in boreal forests. *Am. Nat.* 139:690–705. (12; 24)

PASTOR, J., and W.M. POST. (1986) Influence of climate, soil moisture and succession on forest carbon and nitrogen cycles. *Biogeochemistry* 2:3–27. (Issues; 9; 20)

PASTOR, J., M.A. STILLWELL, and D. TILMAN. (1987) Little bluestem litter dynamics in Minnesota old fields. *Oecologia* 72:327–330. (24)

PATTEN, B.C. (1993) Toward a more holistic ecology, and science: the contribution of H.T. Odum. *Oecologia* 93:597–602. (2)

PEARSON, T.H. (1975) The benthic ecology of Loch Linnhe and Loch Eil, a sea-loch system on the west coast of Scotland. IV. Changes in the benthic fauna attributable to organic enrichment. *J. Exp. Mar. Biol. Ecol.* 20:1–41. (4)

PEARSON, T.H., and R. ROSENBERG. (1978) Macrobenthic succession in relation to organic enrichment and pollution of the marine environment. *Oceanogr. Mar. Biol. Annu. Rev.* 16:229–311. (4)

PEDERSEN, T.F., G.B. SHIMMIELD, and N.B. PRICE. (1992) Lack of enhanced preservation of organic matter in sediments under the oxygen minimum on the Oman Margin. *Geochim. Cosmochim. Acta* 56:545–551. (13)

PERSSON, T. (1989) Role of soil animals in N and C mineralization. *Plant Soil* 81:185–189. (16)

PETERS, R.H. (1976) Tautology in evolution and ecology. *Am. Nat.* 110:1–12. (2)

PETERS, R.H. (1988) Some general problems for ecology illustrated by food web theory. *Ecology* 69:1673–1676. (17)

PETERS, R.H. (1991) *A Critique of Ecology.* Cambridge University Press, New York. (17)

PETERSON, B., B. FRY, and L. DEEGAN. (1993) The trophic significance of epilithic algal production in a fertilized tundra river ecosystem. *Limnol. Oceanogr.* 38:872–878. (30)

PETERSON, C.G., and N.B. GRIMM. (1992) Temporal variation in enrichment effects during periphyton succession in a nitrogen-limited desert stream ecosystem. *J. N. Am. Benthol. Soc.* 11:20–36. (1)

PETERSON, C.H., and S.V. ANDRE. (1980) An experimental analysis of interspecific competition among marine filter feeders in a soft-sediment environment. *Ecology* 61:129–139. (3)

PETERSON, S.R., and J.B. LOW. (1977) Waterfowl use of Uinta Mountain wetlands in Utah. *J. Wildl. Manag.* 41:112–117. (12)

PICKETT, S.T.A., J. KOLASA, and C.G. JONES. (1994) *Ecological Understanding: The Nature of Theory and the Theory of Nature.* Academic Press, Orlando. (Issues)

PILSKALN, C.H. (1991) Biogenic aggregate sedimentation in the Black Sea basin, in *Black Sea Oceanography* (eds. E. Izdar and J.W. Murray), Kluwer, Dordrecht, pp. 200–214. (5)

PILSKALN, C.H., and T.C. JOHNSON. (1991) Seasonal signals in Lake Malawi sediments. *Limnol. Oceanogr.* 36:544–557. (5)

PIMM, S.L. (1982) *Food Webs.* Chapman & Hall, London. (16; 17)

PIMM, S.L. (1984) The complexity and stability of ecosystems. *Nature* 307:321–326. (14)

PIMM, S.L. (1991) *The Balance of Nature? Ecological Issues in the Conservation of Species and Communities.* The University of Chicago Press, Chicago. (7)

PIMM, S.L., and J.H. LAWTON. (1977) The number of trophic levels in ecological communities. *Nature* 268:329–331. (6; 25)

PIMM, S.L., LAWTON, J.H., and J.E. COHEN. (1991) Food web patterns and their consequences. *Nature* 350:669–674. (14; 17)

PLAYLE, R. *The Effects of Aluminum on Aquatic Organisms: 1) Alum Additions to a Small Lake, and 2) Aluminum-26 Tracer Experiments with Minnows.* University of Manitoba, Winnipeg, 1985. Dissertation. (30)

POIER, K.R., and J. RICHTER. (1992) Spatial properties of earthworms and soil properties in an arable loess soil. *Soil Biol. Biochem.* 24:1601–1608. (10)

POLIS, G.A. (1991) Complex Desert food webs: an empirical critique of food web theory. *Am. Nat.* 138:123–155. (17)

POLIS, G.A. (1994) Food webs, trophic cascades and multifactorial analyses of community structure. *Aust. J. Ecol.* 19:121–136. (17)

PORTER, K.G. (1977) The plant–animal interface in freshwater ecosystems. *Am. Sci.* 65:159–170. (6)

POWER, M.E. (1984) Depth distributions of armored catfish: predator-induced resource avoidance? *Ecology* 65:523–528. (6)

POWER, M.E. (1990a) Benthic turfs versus floating mats of algae in river food webs. *Oikos* 58:67–79. **(6)**

POWER, M.E. (1990b) Effects of fish in river food webs. *Science* 250:411–415. **(6; 15; 30)**

POWER, M.E. (1990c) Resource enhancement by indirect effects of grazers: armored catfish, algae, and sediment. *Ecology* 71:897–894. **(8)**

POWER, M.E. (1992) Hydrologic and trophic controls of seasonal algal blooms in northern California rivers. *Arch. Hydrobiol.* 125:385–410. **(6)**

POWER, M.E., T.L. DUDLEY, and S.D. COOPER. (1989) Grazing catfish, fishing birds, and attached algae in a Panamanian stream. *Environ. Biol. Fishes* 26:285–294. **(6)**

POWER, M.E., J.C. MARKS, and M.S. PARKER. (1992) Community-level consequences of variation in prey vulnerability. *Ecology* 73:2218–2223. **(6)**

POWER, M.E., W.J. MATTHEWS, and A.J. STEWART. (1985) Grazing minnows, piscivorous bass and stream algae: dynamics of a strong interaction. *Ecology* 66:1448–1456. **(1; 6)**

POWER, M.E., A.J. STEWART, and W.J. MATTHEWS. (1988) Grazer control of algae in an Ozark Mountain stream: effects of short- term exclusion. *Ecology* 69:1894–1898. **(6)**

PRICE, P.W. (1992) The resource-based organization of communities. *Biotropica* 24:273–282. **(8)**

PROCTOR, L.M., and J.A. FUHRMAN. (1991) Roles of viral infection in organic particle flux. *Mar. Ecol. Prog. Ser.* 69:133–142. **(5)**

PULLIAM, H.R. (1988) Sources, sinks and population regulation. *Am. Nat.* 132:652–661. **(1)**

QUINTON, J.C., and R.W. BLAKE. (1990) The effect of feed cycling and ration level on the compensatory growth response in rainbow trout, *Oncorhynchus mykiss*. *J. Fish Biol.* 37:33–41. **(18)**

RAICH, J.W., E.B. RASTETTER, J.M. MELLILO, D.W. KICKLIGHTER, P.A. STEUDLER, B.J. PETERSON, A.L. GRACE, E. MOORE III, and C.J. VOROSMARTY. (1991) Potential net primary productivity in South America: application of a global model. *Ecol. Appl.* 4:399–429. **(9)**

RAMLAL, P.S., C.A. KELLY, J.W.M. RUDD, and A. FURUTANI. (1993) Sites of methyl mercury production in remote Canadian Shield lakes. *Can. J. Fish. Aquat. Sci.* 50:972–979. **(30)**

RASTETTER, E.B, R.B. KANE, G.R. SHAVER, and J.M. MELILLO. (1992b) Changes in C storage by terrestrial ecosystems: how C–N interactions restrict responses to CO_2 and temperature. *Wat. Air Soil Pollut.* 64:327–344. **(20)**

RASTETTER, E.B., A.W. KING, B.J. COSBY, G.M. HORNBERGER, R.V. O'NEILL, and J.E. HOBBIE. (1992a) Aggregating fine-scale ecological knowledge to model coarser-scale attributes of ecosystems. *Ecol. Appl.* 2:55–70. **(19; 21)**

RASTETTER, E.B., and G.R. SHAVER. (1992) A model of multiple-element limitation for acclimating vegetation. *Ecology* 73:1157– 1174. **(21)**

RASTETTER, E.B., M.G. RYAN, G.R. SHAVER, J.M. MELILLO, K.N. NADELHOFFER, J.E. HOBBIE, and J.D. ABER. (1991) A general biogeochemical model describing the responses of the C and N cycles in terrestrial ecosystems to changes in CO_2, climate, and N deposition. *Tree Physiol.* 9:101–126. **(9)**

REA, A.M. (1983) *Once a River: Bird Life and Habitat Changes on the Middle Gila*. University of Arizona Press, Tuscon, Arizona. **(12)**

REAL, L.A., and J.H. BROWN. (1991) *Foundations of Ecology: Classic Papers with Commentaries*. The University of Chicago Press, Chicago. **(28)**

REICHELT, A.C. (1991) Environmental effects of meiofaunal burrowing. *Symp. Zool. Soc. Lond.* 63:33–52. **(14)**

REICHMAN, O.J., and S. AITCHISON. (1981) Mammal trails on mountain slopes: optimum path in relation to slope angle and body size. *Am. Nat.* 117:416–420. **(8)**

REINERS, W.A. (1986) Complementary models for ecosystems. *Am. Nat.* 127:59–73. **(Issues; 1; 23)**

REISE, K. (1985) *Tidal Flat Ecology: An Experimental Approach to Species Interactions.* Ecological Studies Series 54, Springer-Verlag, Berlin. **(4)**

REMMERT, H. (ed.). (1991) *The Mosaic-Cycle Concept of Ecosystems.* Springer-Verlag, Berlin. **(25)**

RENDELL, W.B., and R.J. ROBERTSON. (1989) Nest-site characteristics, reproductive success and cavity availability for tree swallows breeding in natural cavities. *Condor* 91:875–885. **(12)**

REYNOLDS, C.S. (1984) *The Ecology of Freshwater Phytoplankton.* Cambridge University Press, Cambridge. **(23)**

RHOADS, D.C. (1967) Biogenic reworking of intertidal and subtidal sediments in Buzzards Bay, Massachusetts. *J. Geol.* 75:461–474. **(3)**

RHOADS, D.C., and D.J. STANLEY. (1965) Biogenic graded bedding. *J. Sediment. Petrol.* 35:956–963. **(3)**

RHOADS, D.C., and D.K. YOUNG. (1970) The influence of deposit-feeding organisms on sediment stability and community trophic structure. *J. Mar. Res.* 28:150–178. **(3; 14)**

RHOADS, D.C., and D.K. YOUNG. (1971) Animal sediment relationships in Cape Cod Bay. II. Reworking by *Molpadia oolitica* (Holothuroidea). *Mar. Biol.* 11:255–261. **(3)**

RICE, D.L., and D.C. RHOADS. (1989) Early diagenesis of organic matter and the nutritional value of sediment, in *Ecology of Marine Deposit Feeders* (eds. G.R. Lopez, G.L. Taghon, and J.S. Levinton), Springer-Verlag, New York, pp. 59–97. **(3)**

RICE, W.R. (1984) Disruptive selection of habitat preference and the evolution of reproductive isolation: a simulation study. *Evolution* 38:1251–1260. **(25)**

RICHMAN, S. (1958). The transformation of energy by *Daphnia pulex. Ecol. Monogr.* 28:273–291. **(18)**

RICKLEFS, R.E. (1990) *Ecology.* W.H. Freeman, New York. **(1)**

RIDGE, E.H. (1976) Studies on soil fumigation. II. Effects on bacteria. *Soil Biol. Biochem.* 8:249–253. **(10)**

RIEBESELL, U. (1992) The formation of large marine snow and its sustained residence in surface waters. *Limnol. Oceanogr.* 37:63–76. **(5)**

RIEDMAN, M.L., and J.A. ESTES. (1990) The sea otter (*Enhydra lutris*): behavior, ecology, and natural history. *U.S. Fish and Wildlife Service, Biological Report* 90(14). **(15)**

RITCHIE, J.C. (1986) Climatic change and vegetation response. *Vegetatio* 67:65–74. **(30)**

RITCHIE, M.E., and D. TILMAN. (1992) Interspecific competition among grasshoppers and their effect on plant abundance in experimental field communities. *Oecologia* 89:524–532. **(24)**

RODEN, C.M. (1984) The 1980/81 phytoplankton cycle in the coastal waters of Connemara, Ireland. *Estuarine Costal Shelf Sci.* 18:485–497. **(18)**

RODEN C.M, P.G. RODHOUSE, M.P. HENSEY, T. MCMAHON, T.H. RYAN, and J.P. MERCER. (1987) Hydrography and the distribution of phytoplankton in Killary Harbour: a fjord in western Ireland. *J. Mar. Biol. Assoc U.K.* 67:359–371. **(18)**

ROLAND, J. (1993) Large-scale forest fragmentation increases the duration of tent caterpillar outbreak. *Oecologia* 93:25–30. **(8)**

ROMANOVSKY, Y.E., and L.V. POLISHCHUK. (1982) A theoretical approach to calculation of secondary production at the population level. *Int. Rev. Gesamten. Hydrobiol.* 67:341–359. **(11)**

ROOT, R.B. (1987) The challenge of increasing information and specialization. *Bull. Ecol. Soc. Am.* 68:538–543. **(Issues)**

ROSS A.H., W.S.C. GURNEY, and M.R. HEATH. (1994) A comparative study of the ecosystem dynamics of four fjordic ecosystems. *Limnol. Oceanogr.* 39:318–343. **(18)**

ROSS A.H., W.S.C. GURNEY, M.R. HEATH, S.J. HAY, and E.W. HENDERSON. (1993) A strategic simulation model of a fjord ecosystem. *Limnol. Oceanogr.* 38:128–153. **(18)**

Ross A.H., and R.M. Nisbet. (1990) Dynamic models of growth and reproduction of the mussel *Mytilus edulis* (L). *Funct. Ecol.* 4:777–787 (**18**)

Rosswall, T., R.G. Woodmansee, and P.G. Risser (eds.). (1988) *Scales and Global Change: Spatial and Temporal Variability in Biospheric and Geospheric Processes.* John Wiley & Sons, Chichester. (**28**)

Rovira, A.D. (1976) Studies on soil fumigation. I. Effects on ammonium, nitrate and phosphate in soil and on the growth, nutrition and yield of wheat. *Soil Biol. Biochem.* 8:241–247. (**10**)

Rudd, J.W.M., C.A. Kelly, D.W. Schindler, and M.A. Turner. (1988) Disruption of the nitrogen cycle in acidified lakes. *Science* 240:1515–1517. (**22; 30**)

Rudemann, R., and W.J. Schoonmaker. (1938) Beaver dams as geologic agents. *Science* 88:523–525. (**12**)

Runnegar, B. (1994) Proterozoic eukaryotes: evidence from biology and geology, in *Early Life on Earth* (ed. S. Bengtson), Columbia University Press, New York, pp. 287–297. (**13**)

Ruttenberg, K.C. (1990) *Diagenesis and Burial of Phosphorus in Marine Sediments: Implications for the Marine Phosphorus Budget.* Yale University, New Haven, 1990. Dissertation. (**13**)

Ruttenberg, K.C., and Berner, R.B. (1993) Authigenic apatite formation and burial in sediments from non-upwelling, continental margin environments. *Geochim. Cosmochim. Acta* 57:991–1007. (**13**)

Ryan T.H., P.G. Rodhouse, C.M. Roden, and M.P. Hensey. (1986) Zooplankton and fauna of Killary Harbour: the seasonal cycle in abundance. *J. Mar. Biol. Assoc. U.K.* 66:731–748. (**18**)

Santos, P.F., J. Phillips, and W.G. Whitford. (1981) The role of mites and nematodes in early stages of buried litter decomposition in a desert. *Ecology* 62:664–669. (**11**)

Sarnelle, O. (1992) Contrasting effects of *Daphnia* on ratios of nitrogen to phosphorus in a eutrophic, hardwater lake. *Limnol. Oceanogr.* 37:1527–1542. (**23**)

Scheu, S., and V. Wolters. (1991) Influence of fragmentation and bioturbation on the decomposition of ^{14}C-labelled beech leaf litter. *Soil Biol. Biochem.* 23:1029–1034. (**10**)

Schimel, D.S. (1993) Population and community processes in the response of terrestrial ecosystems to global change in, *Biotic Interactions and Global Change* (eds. P.M. Kareiva, J. Kingsolver, and R.B. Huey), Sinauer Associates, Sunderland, pp. 45–54. (**1**)

Schimel, D.S., T.G.F. Kittel, A.K. Knapp, T.R. Seastedt, W.J. Parton, and V.B. Brown. (1991) Phyiological interactions along resource gradients in a tallgrass prairie. *Ecology* 72:672–684. (**20**)

Schindler, D.W. (1977) Evolution of phosphorus limitation in lakes: natural mechanisms compensate for deficiencies of nitrogen and carbon in eutrophied lakes. *Science* 195:260–262. (**30**)

Schindler, D.W. (1980) Experimental acidification of a whole lake: A test of the oligotrophication hypothesis, in *Ecological Impact of Acid Precipitation: Proceedings of an International Conference* (eds. D. Drablos and E. Tollan), Sandefjord, Norway, March 11–14, 1980, SNSF Project, Oslo, pp. 370–374. (**30**)

Schindler, D.W. (1986) The significance of in-lake production of alkalinity. *Water Air Soil Pollut.* 30:931–944. (**30**)

Schindler, D.W. (1987) Detecting ecosystem responses to anthropogenic stress. *Can. J. Fish. Aquat. Sci.* 44 (Supp. 1):6–25. (**7; 22; 30**)

Schindler, D.W. (1988) Experimental studies of chemical stressors on whole lake ecosystems. Eduardo Baldi Memorial Lecture. *Int. Ver. Theor. Angew. Limnol. Verh.* 23:11–41. (**30**)

Schindler, D.W. (1990a) Experimental perturbations of whole lakes as tests of hypotheses concerning ecosystem structure and function. *Oikos* 57:25–41. (**7; 30**)

Schindler, D.W. (1990b) Natural and anthropogenically imposed limitations to biotic richness in freshwaters, in *The Earth in Transition; Patterns and Processes of Biotic Impoverishment* (ed. G. Woodwell), Cambridge University Press, Cambridge, pp. 425–462. (**30**)

SCHINDLER, D.W., S.E. BAYLEY, P.J. CURTIS, B.R. PARKER, M.P. STAINTON, and C.A. KELLY. (1992) Natural and man-caused factors affecting the abundance and cycling of dissolved organic substances in Precambrian Shield lakes. *Hydrobiologia* 229:1–21. **(30)**

SCHINDLER, D.W., G.J. BRUNSKILL, S. EMERSON, W.S. BROECKER, and T.-H. PENG. (1972) Atmospheric carbon dioxide: its role in maintaining phytoplankton standing crops. *Science* 177:1192–1194. **(30)**

SCHINDLER, D.W., E.J. FEE, and T. RUSZCZYNSKI. (1978) Phosphorus input and its consequences for phytoplankton standing crop and production in the Experimental Lakes Area and in similar lakes. *J. Fish. Res. Board Can.* 35:190–196. **(30)**

SCHINDLER, D.W., T.M. FROST, K.H. MILLS, P.S.S. CHANG, I.J. DAVIES, D.L. FINDLAY, D.F. MALLEY, J.A. SHEARER, M.A. TURNER, P.J. GARRISON, C.J. WATRAS, K. WEBSTER, J.M. GUNN, P.L. BREZONIK, and W.A. SWENSON. (1991) Comparisons between experimentally- and atmospherically-acidified lakes during stress and recovery. *Proc. Roy. Soc. Edinburgh* 97B:193–226. **(30)**

SCHINDLER, D.W., R.H. HESSLEIN, and M.A. TURNER. (1987) Exchange of nutrients between sediments and water after 15 years of experimental eutrophication. *Can. J. Fish. Aquat. Sci.* 44 (Suppl. 1):26–33. **(30)**

SCHINDLER, D.W., K.H. MILLS, D.L. FINDLAY, J.A. SHEARER, I.J. DAVIES, M.A. TURNER, G.A. LINSEY, and D.R. CRUIKSHANK. (1985) Long-term ecosystem stress: the effects of years of experimental acidification on a small lake. *Science* 228:1395–1401. **(7; 30)**

SCHLESINGER, W.H. (1991) *Biogeochemistry: An Analysis of Global Change.* Academic Press, San Diego, California. **(Issues; 28)**

SCHNEIDER, E.D., and J.J. KAY. (1994) Life as a manifestation of the second law of thermodynamics, in *Modelling of Complex Systems* (eds. D. Mikulecky, D. Whitten, and M. Whitten). *J. Adv. Math. Comput. Med.* Special Issue (in press). **(2)**

SCHNEIDER, S.H. (1989) *Global Warming.* Vintage Books, New York, NY. **(28)**

SCHOENER, T.W., and D.A. SPILLER. (1992) Is extinction rate related to temporal variability in population size? An empirical answer for orb spiders. *Am. Nat.* 139:1176–1207. **(7)**

SCHOWALTER, T.D., W.W. HARGROVE, and D.A. CROSSLEY, JR. (1986) Herbivory in forested ecosystems. *Annu. Rev. Entomol.* 31:177–196. **(8)**

SCHULZE, E-D., and MOONEY, H.A. (eds.). 1993. *Biodiversity and Ecosystem Function.* Springer-Verlag, Berlin. **(Issues)**

SCIENTIFIC ASSESSMENT TEAM. (1993) *Viability Assessments and Management Consideration for Species Associated with Late-Successional and Old-Growth Forests of the Pacific Northwest.* USDA Forest Service, Portland. **(31)**

SEALE, D.B. (1980) Influence of amphibian larvae on primary production, nutrient flux, and competition in a pond ecosystem. *Ecology* 61:1531–1550. **(8)**

SEASTEDT, T.R., S.W. JAMES, and T.C. TODD. (1988) Interactions among soil invertebrates, microbes and plant growth in the tallgrass prairie. *Agric. Ecosyst. Environ.* 24:219–228. **(24)**

SEITZINGER, S. (1988) Denitrification in freshwater and coastal marine ecosystems: ecological and geochemical significance. *Limnol. Oceanogr.* 33:702–724. **(4)**

SELLERS, P.J., J.A. BERRY, G.J. COLLATZ, C.B. FIELD, and F.G. HALL. (1993) Canopy reflectance and transpiration III. A reanalysis using improved leaf models and a new canopy integration scheme. *Remote Sens. Environ.* 42:187–216. **(20)**

SETÄLÄ, H., and V. HUHTA. (1991) Soil fauna increase *Betula pendula* growth: laboratory experiments with coniferous forest floor. *Ecology* 72:665–671. **(16)**

SETON, E.T. (1929) *Lives of Game Animals.* Volume 4, Part 2, Rodents, etc. Doubleday, Doran, Garden City, New York. **(12)**

SHACHAK, M. (1980) Energy allocation and life history strategy of the desert isopod, *Hemilepistus reaumuri. Oecologia* 45:404–413. **(27)**

SHACHAK, M., and S. BRAND. (1988) Relationship among settling, demography and habitat selection: an approach and a case study. *Oecologia* 76:620–627. **(27)**

SHACHAK, M., and S. BRAND. (1991) Relationship among spatiotemporal heterogeneity, population abundance and variability in a desert, in *Ecological Heterogeneity. Ecological Studies Series 86* (eds. J. Kolasa and S.T.A. Pickett), Springer-Verlag, New York, pp. 202–223. **(27)**

SHACHAK, M., E.A. CHAPMAN, and Y. STEINBERGER. (1976) Feeding, energy flow and soil turnover in the desert isopod, *Hemilepistus reaumuri*. *Oecologia* 24:57–69. **(27)**

SHACHAK M., C.G. JONES, and S. BRAND. (1994) The role of animals in arid ecosystems: snails and isopods as controllers of soil formation, erosion and desalization. *Catena*. (in press). **(27)**

SHACHAK, M., C.G. JONES, and Y. GRANOT. (1987) Herbivory in rocks and the weathering of a desert. *Science* 236:1098–1099. **(1; 14)**

SHACHAK, M., and P. NEWTON. (1985) The relationship between brood care and environmental unpredictability in the desert isopod, *Hemilepistus reaumuri*. *J. Arid Environ.* 9: 199–209. **(27)**

SHACHAK, M., Y. STEINBERGER, and Y. ORR. (1979) Phenology, activity and regulation of radiation load in the desert isopod, *Hemilepistus reaumuri*. *Oecologia* 40:133–140. **(27)**

SHACHAK, M., and A. YAIR. (1984) Population dynamics and the role of *Hemilepistus reaumuri* in a desert ecosystem. *Symp. Zool. Soc. Lond.* 53:295–314. **(27)**

SHARON, D. (1980) The distribution of hydrologically effective rainfall incident on sloping ground. *J. Hydrol.* 46:165–188. **(27)**

SHARPLEY, A.N., J.K. SYERS, and J.A. SPRINGETT. (1979) Effect of surface casting earthworms on the transport of phosphorus and nitrogen in surface runoff from a pasture. *Soil Biol. Biochem.* 11:459–462. **(10)**

SHEARER, J.A., E.J. FEE, E.R. DeBRUYN, and D.R. DeCLERCQ. (1987) Phytoplankton primary production and light attenuation responses to the experimental acidification of a small Canadian Shield lake. *Can. J. Fish. Aquat. Sci.* 44:83–90. **(30)**

SHERWOOD, B.A., S.L. SAGER, and H.D. HOLLAND. (1987) Phosphorus in foraminiferal sediments from North Atlantic Ridge cores and in pure limestones. *Geochim. Cosmochim. Acta* 51:1861–1866. **(13)**

SHIPITALO, M.J., and R. PROTZ. (1989) Chemistry and micromorphology of aggregartion in earthworm casts. *Geoderma* 45:357–374. **(10)**

SHUGART, H.H. (1984) *A Theory of Forest Dynamics*. Springer-Verlag, New York. **(9)**

SIERSZEN, M.E., and T.M. FROST. (1993) Response of predatory zooplankton populations to the experimental acidification of Little Rock Lake, Wisconsin. *J. Plankton Res.* 15:553–562. **(22)**

SILVER, M.W., and M.M. GOWING. (1991) The "particle" flux: origins and biological components. *Prog. Oceanogr.* 26:75–113. **(5)**

SILVER, M.W., M.M. GOWING, D.C. BROWNLEE, and J.O. CORLISS. (1984) Ciliated protozoa associated with oceanic sinking detritus. *Nature* 309:246–248. **(5)**

SILVER, M.W., M.M. GOWING, and P.J. DAVOLL. (1986) The association of photosynthetic picoplankton and ultraplankton with pelagic detritus through the water column (0–2000 m). *Can. Bull. Fish. Aquat. Sci.* 214:311–341. **(5)**

SILVER, M.W., A.L. SHANKS, and J.D. TRENT. (1978) Marine snow: microplankton habitat and source of small-scale patchiness in pelagic populations. *Science* 201:371–373 **(5)**

SIMON, H.A., and A. ANDO. (1961) Aggregation of variables in dynamic systems. *Econometrica* 29:111–138. **(19)**

SINCLAIR, T.R., D.E. MURPHY, and K.R. KNOERR. (1976) Development and evaluation of simplified models for simulating canopy photosynthesis and transpiration. *J. Appl. Ecol.* 13:813–830. **(19)**

SINGER, F.J., W. SCHREIER, J. OPPENHEIM, and E.O. GARTEN. (1989) Drought, fires, and large mammals. *BioScience* 39:716–722. **(19)**

SLOBODKIN, L.B. (1968) How to be a predator. *Am. Zool.* 8:43–51. (**23**)

SLOBODKIN, L.B. (1985) Breakthrough in ecology, in *The Identification of Progress in Learning* (ed. T. Hägerstrand), Cambridge University Press, Cambridge, pp.187–195. (**Issues**)

SLOBODKIN, L. (1988) Intellectual problems of applied ecology. *Bioscience* 38:337–342. (**29**)

SLOBODKIN, L. (1989) Looking again at blooms—the null case of the paradox of the plankton, in *Novel Phytoplankton Blooms: Causes and Impacts of Recurrent Brown Tides and Other Unusual Blooms* (eds. E. Carpenter, M.Bricelj, and E. Cospar), *Coastal and Estuarine Studies* 35:341–348. Springer-Verlag, Berlin. (**29**)

SLOBODKIN, L.B. (1992a) A summary of the special feature and comments on its theoretical context and importance. *Ecology* 73:1564–1566. (**Issues**)

SLOBODKIN, L.B. (1992b) *Simplicity and Complexity in Games of the Intellect.* Harvard University Press, Cambridge. (**29**)

SLOBODKIN, L.B. (1993) An appreciation: George Evelyn Hutchinson. *J. Anim. Ecol.* 62:390–394. (**29**)

SLOBODKIN, L.B. (1994a) Scientific goals require literal empirical assumptions. *Ecol. Appl.* (in press). (**29**)

SLOBODKIN, L.B. (1994b) The connection between single species and ecosystems, in *Water Quality and Stress Indicators: Linking Levels of Organization* (ed. D.W. Sutcliffe), Freshwater Biological Association, Ambleside, (in press). (**29**)

SMITH, D.C., M. SIMON, A.L. ALLDREDGE, and F. AZAM. (1992) Intense hydrolytic enzyme activity on marine aggregates and implications for rapid particle dissolution. *Nature* 359:139–142. (**5**)

SMITH, M.L., J.N. BRUHN, and J.M. ANDERSON. (1992) The fungus *Armillaria bulbosa* is among the largest and oldest living organisms. *Nature* 356:428–431. (**10**)

SMITS, A.W. (1985) Behavioural and dietary responses to aridity in the chuckwalla, *Sauromalus hispidus. J. Herpetol.* 19:441–449. (**23**)

SMITS, A.W., J. WARD, and H. LILLYWHITE. (1986) Effects of hyperkalemia on thermoregulatory and feeding behaviours of the lizard, *Sauromalus hispidus. Copeia* 1986:518–520. (**23**)

SMOCK, L.A., J.E. GLADDEN, J.L. RIEKENBERG, L.C. SMITH, and C.R. BLACK. (1992) Lotic macroinvertebrate production in 3 dimensions: channel surface, hyporheic, and floodplain environments. *Ecology* 73:876–886. (**1**)

SOLBRIG, O.T. (1981) Adaptive characteristics of leaves with special relevance to violets, in *Evolutionary Physiological Ecology* (ed. P. Calow), Cambridge University Press, Cambridge, England, pp. 127–150. (**25**)

SOHLENIUS, B., and S. BOSTROM. (1984) Colonization, population development and metabolic activity of nematodes in buried barley straw. *Pedobiologia* 27:67–78. (**11**)

SOMMER, U. (1992) Phosphorus-limited *Daphnia*—intraspecific facilitation instead of competition. *Limnol. Oceanogr.* 37:966–973. (**23**)

SORANNO, P.A., S.R. CARPENTER, and M.M. ELSER. (1993) Zooplankton community dynamics, in *The Trophic Cascade in Lakes* (eds. S.R. Carpenter and J.F. Kitchell), Cambridge University Press, Cambridge, pp. 116–152. (**7**)

SPAIN, A.V., P. LAVELLE, and A. MAROITTI. (1992). Stimulation of plant growth by tropical earthworms. *Soil Biol. Biochem.* 24:1629–1634. (**10**)

STEELE, J.H. (1978) Some comments on plankton patches, in *Spatial Pattern in Plankton Communities* (ed. J.H. Steele), Plenum, New York, pp. 1–20. (**28**)

STEELE J.H., and B.W. FROST. (1977) The structure of plankton communities. *Phil. Trans. R. Soc. Lond. Ser. B* 280:485–534 (**18**)

STEELE, R.G.D., and J.H. TORRIE. (1980) *Principles and Procedures of Statistics, A Biometrical Approach*, Second Edition. McGraw-Hill, New York. (**8**)

STEFFEN, W.L., B.H. WALKER, J.S. INGRAM, and G.W. KOCH (eds.). (1992) *Global Change and Terrestrial Ecosystems—The Operational Plan*. IGBP Report No. 21, The International Geosphere-Biosphere Programme, Stockholm. **(Issues)**

STENECK, R.S. (1983) Escalating herbivory and resulting adaptive trends in calcareous algal crusts. *Paleobiology* 9:44–61. **(15)**

STERNER, R.W. (1990) N:P resupply by herbivores: zooplankton and the algal competitive arena. *Am. Nat.* 136:209–229. **(23)**

STERNER, R.W. (1993) *Daphnia* growth on varying quality of *Scenedesmus*: mineral limitation of zooplankton. *Ecology* 74:2351–2360. **(23)**

STERNER, R.W. (1994) Seasonal patterns in phosphorus and trace nutrient limitation. *Limnol. Oceanogr.* 39: (in press). **(23)**

STERNER, R.W., J.J. ELSER, and D.O. HESSEN. (1992) Stoichiometric relationships among producers and consumers in food webs. *Biogeochemistry* 17:49–67. **(1; 7; 23; 24)**

STERNER, R.W., D.D. HAGEMEIER, W.L. SMITH, and R.F. SMITH. (1993) Phytoplankton nutrient limitation and food quality for *Daphnia*. *Limnol. Oceanogr.* 38:857–871. **(23)**

STEWART-OATEN, A., W. MURDOCH, and K. PARKER. (1986) Environmental impact assessments: "pseudoreplication" in time? *Ecology* 67:929–940. **(22)**

STRONG, D.R. (1992) Are trophic cascades all wet? Differentiation and donor-control in speciose ecosystems. *Ecology* 73:747–754. **(15; 17)**

STRONG, D.R. (1988) Special feature editor. Food web theory: a ladder for picking strawberries? *Ecology* 69:1647–1676. **(17)**

STRONG, D.R., J.H. LAWTON, and R. SOUTHWOOD. (1984) *Insects on Plants: Community Patterns and Mechanisms*. Blackwell Scientific, Oxford. **(17)**

SUNDQUIST, E.T. (1985) Geological perspectives on carbon dioxide and the carbon cycle, in *The Carbon Cycle and Atmospheric CO_2: Natural Variations Archean to Present* (eds. E.T. Sundquist and W.S. Broeker), Geophysical Monograph 32, American Geophysical Union, Washington, DC, pp. 5–59. **(13)**

SUNDQUIST, E.T. (1993) The global carbon dioxide budget. *Science* 259:934–941. **(13)**

SUTHERLAND, J.P. (1974) Multiple stable points in natural communities. *Am. Nat.* 108:859–873. **(23)**

SUTHERLAND, J.P. (1990) Perturbations, resistance, and alternative views of the existence of multiple stable states in nature. *Am. Nat.* 136:270–275. **(23)**

SWAIN, E.B. (1973) A history of fire and vegetation in northeastern Minnesota as recorded in lake sediments. *Quat. Res.* 3:383–396. **(30)**

SWAIN, E.B. (1985) Measurement and interpretation of sedimentary pigments. *Freshwat. Biol.* 15:53–75. **(30)**

SWANK, W.T., and J.E. DOUGLASS. (1974) Streamflow greatly reduced by converting deciduous hardwood stands to white pine. *Science* 185:857–859. **(9)**

SWANK, W.T., L.W. SWIFT, JR., and J.E. DOUGLASS. (1988) Streamflow changes associated with forest cutting, species conversions, and natural disturbances, in *Forest Hydrology and Ecology at Coweeta* (eds. W.T. Swank and D.A. Crossley Jr.), Springer-Verlag, New York, pp. 297–312. **(9)**

SWANSON, D.R. (1987) Two medical literatures that are logically but not bibliographically connected. *J. Am. Soc. Inform. Sci.* 38:228–233. **(Issues)**

TANSLEY, A.G. (1935) The use and abuse of vegetational concepts and terms. *Ecology* 16:284–307. **(Issues; 17; 27)**

TANSLEY, A.G. (1949) *Britain's Green Mantle*. George Allen and Unwin, London. **(14)**

TAUBE, M. (1985) *Evolution of Matter and Energy on a Cosmic and Planetary Scale*. Springer-Verlag, New York. **(25)**

TAYLOR A.H., A.J. WATSON, and J.E. ROBERTSON. (1992) The influence of the spring phytoplankton bloom on carbon dioxide and oxygen concentrations in the surface waters of the north-east Atlantic during 1989. *Deep Sea Res.* 39:137–152. (18)

TAYLOR, B.E. (1985) Effects of food limitation on growth and reproduction of *Daphnia*. *Arch. Hydrobiol. Beihefte* 21:285–296. (18)

TAYLOR, B.E. and W. GABRIEL. (1985) Reproductive strategies of two similar *Daphnia* species. *Proc. Int. Assoc. Theoret. Appl. Limnol.* 22:3047–3050. (18)

TETT, P.A., A. EDWARDS, and K. JONES. (1986) A model for the growth of shelf-sea phytoplankton in summer. *Estuarine, Coastal & Shelf Sci.* 23:641–672. (18)

THAYER, C.W. (1979) Biological bulldozers and the evolution of marine benthic communities. *Science* 203:458–461. (14)

THOMAS, J.W., E.D. FORSMAN, J.B. LINT, E.C. MESLOW, B.R. NOON, and J. VERNER. (1990) *A Conservation Strategy for the Northern Spotted Owl.* A report of the Interagency Scientific Committee to address the conservation of the northern spotted owl. USDA Forest Service, USDI Bureau of Land Management, Fish and Wildlife Service, and National Park Service, Portland. (31)

TIEDJE, J.M., R.K. COLWELL, Y.L. GROSSMAN, R.E. HODSON, R.E. LENSKI, R.N. MACK, and P.J. REGAL. (1989) The planned introduction of genetically engineered organisms: ecological considerations and recommendations. *Ecology* 70:298–315. (28)

TILMAN, D. (1982) *Resource Competition and Community Structure.* Princeton University Press, Princeton, New Jersey. (25)

TILMAN, D. (1988) *Plant Strategies and the Dynamics and Structure of Plant Communities.* Princeton University Press, Princeton, New Jersey. (24; 26)

TILMAN, D., and D. WEDIN. (1991) Plant traits and resource reduction for five grasses growing on a nitrogen gradient. *Ecology* 72:685–700. (24)

TOFT, C.A., and P.J. SHEA. (1983) Detecting community-wide patterns: estimating power strengthens statistical inference. *Am. Nat.* 122:618–625. (8)

TOMLIN, A.D., D. MCCABE, and R. PROTZ. (1992) Species composition and seasonal variation of earthworms and their effect on soil properties in southern Ontario, Canada. *Soil Biol. Biochem.* 24:1451–1458. (10)

TONGWAY, D.J., J.A. LUDWIG, and W.G. WHITFORD. (1989) Mulga log mounds: fertile patches in the semi-arid woodlands of eastern Australia. *Aust. J. Ecol.* 14:263–268. (10)

TONN, W.M. (1985) Density compensation in Umbra-Perca fish assemblages of northern Wisconsin Lakes. *Ecology* 66:415–429. (22)

TRITTON, L.M., and J.W. HORNBECK. (1982) Biomass equations for major tree species of the northeast. *USDA Forest. Service Gen. Tech. Report NE-69.* (9)

TURNER, M.A., E.T. HOWELL, M. SUMMERBY, R.H. HESSLEIN, D.L. FINDLAY, and M.B. JACKSON. (1991) Changes in epilithon and epiphyton associated with experimental acidification of a lake to pH 5. *Limnol. Oceanogr.* 36:135–149. (30)

TURNER, M.G., V.H. DALE, and R.H. GARDNER. (1989a) Predicting across scales: theory development and testing. *Landscape Ecol.* 3:245–252. (19)

TURNER, M.G., R.H. GARDNER, V.H. DALE, and R.V. O'NEILL. (1989b) Predicting the spread of disturbance across heterogeneous landscapes. *Oikos* 55:121–129. (19)

TURNER, M.G., R.V. O'NEILL, R.H. GARDNER, and B.T. MILNE. (1989c) Effects of changing spatial scale on the analysis of landscape pattern. *Landscape Ecol.* 3:153–162. (19)

TURNER, M.G., Y. WU, W.H. ROMME, L.L. WALLACE, and A. BRENKERT. (1994) Simulating interactions between ungulates, vegetation and fire in northern Yellowstone National Park during winter. *Ecol. Appl.* (in press). (19)

ULANOWICZ, R.E. (1986) *Growth and Development of Ecosystems and Communities.* Springer-Verlag, New York. (26)

UNITED STATES DEPARTMENT OF THE INTERIOR. (1992) *Recovery Plan for the Northern Spotted Owl—Final Draft*. USDI Fish and Wildlife Service, Portland. (**31**)

UNITED STATES ENVIRONMENTAL PROTECTION AGENCY. (1990) *Reducing Risk: Setting Priorities and Strategies for Environmental Protection*. The Report of the Science Advisory Board: Relative Risk Reduction Strategies Committee to William K. Reilly, Administrator, United States Environmental Protection Agency. SAB-EC-90–021. (**28**)

VAN HOOFF, P. (1982) Earthworm activity as a cause of splash erosion in a Luxembourg forest. *Geoderma* 31:195–204. (**10**)

VAN SOEST, P.J. (1982) *Nutritional Ecology of the Ruminant*. Cornell University Press, Ithaca, New York. (**24**)

VANBLARICOM, G.R., and J.A. ESTES (eds.). (1988) *The Community Ecology of Sea Otters*. Springer-Verlag, Berlin. (**15**)

VERHOEF, H.A., and L. BRUSSAARD. (1990) Decomposition and nitrogen mineralization in natural and agro-ecosystems: the contribution of soil animals. *Biogeochemistry* 11:175–211. (**10; 16**)

VERMEIJ, G.J. (1977) The mesozoic marine revolution: evidence from snails, predators and grazers. *Paleobiology* 3:245–258. (**15**)

VERMEIJ, G.J. (1978) *Biogeography and adaptation: Patterns of Marine Life*. Harvard University Press, Cambridge, Massachusetts. (**15**)

VITOUSEK, P.M. (1986) Biological invasions and ecosystem properties: Can species make a difference? in *Ecology of Biological Invasions of North America and Hawaii* (eds. H.A. Mooney and J.A. Drake), Springer-Verlag, New York, pp. 163–176. (**22**)

VITOUSEK, P.M. (1990) Biological invasions and ecosystems processes: towards an integration of population biology and ecosystem studies. *Oikos* 57:7–13. (**22**)

VITOUSEK, P.M., and J. LUBCHENCO. (1994) Limits to sustainable use of resources: from local effects to global change, in *Defining and Measuring Sustainability: Biological and Physical Foundations* (eds. M. Munasinghe, W. Shearer, and T. Lovejoy), World Bank and United Nations University, Washington, D.C. (in press). (**28**)

VITOUSEK, P.M., and L.R. WALKER. (1989) Biological invasion by *Myrica faya* in Hawai'i: plant demography, nitrogen fixation and ecosystem effects. *Ecol. Monogr.* 59:247–265. (**20; 22**)

VOLLENWEIDER, R.A. (1968) Scientific fundamentals of the eutrophication of lakes and flowing waters, with particular reference to phosphorus and nitrogen as factors in eutrophication. *OECD Technical Report* DAS/CSI/68.27. Revised 1971. (**30**)

VOLLENWEIDER, R.A. (1976) Advances in defining critical loading levels for phosphorus in lake eutrophication. *Mem. Ist. Ital. Idrobiol.* 33:53–83. (**30**)

WALDROP, M.M. (1992) *Complexity: The Emerging Science at the Edge of Order and Chaos*. Simon and Schuster, New York. (**2**)

WALLACE, L.L., M.G. TURNER, W.H. ROMME, R.V. O'NEILL, and Y. WU. (1994) Scale of heterogeneity of forage production and winter foraging by elk and bison. *Landscape Ecol.* (in press). (**19**)

WARDLE, D.A., and G.W. YEATES. (1993) The dual importance of competition and predation as regulatory forces in terrestrial ecosystems: evidence from decomposer food-webs. *Oecologia* 93:303–306. (**16**)

WARING, R.H. (1989) Ecosystems: fluxes of matter and energy, in *Ecological Concepts: The Contributions of Ecology to the Understanding of the Natural World* (ed. J.M. Cherrett), Blackwell Scientific Publications, Oxford, pp. 17–42. (**26**)

WASILEWSKA, L., H. JAKUBCZYK, and E. PAPLINSKA. (1975) Production of *Aphelenchus avenae* Bastian (Nematoda) and reduction of mycelium of saprophytic fungi by them. *Pol. Ecol. Stud.* 1:61–73. (**11**)

WATRAS, C.J., and T.M. FROST. (l989) Little Rock Lake (Wisconsin): perspectives on an experimental ecosystem approach to seepage lake acidification. *Arch. Environ. Contam. Toxicol.* 18:157–165. (**22**)

WATSON, A., J.E. LOVELOCK, and L. MARGULIS. (1978) Methanogenesis, fires, and the regulation of atmospheric oxygen. *Biosystems* 10:293–298. (**13**)

WATSON, J.C. (1993) *The Effects of Sea Otter (Enhydra lutris) Foraging on Rocky Subtidal Communities off Northwestern Vancouver Island, British Columbia.* University of California, Santa Cruz, 1993. Dissertation. (**15**)

WEATHERLY, A.H., and R.W. GILL. (1991) Recovery growth following periods of restricted rations and starvation in rainbow trout, *Salmo gairdneri* Richardson. *J. Fish Biol.* 18:195–208. (**18**)

WEBSTER, J.R., M.E. GURTZ, J.J. HAINS, J.L. MEYER, W.T. SWANK, J.B. WAIDE, and J.B. WALLACE. (1983) Stability of stream ecosystems, in *Stream Ecology* (eds. J.R. Barnes and G.W. Minshall), Plenum Press, New York, pp. 355–395. (**1**)

WEBSTER, J.R., J.B. WAIDE, and B.C. PATTEN. (1975) Nutrient cycling and the stability of ecosystems, in *Mineral Cycling in Southeastern Ecosystems* (eds. F.G. Howell, J.B. Gentry, and H.M. Smith), Energy Research and Development Administration Symposium Series CONF-740513, Oak Ridge, Tennessee, pp. 1–27. (**12**)

WEBSTER, K.E., T.M. FROST, C.J. WATRAS, W.A. SWENSON, M. GONZALEZ, and P.J. GARRISON. (1992) Complex biological responses to the experimental acidification of Little Rock Lake, Wisconsin, USA. *Environ. Pollut.* 78:73–78. (**22**)

WEDIN, D.A., and J. PASTOR. (1993) Nitrogen mineralization dynamics in grass monocultures. *Oecologia* 96:186–192. (**24**)

WEDIN, D.A., and D. TILMAN. (1993) Competition among grasses along a nitrogen gradient: initial conditions and mechanisms of competition. *Ecol. Monogr.* 63:199–229. (**24**)

WEDIN, D.A., and D. TILMAN. (1990) Species effects on nitrogen cycling: a test with perennial grasses. *Oecologia* 84:433–441. (**1; 23; 24**)

WEILENMANN, U., C.R. O'MELIA, and W. STUMM. (1989) Particle transport in lakes: models and measurements. *Limnol. Oceanogr.* 34:1–18. (**5**)

WHITTAKER, R.H. (1973) *Communities and Ecosystems.* Second Edition. Harper & Row, New York. (**26**)

WHITTAKER, R.H., and S.A. LEVIN. (1977) The role of mosaic phenomena in natural communities. *Theor. Pop. Biol.* 12:117–139. (**25**)

WIEGERT, R.G. (1988) The past, present, and future of ecological energetics, in *Concepts of Ecosystem Ecology* (eds. L.R. Pomeroy and J.J. Alberts), Springer-Verlag, New York, pp. 29–55. (**25**)

WILLIAMS, G.C. (1966) *Adaptation and Natural Selection.* Princeton University Press, Princeton. (**1**)

WILLIAMS, G.C. (1992) *Natural Selection: Domains, Levels, and Challenges.* Oxford University Press, Oxford. (**26**)

WILLIG, M.R., and T.E. LACHER JR. (1991) Food selection of a tropical mammalian folivore in relation to leaf-nutrient content. *J. Mammal.* 72:314–321. (**23**)

WINER, B.J. (1971) *Statistical Principles in Experimental Design.* Second Edition. McGraw-Hill, New York. (**8**)

WOLTERS, V. (1991) Soil invertebrates—Effects on nutrient turnover and soil structure—A review. *Z. Pflanzenernähr. Bodenk.* 154:389–402. (**16**)

WOOD, T.G. (1991) Termites in Ethiopia: the environmental implications of their damage and resultant control measures. *Ambio* 20:136–138. (**10**)

WOODIN, S.A. (1974) Polychaete abundance patterns in a marine soft-sediment environment: the importance of biological interactions. *Ecol. Monogr.* 44:171–187. (**3**)

WOODWARD, I.F. (1987) *Climate and Plant Distribution.* Cambridge University Press, Cambridge. (**26**)

WOOTTON, J.T. (1990) *Direct and Indirect Effects of Bird Predation and Excretion on the Spatial and Temporal Patterns of Intertidal Species.* University of Washington, Seattle, 1990. Dissertation. (**6**)

WRIGHT, D.H. (1983) Species-energy theory: an extension of species area theory. *Oikos* 41:496–506. **(25)**

XIA, X., and R. BOONSTRA. (1992) Measuring temporal variability of population density: a critique. *Am. Nat.* 140:883–892. **(7)**

YAIR, A. (1985) Runoff generation in arid and semi-arid zones, in *Hydrological Forecasting* (eds. M.G. Anderson and T.P. Burt), John Wiley & Sons, New York, pp. 183–220. **(27)**

YAIR, A., and A. DANIN. (1980) Spatial variation in vegetation as related to soil moisture regime over an arid limestone hillside, Northern Negev Israel. *Oecologia* 47:83–88. **(27)**

YAIR, A., and J. RUTIN. (1981) Some aspects of the regional variation in the amount of available sediment produced by isopods and porcupines, northern Negev, Israel. *Earth Surface Process. Landforms* 6:221–234. **(14)**

YAIR, A., and M. SHACHAK. (1982) A case study of energy, water and soil flow chains in an arid ecosystem. *Oecologia* 54:389–397. **(27)**

YAIR, A., and M. SHACHAK. (1987) Studies in watershed ecology of an arid area, in *Progress in Desert Research* (eds. M.O. Wurtele and L. Berkofsky), Rowman and Littlefield Publishers, Totowa, pp. 146–193. **(27)**

YAIR A., D. SHARON, and H. LAVEE. (1978) An instrumented watershed for the study of partial area contribution of runoff in arid zone. *Zeitsch. Geom. Supp.* 29:71–82. **(27)**

YOCCOZ, N.G. (1991) Use, overuse, and misuse of significance tests in evolutionary biology and ecology. *Bull. Ecol. Soc. Am.* 72:106–111. **(8)**

YODZIS, P. (1993) Environment and trophodiversity, in *Species Diversity in Ecological Communities* (eds. R.E. Ricklefs and D. Schluter), University of Chicago Press, Chicago, pp. 26–38. **(17)**.

YODZIS, P. (1984) Energy flow and the vertical structure of real ecosystems. *Oecologia* 65:86–88. **(25)**

YODZIS, P. (1988) The indeterminacy of ecological interactions as perceived through perturbation experiments. *Ecology* 69:508–515. **(7)**

ZARET, T.M. (1980) *Predation and Freshwater Communities*. Yale University Press, New Haven. **(7)**

ZARET, T.M. and R.T. PAINE. (1973) Species introduction in a tropical lake. *Science* 182:449–455. **(15)**

ZEIGLER, B.P. (1976) The aggregation problem, in *Systems Analysis and Simulation in Ecology*, Volume IV (ed. B.C. Patten), Academic Press, New York, pp. 299–311. **(19)**

ZHENG, D.W. (1993) *Theoretical Investigations on the Influence of Soil Food Web Structure on Decomposition*. Swedish University of Agricultural Science, Uppsala, 1993. Dissertation. **(16)**

ZIMAN, J. (1985) Pushing back frontiers—or redrawing maps! in *The Identification of Progress in Learning* (ed. T. Hägerstrand), Cambridge University Press, Cambridge, pp. 1–12. **(Issues)**

INDEX